We live in a world in which inconsistency is the rule rather than the exception, and this is particularly true for rewards and frustrations. In some cases, rewards and frustrative nonrewards appear randomly for what seems to be the same behavior; in others, a sequence of rewards is suddenly followed by nonrewards, or large rewards by small rewards. Sometimes we are rewarded for responding quickly, other times for responding slowly. The important common factor in these and other cases is frustration, how we learn about it and how we respond to it. Without our awareness, our long-term dispositions are shaped from infancy and early childhood by such inconsistency of reward and by our reactions to discrepancy, and they are marked by changes in arousal, suppression, persistence, and regression. This book provides a basis in learning theory, and particularly in an animal-based model of frustration theory, for a comprehension not only of the mechanisms controlling these dispositions, but also of their order of appearance in early development and, to an approximation at least, their neural underpinnings.

The explanatory domain of frustration theory covers an area of experimental research that has evolved over some 40 years. Written by the originator of the theory, the book provides an integrated survey of the theory's history and the experimental particulars on which it is based, tracing its development and the experimental research it has stimulated and organized.

Problems in the Behavioural Sciences
GENERAL EDITOR: Jeffrey Gray
EDITORIAL BOARD: Michael Gelder, Richard Gregory, Robert Hinde, Christopher Longuet-Higgins

Frustration theory

Problems in the Behavioural Sciences
1. Contemporary animal learning theory
   A. DICKINSON
2. Thirst
   B. J. ROLLS & E. T. ROLLS
3. Hunger
   J. LE MAGNEN
4. Motivational systems
   F. TOATES
5. The psychology of fear and stress
   J. A. GRAY
6. Computer models of mind
   M. A. BODEN
7. Human organic memory disorders
   A. R. MAYES
8. Biology and emotion
   N. McNAUGHTON
9. Latent inhibition and conditioned attention theory
   R. E. LUBOW
10. Psychobiology of personality
    M. ZUCKERMAN
11. Frustration theory
    A. AMSEL

Frustration theory

An analysis of dispositional learning and memory

Abram Amsel
University of Texas at Austin

Published by the Press Syndicate of the University of Cambridge
The Pitt Building, Trumpington Street, Cambridge CB2 1RP
40 West 20th Street, New York, NY 10011-4211, USA
10 Stamford Road, Oakleigh, Victoria 3166, Australia

© Cambridge University Press 1992

First published 1992

Printed in Canada

Library of Congress Cataloging-in-Publication Data
Amsel, Abram.
Frustration theory : an analysis of dispositional learning and memory / Abram Amsel.
p. cm. – (Problems in the behavioural sciences : 11)
Includes bibliographical references and index.
ISBN 0-521-24784-5 (hc).
1. Frustration. 2. Reward (Psychology). 3. Psychobiology, Experimental. 4. Human behavior – Animal models. I. Title. II. Series.
BF575.F7A553 1992
153.1′53 – dc20   91-42414
                                                       CIP

A catalog record for this book is available from the British Library.

ISBN 0-521-24784-5 hardback

Contents

Preface		*page* vii
List of abbreviations		xi
1	Introduction: reward-schedule effects and dispositional learning	1
2	Motivational and associative mechanisms of behavior	12
3	Frustration theory: an overview of its experimental basis	34
4	Survival, durability, and transfer of persistence	61
5	Discrimination learning and prediscrimination effects	94
6	Alternatives and additions to frustration theory	122
7	Ontogeny of dispositional learning and the reward-schedule effects	137
8	Toward a developmental psychobiology of dispositional learning and memory	174
9	Summing up: steps in the psychobiological study of related behavioral effects	205
10	Applications to humans: a recapitulation and an addendum	216
Appendix: some phenomena predicted or explained by frustration theory		233
References		238
Name index		267
Subject index		274

Preface

There are two kinds of biologically oriented scientists who study behavior – nowadays we call them behavioral neuroscientists. One of these is interested in animals in the way a naturalist is, and often the question is, What can the animal do and how well can the animal do it? Or how intelligent is the animal, in terms of the way we understand cognitive abilities in humans? The other kind of investigator of behavior, or of brain–behavior relationships, uses one of a few animal species, in biological psychology predominantly the rat, pigeon, rabbit, or monkey, as an animal model for general – and specifically human – function and is not much interested in the animal in the naturalist's or cognitivist's way, except insofar as this kind of interest or knowledge affects his or her research and its interpretation. It will become obvious that the orientation of this book is of the latter kind: The experimental animal of choice is overwhelmingly *Rattus norvegicus,* and its study is designed to further the comprehension of what I have called dispositional learning and memory – systems that ordinarily have a long-term historical etiology and in which learning is relatively reflexive and memory implicit and not strongly episodic.

This is a book on frustration theory and not a book on frustration *theories;* I try to underscore this in Chapter 3, where I provide a brief review of theories of frustration. Although theories other than my own, particularly of the partial-reinforcement extinction effect, are referred to throughout, and in particular in Chapter 6, the main purpose of this monograph is to review a line of theorizing and experimental research that has evolved over some 40 years. Frustration theory does not necessarily include in its explanatory domain all of the myriad human manifestations of nonreward, thwarting, failures, and reactions to physical and emotional insult that some have defined as frustration-instigated. It is, however, an attempt to present in some detail an animal-based model of frustration as it is applied to a limited – but still large – number of experimentally established phenomena (perhaps the largest number of phenomena organized by any one such theory) that bear some resemblance to equivalent phenomena in humans to which terms like *arousal, suppression, persistence,* and *regression* have been applied.

The courage even to contemplate writing a book such as this one – to spread "old-fashioned" *S*s and *R*s across the pages of a volume in the face of the steamrolling cognitive revolution – came to me in the serenity of the rustic setting of the Center for Advanced Study in the Behavioral

Sciences at Stanford. (A major reason for proceeding with this task was the hope that it might strengthen the resolve of, or at least express solidarity with, those who think this level of analysis of behavior is useful and even important.) It was at the Center in 1986–7 that the book was conceived and planned and where rough drafts of some of the chapters were written. For this I am grateful for the ambience of the Center, which provided the freedom and the atmosphere in which to think and do such things, to the staff who made it technically possible, and to the National Science Foundation for financial support of my fellowship under Grant BNS 84–11738.

The more direct impetus behind this book was the friendly nagging over many years of a companion in frustration theory and the editor of this series, Jeffrey Gray. I am deeply indebted to him, if not for his persistence, then certainly for his personal guidance and advice throughout the project and for his patient and insightful editing of the manuscript. I would also like to acknowledge the helpful comments I received from two of his associates, Dr. Alan D. Pickering and E. Clea Warburton.

The collaborators in the research that makes up a large part of this book are too numerous to acknowledge here; however, the contributions of all of these colleagues, former and present students and associates, are acknowledged in specific references throughout the book. They contributed to this research over a period of four decades, the first at Tulane University, the second at the University of Toronto, and the third and fourth at the University of Texas at Austin.

The reader will soon note that most of the figures in this book are from publications from my own laboratory. In some cases, particularly in Chapters 5, 7, and the last part of Chapter 8, these are close to being the only available data in the specific areas under discussion. Whenever experimental results from similar experiments are available, they are cited in the text or in the Appendix.

An early draft of this book was the main subject matter of a graduate seminar at the University of Texas at Austin in the spring of 1989. The students in this seminar were Jaime Diaz-Granados, Paul Greene, Michael Lilliquist, Erika Pollock, Lisa Shapiro, and Carlos Zalaquett. Nancy Lobaugh, a postdoctoral research associate in my laboratory who was a consultant on this book in a number of important ways, was also a participant in the seminar. Each of these people went over the early draft chapter by chapter and provided critiques and discussion that have made the end product much better than it would otherwise have been.

I have been very fortunate to have had federal research support through most of the years I have worked on the subject matter of this book: from the National Science Foundation, 1954 to 1978 and 1987 to 1991; from the National Research Council of Canada, 1963 to 1969; from the National Institute for Mental Health, 1978 to 1983; from the National Institute of Child Health and Human Development, 1983 to 1986; and from the National Institute on Alcohol Abuse and Alcoholism, 1987 to the present. Obviously this monograph would not have been written without this support. The most recent of these grants contributed, at least indirectly, to the writing and the final preparation of the manuscript.

Early drafts of portions of the book were prepared by the word processing staff at the Center for Advanced Study in the Behavioral Sciences. Sandra Foster assisted me greatly in every phase of preparing an early version of the complete manuscript, as did Jo Ann Wolf and Anna Tapsak in the final version. I am most grateful to them for their invaluable assistance and for their care and dedication.

Finally, it is a pleasure to express my thanks to Mary Racine, who, in the last stages of the project, guided it over some rough spots and gently to its conclusion.

This book is obviously a kind of scientific autobiography, covering that part of the experimental and theoretical work that has been the constant in my academic career. The other important and parallel constant has been the loving care and inspiration provided by my three sons, Steve, Andrew, and Geoff, and particularly by my wife, Tess. To them, I dedicate this book.

Abbreviations

Frequently used theoretical terms

CS The conditioned stimulus in classical (Pavlovian) conditioning

US (or UCS) The unconditioned stimulus in classical conditioning

CR The conditioned response in classical conditioning

UR (or UCR) The unconditioned response in classical conditioning

S_C **(or S_A)** The conditioned stimulus (or conditioned stimulation arising out of the apparatus) in instrumental conditioning

S_U **(or S_G)** The unconditioned stimulus or goal event (general) in instrumental conditioning

S_R The unconditioned stimulus or rewarding (positive reinforcing) goal event in instrumental conditioning

S_{nonR} The absence of the unconditioned rewarding stimulus in instrumental conditioning

R_G The unconditioned (consummatory) response to a goal event (general) in instrumental conditioning

R_R The unconditioned (consummatory) response to a rewarding goal event in instrumental conditioning

R_F The unconditioned response to a nonreward (S_{nonR}) when reward is anticipated (primary frustration) in instrumental conditioning

r_G The classically conditioned form of R_G; Hull's anticipatory goal response (general)

r_R The classically conditioned form of R_R; anticipatory reward

r_F The classically conditioned form of R_F; anticipatory frustration

s_G The feedback stimulus from Hull's r_G (general)

s_R The feedback stimulus from r_R

s_F The feedback stimulus from r_F

$s_R \rightarrow$ **approach** The instrumental conditioning of approach to the feedback stimulus, s_R

$s_F \rightarrow$ **avoidance** The instrumental conditioning of avoidance to the feedback stimulus, s_F

xii *Abbreviations*

s_F → approach The counterconditioning of approach to the feedback stimulus, s_F; the mechanism of learned persistence

S^HR Hull's designation for habit: strength of association between stimulus and response

D Hull's generalized, undifferentiated drive; strength of the nonassociative component of behavior derived from primary and secondary need states

K Hull's equivalent of generalized drive, derived from tertiary need states; the nonassociative component of incentive motivation

Reward-schedule effects

PA Patterned alternation: a discrimination based on the single alternation of rewarded and nonrewarded trials, usually in a runway

PRAE Partial-reinforcement acquisition effect: a paradoxical effect in which speeds are greater in the early segments of a runway and slower in the goal segment under partial-reinforcement than under continuous-reinforcement conditions

PREE Partial-reinforcement extinction effect: paradoxical greater response persistence (slower extinction) following partial-reinforcement than continuous-reinforcement acquisition

VMREE Variable magnitude of reinforcement extinction effect: paradoxical greater persistence following intermittent high and low magnitudes of reinforcement than following continuous high magnitudes of reinforcement

PDREE Partial delay of reinforcement extinction effect: paradoxical greater persistence if, in acquisition, all responses are rewarded but some rewards are delayed than if all responses are rewarded immediately

MREE Magnitude of reinforcement extinction effect: faster extinction (less persistence) after large than after small magnitudes of reward in acquisition

OEE Overtraining extinction effect: faster extinction after a very large number of acquisition trials than after a smaller number

SNC Successive negative contrast: lower terminal levels of performance to a goal if magnitude of reward is shifted from higher to lower levels than if the lower level of reward is maintained throughout acquisition

DNC Discontinuously negatively correlated reinforcement: a schedule in which the animal is required to take time (e.g., 5 seconds) to complete an appetitive-approach response that normally can be completed much faster (when speed and reward/nonreward are not correlated); a condition in which, in order to take time, each animal learns an idiosyncratic ritualized response that is reinforced on a partially reinforced schedule, the percentage of reinforcement depending on the time cutoff point

1 Introduction: reward-schedule effects and dispositional learning

In the history of the scientific study of learning and memory, a number of terms have been used to describe abstract experimental paradigms that purport to represent different kinds of learning and to reflect the distinct associative mechanisms that underlie them. We will have occasion in this book to refer to several of these experimental paradigms, but the most inclusive ones are classical (Pavlovian) conditioning and instrumental (Thorndikian) learning. As we shall see, these two paradigms are central to understanding the phenomena I call the *reward-schedule effects*. These phenomena of learning and memory depend on a variety of sequences of reward and nonreward, and are the basis for a family of generalizations known as frustration theory.

My major thesis, in general terms, is that inherent in such reward schedules is the buildup of *primary frustration,* defined simply as a temporary state that results when a response is nonreinforced (or nonrewarded in more neutral language in the appetitive case) in the presence of a reward expectancy; that this temporally labile state of frustration is subject to Pavlovian (or classical) conditioning, which is to say that a learned or anticipatory form can be elicited by an originally indifferent or neutral cue; that this conditional form of frustration, like other learned states, is therefore permanent, at least for that situation; and that, together, the primary (unlearned) form and the secondary (learned) form can account for a number of important processes in the dynamics of instrumental behavior. These can be summarized by four descriptive behavioral concepts: invigoration, suppression, persistence, and regression, and these four concepts define what I regard as the endpoints of *dispositional learning*.

The term *dispositional learning* is borrowed from, and bears a strong resemblance to, the term *dispositional memory,* proposed by Thomas (1984), who differentiates it from *representational memory,* as follows:

Dispositional memory is most generally characterized by a memory-indicating discrimination that is made with regard to crucial sensory events present at the organism's sensorium at the time of choice. The concept says nothing about the distinction between classical and instrumental conditioning. In both kinds of conditioning, past experiences with the conditioned stimulus (CS) and unconditioned stimulus (UCS) have allowed the organism to extract from its continuous flow of sensory inputs a "correlation" between a cue stimulus and its contingent consequences, stimulus–stimulus (S–S) or stimulus–response (S–R).

Such "learnings" are usually slow, require many repetitions (trials), and are built up gradually through cumulative memories of past contingent relationships. One supposes (speculatively) that such memory capabilities are very old in evolution – probably as old as a synaptic nervous system that permits "optional" transmission (Sherrington's term). Even lowly invertebrates with some degree of plasticity in their behavior can habituate (i.e., eliminate responses to nonnoxious stimuli) when they are repeated frequently: For example, the sea cucumber can "learn and remember" *not* to withdraw its siphon in response to a light touch! (pp. 374–5)

The Random House dictionary defines dispositions, in my sense, as "aspects and habits of mind that one displays over a long time." A synonym provided in most dictionaries is *temperament,* and references are always made to *personality.* An important mechanism in what I have called dispositional learning is in the interplay between rewards and frustrative nonrewards, particularly when these two consequences result from what is essentially the same behavior. Added to what may be the effects of inherited behavioral tendencies, dispositional learning, as I define it, determines the general, long-term temperamental characteristics of individuals, such as tendencies to approach and to avoid, to persist or to desist, to be aggressive or defensive, to overreact or not to overreact, and to alter or not to alter the direction and/or intensity of behavior when rewards are withheld or reduced. As we shall see in the chapters that follow, in the context of a number of the reward-schedule effects exemplars of dispositional learning occur in all mammalian species that have been studied, including humans.

Finally, there is a feature inherent in much of dispositional learning that is implicit in Thomas's definition and my own and is more the focus in the recent work of neuropsychologists of human memory. It has taken a number of forms; however, the common thread that runs among them is the recognition that there are two fundamental kinds of memory that depend on different kinds of "encoding," that is to say, on the different ways in which these memories are formed or learned. Table 1.1, reproduced from a book by Squire (1987), lists a number of the distinctions, including Thomas's, between these two kinds of memory. The first five items, among which there is little if any difference, are fact versus skill, declarative versus procedural, memory versus habit, explicit versus implicit, knowing that versus knowing how. Among these distinctions, the common theme is that the first item in each case refers to the encoding in memory of an event or episode, usually on the basis of a single exposure of which the "learner" is fully, cognitively aware – memory with record, in terms of the distinction in the ninth pair in the table. The second item in each case refers to a kind of memory that is the product of dispositional learning; that is to say, it is the product of learning requiring many exposures or trials and in which cognitive awareness is not a dominant characteristic – memory without record, in the words of the second part of the ninth pair in the table.

In making the memory versus habit, or "knowing that" versus "knowing

Table 1.1. *Two kinds of memory*

Fact memory	Skill memory
Declarative	Procedural
Memory	Habit
Explicit	Implicit
Knowing that	Knowing how
Cognitive mediation	Semantic
Conscious recollection	Skills
Elaboration	Integration
Memory with record	Memory without record
Autobiographical memory	Perceptual memory
Representational memory	Dispositional memory
Vertical association	Horizontal association
Locale	Taxon
Episodic	Semantic
Working	Reference

Source: Squire (1987, p. 169).

how," distinction, Mishkin and Petri (1984) actually characterize the left-hand item in each case as Tolmanian and the right-hand item as Hullian, a distinction, presumably, between Tolman's cognitive maps and Hull's habits. A further characterization of the metatheoretical orientation of this book, then, is that, dealing as it does with dispositional learning, it is very much more to the right-hand side than the left-hand side of the items in Table 1.1; it is much more a model of learning that is implicit and without awareness than of encoding into memory based on single cognition-engaging episodes.

Behaviorism and neobehaviorism

It is important to begin this book with some statement of metatheoretical position because, in the face of the "cognitive revolution" that has occurred in psychology, and even in the psychology of animal learning from which frustration theory is derived, the stimulus–response (S–R) framework in which the work of this book is cast will seem not to have changed much in some 30 years. Although the theorizing is about mechanisms like anticipation, expectancy, and memory, which can obviously be characterized as cognitive, the constructs are those associated with neobehaviorism. I have argued (Amsel, 1989) that this approach is at the same time more analytic and more constraining than the more mentalistic cognitive approach, and that a little constraint goes a long way in theorizing about behavior and its determinants. In light of the neuroscientific emphasis of our recent work, the position taken in this book can be characterized as a kind of "liberalized" S–R neobehaviorism, the term N. E. Miller (1959) used to characterize his brand of S–R psychology. How does this differ from other forms of behaviorism?

There have been several reasonably distinct versions of behaviorism. The seminal one is, of course, Watson's early (1913/1919) version, which was modified and hardened in 1925, to its detriment, in the opinion of many. There followed Tolman's purposive or molar behaviorism of the 1920s and 1930s; the operational behaviorism of the 1930s and 1940s – that of Pratt, Stevens, early Skinner, Hull, Bergmann, Spence, and others – which appears to have been anticipated by Watson in 1919 and in Lashley's later articles; and, most recently, the radical (descriptive) behaviorism of later Skinner, which can be dated from the famous article "Are Theories of Learning Necessary?" (1950). This last version, from which Skinner seems to depart in some of his more recent writings, most resembles the later, more doctrinaire Watsonian behaviorism. (And, like Watson's, Skinner's earlier position, exemplified by his *Behavior of Organisms* [1938], was not as radical a behaviorism as his later one.) In many respects, the neobehaviorism of Hull – with its emphasis on learned and unlearned S–R associations, adaptiveness of behavior, and concepts derived from physiology; its admonitions to guard against subjectivism and anthropomorphism; and its preference for S–R analyses of such terms as *knowledge, anxiety, purpose,* and *anticipation* – is the kind of behaviorism I hold to be most like my own.

If there is nowadays a focus in the disagreement between the positions of the neobehaviorists and the animal cognitivists who work in animal learning, it is that the former invent constructs to explain behavior, whereas for the latter behavior is in itself unimportant except as a "window on the mind." If we set aside the radical behaviorism of Skinner's (1950) article, then the disagreement between neobehaviorism and what I call "animal cognitivism" is, on closer examination, not only between the S–R and the cognitive explanatory languages but also to a large extent between the experimental topics that each addresses (Amsel, 1989). These differences are discussed in the next sections so as to define further the characteristics of my own approach to theorizing and the subject matter it encompasses.

Neobehaviorism: habits, needs, incentives, and behavior

The first shots in the "cognitive revolution" in psychology, some 30 years ago, were fired by people who wanted to draw firmer lines between investigations involving humans and those involving animals as subjects. The neocognitivists were, of course, concerned with processes they regarded as primarily human – remembering and forgetting of verbal material, processing of information, perception, simulation of human problem solving by computers, formation and identification of concepts, reading, language, and so on. The neocognitivists of the 1960s seemed not to understand that many if not most of the investigators who were identified as learning theorists, and who worked with animal models, were less interested in the cognitive and intellectual abilities of the animals they chose to study than in basic learning processes such as occur in Pavlovian conditioning and instrumental (Thorndikian) learning, and in basic motivational or need systems. In short, they were involved in the formation of animal models of the perhaps more primitive learning, ontogenetically and phylogenetically, that operates not only in people, but in animals as well. Both Hull and Tolman, two of the great learning theorists of the 1930s and 1940s, introduced into psychology theoretical systems the main concern of which was the identification of the factors that contributed to the formation of habits and goal expectancies, in their interactions with the complementary concepts of need, drive, and demand. Hull, for example, was tremendously preoccupied with the mechanisms of adaptation and survival (Amsel & Rashotte, 1984). This preoccupation can be found even in his most formal theorizing – for example, in *Principles of Behavior* (1943) – and his earlier seminal papers showed this interest even more clearly, as did his last book, *A Behavior System* (1952).

The next generation of learning theorists – H. F. Harlow, N. E. Miller, O. H. Mowrer, B. F. Skinner, and K. W. Spence – carried on a tradition of experimental research and theory in which the aim was to understand the basic associative and motivational processes of humans by studying and understanding these same processes in lower animals (or so, at least, it has seemed to me). True, Skinner's work with pigeons has tremendously increased our understanding of the ways of the pigeon; but obviously Skinner has been interested in pigeons more abstractly, as an animal model for the experimental analysis of behavior, and not in the manner in which an ethologist might be interested in pigeons. Harlow's later work was primarily with monkeys and other nonhuman primates, and this work has contributed greatly to the knowledge we have about emotional-affectional systems of primates. But, clearly, this was not Harlow's only, or even primary, intention; the purpose of Harlow's work was to understand affectional (and other) systems in people. As I suggested about three decades ago (Amsel, 1961), at around the time the cognitive revolution began, one of the characteristics of many if not most of the learning theorists who had worked with animals in the preceding quarter-century (or more) was that they had not been interested in the animals they were studying in the way a naturalist is, but rather had been interested in these animals as "preparations" from which it would be possible to develop hypotheses about associative and motivational-emotional processes of higher animals in general and humans in particular. It was for this reason that applications of learning theory to personality and psychopathology came more from research in animals than from work in human learning (e.g., Dollard &

Miller, 1950). These theorists of learning whose subjects were animals were interested in the level of behavioral analysis that I earlier referred to as my own interest in "dispositional learning."

Up to about the late 1960s, it would have been fair to say that learning theorists who based their ideas on work with animals (usually, but not exclusively, the pigeon, rat, and monkey) were at least as interested in the conative and affective aspects of behavior as in the cognitive aspect, *and they were seldom interested in the pigeon, rat, or monkey.* They tended to be functionalist in outlook and to be influenced by Darwin, Pavlov, and Freud in their emphasis on adaptiveness, motivation and reinforcement, and the *nonintentional nature of an important part of learning.* In contrast, the neocognitivists showed very little interest in the conative or affective categories – and they still show very little interest in them.

Animal cognitivism: behavior as a window on the mind

Unlike their natural predecessor, Tolman, the neocognitivists who work with animals appear to regard behavior simply as a vehicle for understanding thinking in animals. Two of the interesting features of this approach, to which I do not subscribe, are that behavior per se is not considered important and that, again unlike Tolman, emotion and motivation are virtually ignored (see *Animal Cognition* by Roitblat, Bever, & Terrace, 1984). The fathers of our troubled science, following more or less standard philosophical practice, were comfortable with a definition of psychology that included the cognitive, the conative, and the affective. Now it appears that, for most psychologists, even for the *animal* cognitivists, only the cognitive remains. This direction, taken by the field of animal learning, is not taken in this book.

As we shall see, frustration theory, and the experiments from which it was derived, and which were deduced as particulars from it, is made up of at least as many motivational as associative (or cognitive) constructs. These are an amalgam of the basic Hullian motivational theory and the related experimental and theoretical work of such theorists as K. W. Spence, N. E. Miller, and O. H. Mowrer. *The Motivation of Behavior* by J. S. Brown (1961) is perhaps the most comprehensive treatment of this tradition in motivation theory. The interpretation and use in the present book of this kind of motivation theory is the subject of Chapter 2.

Developmental psychobiology and levels of functioning in dispositional learning

One direction in which animal-based learning theory has moved (the other being toward animal cognition) is toward behavioral neuroscience or psychobiology, and it will be obvious in the later chapters of this book that this is the direction in which some of the work on dispositional learning appears to have gone. If we add a comparative (phylogenetic) influence and a developmental (ontogenetic) dimension to this psychobiological approach, we have a summary description of the way my own thinking and the work in my laboratory have gone in recent years (see Chapters 7 and 8).

If one studies learning and memory from an ontogenetic or evolutionary perspective, or from the perspective of brain-damaged animals and people, it becomes relatively easy to accept the view that there are "lower" and "higher" levels of processing in learning and memory and that both (or all) levels reside in the intact human adult.

Using simple behavioral techniques, we have been able to provide developmental data showing the order of emergence of a number of reward-schedule effects in ontogeny, and the sequence of emergence of these effects has been taken to represent stages in the development of simple memory and dispositional learning. In Chapter 7 we shall see that, at 11 days of age, rat pups can learn to discriminate the presence or absence of a reward by approaching on odd-numbered, rewarded trials and suppressing approach on even-numbered, nonrewarded trials on a single-alternating reward–nonreward schedule. They can do this if the time interval between rewards and nonrewards is brief (8 seconds) but not if it is longer (30 to 60 seconds). As animals grow older, their developing memory allows them to span longer and longer intervals of time. By 14 days of age, pups on a randomly ordered training schedule of intermittent rewards and nonrewards are relatively more persistent in approach to a goal than pups rewarded on all trials when reward is no longer present at the goal (experimental extinction). At 18 days of age, extinction is faster, with concomitant heightened frustration, after training with large rather than with small rewards. At about 25 days, shifting from a large to a small reward produces an emotionally related depression in performance (called successive negative contrast, SNC) relative to animals that have received the small reward throughout training.

We know that, in adult rats, these behavioral effects and others depend on the integrity of the limbic system, a region of the brain underlying and bordering the neocortex that includes the hippocampal formation, the septum, and the amygdala. So it is reasonable to examine (as we do in Chapter 8) the parallel development of these aspects of brain function and the behavior they appear to control. We shall see that a portion of the limbic system that appears to be important in learning and memory, the septo-hippocampal system, shows a particularly rapid rate of development in late prenatal and early postnatal stages. We can therefore ask such questions as: What developmental changes in this brain system are related to the demonstrated stages of development of learning and memory? How do lesions in this system affect the emergence of the behavioral effect from which learning and memory are inferred? And if lesions eliminate effects that occur early in infancy, is there a recovery of these functions later in development? Conversely, do the kinds of training that are required for the earliest manifestation of these effects on learning and memory affect neuronal plasticity? Do they induce acceleration or retardation in some of the structural landmarks of brain development – cell size and number, dendritic branching, neurotransmitter release, synapse formation – and how do these changes in turn relate to an accelerated or a retarded appearance of the later behavioral effects? How do teratogenic treatments, such as exposure to alcohol in utero by the feeding of alcohol to pregnant rats, or the exposure of infant rats to X-irradiation of particular parts of the brain affect such structural changes in their offspring, and how do these brain changes affect the ages of appearance of normal capacity to learn and remember? It should be clear, however, that none of these questions could be asked until the theory and research on what I have called dispositional learning had reached a certain stage of maturity.

This book deals with the psychodynamics and developmental psychobiology of dispositional learning from a neobehavioristic orientation: Frustration theory is an attempt to explain and to predict – to integrate – a number of phenomena of instrumental learning that have in common their dependence on the properties of frustrative reward, which in turn depends on conditioned expectancies and the failure of their confirmation. It is an animal model that may lead to a more general understanding of the development in humans of such specific problems as abnormalities in emotional affect, attention deficits with hyperactivity, deficits in the capacity to suppress behavior and form discriminations, and deficits or excesses of persistent or perseverative behavior.

As we discuss the characteristics of dispositional learning and the various reward schedules under which it occurs, it should become plain that the environmental control of this kind of learning is both external and internal to the learner, and that in most of the cases to be considered, the internal environment is much the more important of the two. The exception to this assertion is the case in which there are differential external stimuli controlling differential responding, the case of discrimination learning; in the majority of cases we consider, however, there are no differential external cues at the point of the goal-oriented responding. The interoceptive control that then comes to the fore arises, in these reward-schedule cases, entirely out of the nature of the rewards, nonrewards, reduced rewards, and delayed rewards that occur at the termination of an instrumental response. These are of two kinds. The first is a direct consequence of what has happened on the last learning trial; it is a direct carryover from Trial N to Trial $N + 1$: a short- or intermediate-term trace or memory of the last learning experience or trial. This has been the interoceptive-stimulus mechanism featured in many of Capaldi's (e.g., 1967) explanations of reward-schedule learning, and in his view it can account even for relatively long-term memorial consequences of reinforcement and nonreinforcement. In this case the reinforcer or nonreinforcer (the reward or nonreward) on one trial gives rise to the internal stimulation (the carried-over stimulus trace or the memory of the previous trial) that controls the behavior on the next trial.

The second kind of guiding internal stimulation also depends on the goal event, the reward or the nonreward, but in a different way. In this case the internal stimuli that guide behavior arise out of classically conditioned (Pavlovian) internal responses based on the goal events of all preceding trials. These goal events serve as the unconditioned stimuli (e.g., food) and unconditioned responses (eating) that are the basis for the Pavlovian conditioned responses. These are the kinds of internal responses that Hull (1931) called *fractional anticipatory goal responses* (r_G–s_G). They are essentially internal responses whose function is to produce internal feedback stimulation ("pure stimulus acts") that then control ongoing behavior. The kind of theorizing that has involved such interactions has been called *conditioning model theory* (Lachman, 1960), a term I have adopted and used in my own theorizing (Amsel, 1962). Others have investigated such interactions under the headings of *two-process theory* (Gray, 1975; Rescorla & Solomon, 1967) and the *second learning process* (Trapold & Overmier, 1972). In this book, the notation I employ is the Hullian one, but I make no distinctions and see no important differences between this one and the others.

Summary of the metatheory

The metatheoretical position of this book can be summarized as follows: It is a neobehaviorism involving stimuli and responses at both the descriptive and explanatory levels, the S–R theoretical language being a "liberalized" one in N. E. Miller's (1959) meaning. There is no attempt to avoid coordination of such constructs with physiological states and processes. In this sense, the position is not fundamentally different from early Watson or even from the Skinner of *Behavior of Organisms* (1938), in which intermediary constructs like drive, emotion, and reflex reserve were acceptable and the strength of this latter state was expressed in a series of "dynamic laws" that were clearly and admittedly Sherringtonian. The present position, in this respect, is also close to Hull's in *Principles of Behavior,* in which such allusions to physiology and uses of physiological-sounding constructs (e.g., "afferent neural interaction," "receptor–effector connection") can be found on every other page. Hull stated his position on the role of neurophysiology in the study of behavior as follows:

There can hardly be any doubt that a theory of molar behavior founded upon an adequate knowledge of both molecular and molar principles would in general be more satisfactory than one founded upon molar considerations alone. But here again the history of physical science is suggestive. Owing to the fact that Galileo and Newton carried out their molar investigations, the world has had the use of a theory which was in very close approximation to observations at the molar level for nearly three hundred years before the development of the molecular science of modern relativity and quantum theory. Moreover, it is to be remembered that science proceeds by a series of successive approximations; it may very well be that had Newton's system not been worked out when it was there would have been no Einstein and no Planck, no relativity and no quantum theory at all. It is conceivable that the elaboration of a systematic science of behavior at a molar level may aid in the development of an adequate neurophysiology and thus lead in the end to a truly molecular theory of behavior firmly based on physiology. (1943, p. 20)

In this regard, my position is more like Hull's and less like Spence's (e.g., 1956), which, in line with his well-known disagreements with Hull on "physiologizing," tended to avoid physiological-sounding constructs (see Amsel & Rashotte, 1984). Indeed, in what may have been the last statement of his position on this matter, in a "Pavlovian Conference on Higher Nervous Activity," Spence (1961) not only repeated and confirmed his commitment to the discovery of a "set of abstract *theoretical concepts*" (p. 1188, italics his), but pointed out, with some justification, that on close examination Pavlov's concepts, such as internal inhibition, were of the same sort as his own and did not refer to any identifiable neurophysiology.

Another, and perhaps more important, aspect of the metatheoretical orientation of this book is that, while it is Hull–Spence in lineage, it is clearly more in the spirit of the Hull and Spence of the 1930s than of their more formalized, sometimes mathematical treatments. For example, it borrows a great deal more from Hull's earlier theoretical papers (Amsel & Rashotte, 1984) than from his more formalized treatment of learning in *Principles,* and more from Spence's famous papers on discrimination learning (1936, 1937, 1940) than from his more formal treatments (1956). As in Hull's early papers and in the work of those who followed in this tradition (e.g., Berlyne's [1960/1964] analyses of curiosity and thinking; Miller's [1944/1948] analyses of conflict and displacement; Mowrer's [1939] analysis of anxiety), terms such as *anticipation* and *expectancy* are not avoided; but when they enter into theoretical statements, they are reduced to their meaning as classically conditioned responses and to the S–R notation, so that they can find integration with other constructs of the theory.

Outline of the book

There are ten chapters in this book. Chapter 2 reviews the notation of the motivational and associative constructs of S–R neobehaviorism in which

Table 1.2. *Steps in the psychobiological study of related behavioral effects*

1. Observe and describe a number of apparently related behavioral effects.
2. Develop a conceptualization of these effects in terms of empirical-construct theory.
3. Study these effects ontogenetically for their presence or absence at various developmental stages, and for the order of their first appearance.
4. Study these effects for their presence or absence in relation to the presence or absence of portions of, or activities of, their presumed neural substrate.
5. Relate the order of appearance of the effects to the developing neural substrate.
6. Modify empirical-construct theory on the basis of findings from 4 and 5.

the content of this book is cast. Chapter 3 introduces frustration theory and provides an overview of its explanatory scope. Chapter 4 sets the stage for the later chapters on development by providing evidence from a number of experiments that what is learned during a specific series of intermittent rewards and frustrations may become a quite general and permanent disposition to control the animal's behavior much later in life. As we shall see, perhaps the crucial mechanism in this dispositional learning can be conceptualized as the acquisition of conditioned or anticipated frustration and its association with some specific mode of responding – persistent or desistant, aggressive or regressive. Chapter 5 provides a frustration theory account of mechanisms involved in differential external-cue-based responding: How are discriminations formed? How can their formation be facilitated or retarded by prior dispositional learning experience? In Chapter 6 we summarize what has gone before, placing special emphasis on alternative analyses of theories of resistance to extinction following intermittent reinforcement. In Chapter 7, in which we move to a developmental analysis of dispositional learning, beginning with the infant animal, the major focus is on the sequence of appearance in ontogeny of a number of reward-schedule effects. This leads quite naturally, in Chapter 8, to a discussion of the parallel development of the putative neuroanatomical and neurophysiological substrates of this sequence of behavioral effects.

Some years ago, these relationships between brain and behavior, in the context of development, suggested the strategy of research that is summarized in Table 1.2 (revised from Amsel & Stanton, 1980). In the chapters that follow it should become clear that, not entirely by design, we have appeared to follow this set of guiding principles. In Chapter 9, a reprise of the content of this book, we examine the manner and extent to which the theme and content of Table 1.2 have been developed in the previous eight chapters. And in Chapter 10, after a brief recapitulation of applications to "human indispositions," I conclude with an application to attention deficit–hyperactivity disorders in children.

2 Motivational and associative mechanisms of behavior

In Chapter 1, the methodological and metatheoretical positions on learning theory of this book were outlined. The present chapter is a more specific treatment of these positions, particularly with respect to the role of the interaction of motivational and associative mechanisms. This interaction is a critical feature of dispositional learning and the four characteristics of behavior it encompasses: invigoration, suppression, persistence, and regression.

A brief history of the motivation concept

Throughout history there have been a variety of conceptions of motivation (animistic, religious, rationalistic, teleological-adaptive, instinctive), all of which are, to some extent, still a part of our everyday explanatory language. However, scientific conceptions of motivation have moved increasingly toward the identification of specific mechanisms.

One of the most important movements in the direction of a more mechanistic definition of motivation emerged from Darwin's theory of natural selection – the theory of evolution. Darwin's *Origin of Species* (1859) emphasized the role of adaptation and survival as factors of significance in behavior. As a biologically based conception of an important determinant of behavior, it was, as we shall see, the forerunner of learning theories such as C. L. Hull's (1943), in which adaptation played a principal role.

After Darwin, it seemed reasonable to make a distinction among proximal, developmental or historical, and evolutionary mechanisms determining behavior. Proximal events, in this trilogy, are those that immediately precede behavior, the instigating stimulus. Developmental or historical mechanisms include those we normally call learning, the residue that is laid down in the organism as a function of its own previous experience. And, of course, evolutionary mechanisms refer to what is laid down in the gene pool, a quite different kind of historical explanation of behavior. The history of explanations of behavior is written at these three levels.

In what sense do the past, present, and future determine a present behavior? It is important in the context of a theory such as the present one, which employs constructs connoting anticipation, to answer this question.

The past, of course, determines behavior in terms of some kind of trace

– memory, habit – laid down and existing in some form in the nervous system. The present determines behavior as an immediate instigating effect. In scientific terms, the future cannot determine present behavior. However, the future determines the present in the same way the past determines the present. Anticipation of the future can determine present behavior, and anticipation or expectation is defined by conditioning, a kind of learning – a past event: As a present event controlled by the past, in scientific terms, anticipation has no teleological connotation. This is important as we examine the next major attempt to introduce motivational conceptions, which was the advent of the use of the terms *instinct* and *propensity,* early mechanistic explanations that were, in some respects, influenced by Darwinism.

Perhaps the most frequently cited theorists of motivation, following Darwin, were the social psychologist William McDougall (1871–1938) and the psychoanalyst Sigmund Freud (1856–1939). The instinct doctrine flowered particularly in the writings of McDougall, who proposed an instinct to account for virtually every kind of behavior; he juxtaposed long lists of behaviors and long lists of instincts that explained those behaviors. An influential instance of the instinct doctrine could also be found in Freud's theorizing, the life and death instincts, among others, playing a prominent part. The use of instincts as explanatory rather than as descriptive terms came to be criticized as an example of the fallacy of *hypostatization* – nominal explanation or explaining by naming. Explaining things by naming them – or renaming them – was seen to offer no particular theoretical advantage, since there were as many degrees of freedom in the explanatory language as there were particulars to be explained.

A major advance in the move toward mechanism in motivation theory was the homeostatic approach, which was in some ways a derivative of the earlier approaches but was, in major respects, a result of increased knowledge about "The wisdom of the body," the title of W. B. Cannon's (1932) important book. Major contributors to the homeostatic approach were Claude Bernard (1813–1878), W. B. Cannon (1871–1945), and Curt Richter (1895–1988).

The homeostatic approach brings with it such concepts as need, drive, and arousal, circadian rhythms, and diurnal effects. It is, in general, a physiological approach to the study of motivation – or at least to its basic mechanisms. It is not only mechanistic; it is deterministic, and it is organismic. It emphasizes antecedent conditions, such as deprivation and noxious stimulation, and it leads naturally to a distinction, within the motivational construct, between nonassociative and associative mechanisms determining behavior. This distinction, which became important in a number of learning theories, is not found in the theoretical writings of most recent animal cognitivists. Like a thermostat, a homeostat is constantly being adjusted to a particular set point, and returning it to that set point involves some kind of behavior. In some reinforcement theories (e.g., that of Berlyne, 1960, 1967), the reinstatement of some "optimal" homeostatic condition defines the concept of reinforcement and the optimal conditions for learning.

The survival of older nonmechanistic explanations

As I asserted earlier, to some extent in the scientific study of behavior, and to a major extent in its nonscientific explanations, all of the concepts in the history of the motivation idea are still with us: animism, religious conceptions, rationalism and idealism, evolutionary adaptive thinking, the instinct doctrine, homeostatic determinism. The homunculus still exists in some forms: Early and late, it has existed in Freud's censor, which guards the entry to the conscious. A more recent example is Chomsky's (1959) notion of linguistic competence. Rationalism and idealism are very much alive in cognitive psychology, but in this case there appears to be very little differentiation between the motivational and the nonmotivational: Cognitive psychology is by definition not conative or affective, and the explanations are heavily rationalistic.

There is still a great deal of nominal explanation. In social psychology, for example, there is achievement motivation, which defines achievement. In animal learning psychology, for example, there have been the curiosity and exploratory drives that explain the tendencies to be curious and to explore. The manipulation drive explains manipulatory behavior. And more recently, the concepts of learned helplessness, learned safety, learned laziness, and learned indolence have been used to explain helplessness, laziness, and so on. This kind of thinking does not die or even fade away; it just takes changing forms.

A newer quasi-motivational explanation is Seligman's (1970) concept of preparedness. If animals or humans are prepared to do or learn things, they do or learn them easily; if they are unprepared, they do or learn with more difficulty; if contraprepared, the doing or learning is even harder. This is a form of evolutionary-adaptive thinking, which is substituted for what might be regarded as a simpler approach: identifying species-typical behaviors and asserting that, as far as is known, they are characteristics of the species at given stages of development under normal conditions. Hull (1943) introduced a concept of unlearned habit ($_sU_R$) that defined such behaviors, as did Tolman (e.g., 1938). This is the way these two great systematists acknowledged "preparedness" – that not everything organisms do requires the same degree of learning, and that some behaviors are more inborn than others, depending on the species.

It is not unreasonable to refer to transitional, sensitive, or critical periods in ontogeny for the emergence of certain behaviors and then to relate these behaviors to some kind of explanatory network that involves learning and motivation. But if the word *drive* refers to some kind of inherited predisposition, and nothing more, it is obviously not as tightly defined an explanatory concept as it could be if it were related to specific defining events in the life of the organism.

Associative and nonassociative factors in behavior: a brief history

The distinction between associative and nonassociative factors, as we saw in Chapter 1, is an undervalued currency in these days of cognitive psychology. There was a time, starting in about the late 1920s, when a group of theorists found themselves in general agreement that behavior was a joint function of associative and nonassociative variables. And this idea took a variety of forms. Woodworth was one of the first to make this explicit, in his distinction between *mechanism* and *drive*. This was a machine analogy: The engine and the steering mechanism determined the manner and direction the machine would move; the drive was the fuel. Tolman's distinction was between *cognitions* and *demands*. Hull's distinction, between *habit-strength* ($_sH_R$) and *drive* (D), added the explicit assumption that the relationship was multiplicative; it was not simply that both factors were necessary for behavior, but that some of both were necessary: If either factor was zero, the result was zero. For Lewin (1943), the associative factor was the *psychological environment, the nonassociative, the person*.

Other versions of this distinction dealt with "emotion" rather than "response" or "behavior." Social psychologists Schachter and Singer (1962) held emotion to be a function of the *situation as perceived* (associative) and *arousal* (nonassociative). The Pavlovian Simonov's (1969) theory of emotion made it a function of *information necessary minus information available* (associative) and *need* (nonassociative). These two versions, like Hull's, involved multiplicative relationships. In Bindra's (1969) case the two factors are *selective attention* and the *central motive state* (CMS), although the CMS is not, as we shall see, a purely nonassociative construct.

All of these ideas involve the same two basic distinctions. First, there is the distinction between behavior, on the one hand, and the determinants of behavior, on the other. And the determinants are of two sorts: a learning or associative factor and a motivational (need or drive or demand) factor. Table 2.1 identifies the theorists and provides the general equations defining their basic positions, which in every case, as we have seen, makes behavior a function of associative and nonassociative variables. What we also have in each of these cases is an explicit acknowledgment of what came to be known as the "learning–performance distinction," a very im-

Table 2.1. *The learning–performance distinction*

Woodworth	Behavior = f(Mechanism, Drive)
Tolman	Behavior = f(Cognitions, Demands)
Hull–Spence	Response strength = $f[\ (_sH_R \times (D + K) - I_R - {_sO_R}]$
Lewin	Behavior = f(Psychological environment, Person)
Schachter	Emotion = f(Situation as perceived $\times$ Arousal)
Simonov	Emotion = $-$Need($I_n - I_a$)
Bindra	Response = f(Selective attention, CMS)

portant distinction in frustration theory. The hypothetical associative factor (habit, information, etc.) is learning; the behavior or response consequent upon the combined action of the associative and nonassociative factors is performance.

It is important to understand that in each of the cases outlined in Table 2.1, and in the present context, the nonassociative factor is in the first instance a general or a generalized one. For example, in Schachter's theory, in which emotion is conceived to be a joint function of the situation as perceived and arousal, and in which in a social-psychological experiment the situation was controlled by, say, the behavior of a "confederate," or by a cover story of some kind, the arousal was defined by a drug, usually adrenaline, which was employed to induce nonspecific arousal. The nature of the arousal (the "emotion") was interpreted by the subject in terms of the situation. So if a fearful situation was staged, the arousal was interpreted as fear. If the situation was perceived as frustrating, the arousal might be interpreted as anger, and so on. As we shall shortly see, to the extent that an arousal factor has specific stimulus or cue characteristics, it can enter into associative relationships with behavior, providing motivation with an associative as well as a nonassociative component.

An interesting personal sidelight, in this connection, is that the first time I heard about Simonov's theory of emotion was in 1967, in Moscow, where he told me about it and jotted down the equation shown in Table 2.1. I remarked that his equation resembled the Hullian multiplicative function (of which he appeared to be unaware), with emotion substituted for response. In his case, E is emotion and N is need, which is negative in the equation, I_n is information necessary, and I_a is information available. And this statement about information, of course, is the associative factor. Need is the basic generalized motivational condition. The implications are these: If need is zero, emotion is necessarily zero (which is just like the Hullian formula). If I_n equals I_a, then, of course, emotion is zero as well. So if you have the information you need, no emotion is involved. If I_a is greater than I_n, the information available is more than needed, and emotion is positive. And if the information needed is less than the information available, E is negative.

The point of the examples provided in this section is that the same ideas emerge in a variety of theoretical, and even national, contexts; and they emerge in the contexts of explaining behavior, including emotional behavior. Until recently, then, the ascription of behavior to the factors of learning and drive (or need) had been quite general. In this book I adopt the Hullian version of this general explanatory strategy.

Cue and drive function in motivation

Our next task is to break down the motivational concept into the components we briefly identified, to deal with *interpretations* of the motivational concept, particularly as they have been employed in the context of learning theory. Our earlier treatment involved the nonassociative mechanism in behavior, the general arousal or drive-strength function. A more complete treatment of motivation includes an associative aspect, a consideration of the cue function as distinguished from the drive function of motivation.

Hullian theories

We begin with Hull's (1943) own theory of drive (D) and cue (S_D) function (Figure 2.1). Hull's was a biologically based homeostatic theory of motivation, as were so many others in those days. In Hull's terms, deprivation or noxious stimulation or any other kind of instigation to departure from homeostasis produces a condition of need, which has two characteristics. It has the characteristic of contributing to the generalized, nonspecific pool of drive strength (D), which refers to the activating or arousal factor, much like Schachter's administration of adrenaline. But each need system also has connected with it a characteristic drive stimulus (S_D).

It is therefore the case, in Hullian theory, as in the present treatment, that the motivation construct has an associative as well as a nonassociative component; it can serve either an associative or a nonassociative function, the first through the drive stimulus (S_D), the second through generalized drive strength (D). The drive stimulus, like any other stimulus, can acquire associative control over behavior. If two or more needs exist simultaneously, all contribute to the pool of generalized drive strength (D). However, the contribution of the relevant need ($\overline{D}$) is greater than the contribution of the irrelevant ($\dot{D}$), and each retains its individuality by virtue of its own drive stimulus S_D, which can enter into associative relationships quite differentially. The generalized drive factor simply multiplies whatever habit is dominant at a given moment. This is the concept of the habit-family hierarchy (Hull, 1934), a hierarchy of associative strengths in

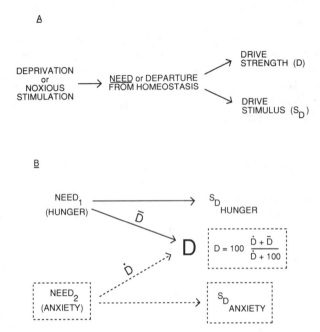

Figure 2.1. Hull's theory of drive and cue function in motivation. (A) Two motivational factors resulting from need. (B) Drive strength and drive stimuli resulting from relevant (1) and irrelevant (2) needs. The equation to the right reflects the assumption that $\bar{D}$, the contribution of the relevant need, carries more weight in the total drive strength (D) than $\dot{D}$, the contribution of the irrelevant need.

relation to a given compound of external and drive stimuli. The response that has the greatest "habit-loading" for that stimulus complex is activated by the total pool of existing drive strength. As we shall see, all of these conceptions in Hull's motivation theory are premises in frustration theory.

A view of cue and drive functions different from Hull's but in the same family is depicted in Figure 2.2. According to Miller and Dollard (1941), the cue–drive distinction depends on intensity. Any departure from homeostasis produces a drive stimulus; and when this stimulus reaches some superthreshold level of intensity, energizing drive strength is the result. This is depicted in the bottom portion of the figure in the case of the intensity of any internal or external stimulus. This version of cue and drive functions is different from Hull's in that, here, both functions do not emerge independently out of conditions of need.

The reticular activating system and arousal theories

In *The Organization of Behavior,* Hebb (1949) asserted that there was little support in neurophysiology for a concept of generalized drive. Others (e.g.,

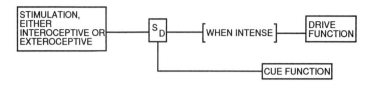

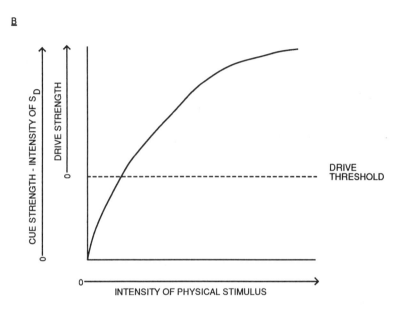

Figure 2.2. Miller and Dollard's (1941) view of cue and drive functions. (A) From Kimble (1961, p. 397). (B) Graphic representation of the hypothesis.

Bolles, 1958) wrote that there was nothing about the concept of generalized drive that increased our understanding of behavior. In all such cases, whether sole importance was attached to the firing of cell assemblies and phase sequences, as in Hebb's case, or to the hypothetical association mechanism, as in Bolles's case, the energizing or arousing aspects of the account were considered insignificant.

In the year Hebb's book appeared, an important paper was published by two neurophysiologists (Moruzzi & Magoun, 1949). The title of this paper was "Brain Stem Reticular Formation and Activation of the EEG," and it reported the identification of a nonspecific nerve network, the "reticular formation," that passed through the central core of the brain stem, the medulla, the pons, midbrain, hypothalamus, and thalamus, to the cortex, where it exerted a general activating or arousing function. It described

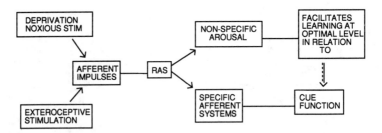

Figure 2.3. Schematic of arousal and the reticular activating system. (See Figure 2.4 for the details of the connection denoted by the dashed arrow.)

a more diffuse sensory pathway than in the more direct afferent sensory nerves. A diffuse network, a reticulum, it was first called the ARAS (ascending reticular activating system) and was later referred to as the RAS, because there appeared to be descending as well as ascending components. Activation of this pathway was said to cause increases in arousal, detected as desynchronization of the electroencephalogram (EEG). Severing of this pathway, but not the classical afferent pathway, caused permanent sleep and a sleeping EEG. As we have seen, the report of this alerting or arousal system emerged in the literature at the same time as Hebb's famous book. About five years later Hebb (1955) changed his position and acknowledged that the arousal system was very important in performance.

While recent evidence provides a more specific basis for brain stem origins of the changes in activation attributed to the RAS, implicating the monoaminergic nuclei (raphe, locus coeruleus) and the noradrenergic system (e.g., Aston-Jones & Bloom, 1981; Brodal, 1981), the RAS, as it was conceptualized in the 1950s, was as shown in the block diagram in Figure 2.3. According to this schema, deprivation or exteroceptive (including noxious) stimulation, particularly if intense, results in the activation of afferents in the reticulum and higher brain centers, which produce cortical activation and desynchronization of the EEG. These are related, in turn, to nonspecific arousal. It was hypothesized later that, not only did the RAS produce nonspecific arousal, it also generated specific afferent components – specific stimulus components that could be associated selectively with behavior. The correspondence of RAS thinking to Hull's behavioral constructs D and S_D was remarkable, but the emergence of these ideas led also to some divergences from Hull's theorizing, particularly as they related to the role of need and its reduction and to the concept of reinforcement in learning.

In arousal theory (Berlyne, 1960; Hebb, 1955) an intermediate level of activation is held to be optimal for learned performance. This differs in some respects from Hull's position, because Hull's formula is $R = f(E) = f(H \times D)$. And in that case, the stronger the drive or arousal, the

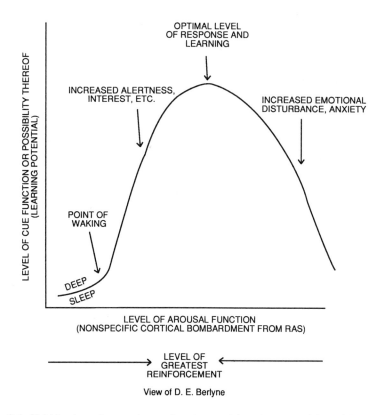

Figure 2.4. Hebb's view of arousal, cue function, and learning potential, and Berlyne's hypothesis of its relation to reinforcement and learning. (Adapted from Hebb, 1955.)

greater the excitatory potential (E) for the specific response. Of course, that does not mean that habit strength (H) for that same response is greater. As we shall see, increasing D may bring with it a changed complex of drive stimuli (S_Ds). Consequently, the habit-family hierarchy might also be changed, in which case a different response would become dominant. Whereas arousal theory was regarded as contradictory to aspects of Hull's position, it really was not, as we shall see later in some detail.

To return to Hebb's (1955) arousal function (Figure 2.4), the ordinate expresses level of cue function, or possibility thereof, which is the learning potential; the abscissa, level of arousal function. And marked off on Hebb's curve are the various states of performance related to learning potential and the level of arousal: from deep sleep to point of awakening, to increased alertness and interest, and so on to an optimal level of responding and learning; and then to increasing emotionality, and disturbance and anxiety, which would presumably result in a decreased efficiency in learning and performance. But the conceptualization of factors affecting increases and decreases in the efficiency of learning is a very general one: For every level of arousal, for every level of disturbance, different kinds of learned performance may emerge. So this statement about learned performance is unspecific; it refers to some hypothetical kind of learning that is affected differentially by the level of arousal. In a sense it is a restatement of the Yerkes–Dodson law, in this case with respect to the effects of general arousal rather than the effects of the strength of an external motivating stimulus, such as electric shock: The Yerkes–Dodson law holds that, if the strength of shock for an incorrect response is increased beyond an optimum, speed of learning is decreased rather than increased.

The specific contribution of Berlyne (1960, 1967), in this context, was to redefine *reinforcement* as a return to some optimal (intermediate) level of arousal from either a lower or higher level. This was a departure from Hull's position in the sense that, for Berlyne, the greatest level of reinforcement did not correspond to the greatest degree of drive or need reduction. In Berlyne's thinking, not only was drive reduction reinforcing; drive induction was also reinforcing (see also F. D. Sheffield, 1965): It could result from an increase in arousal if one was "bored," or a decrease if one was highly aroused. In homeostatic terms, Berlyne's interpretation of reinforcement was that homeostasis did not mean reducing the level of need to zero, but returning it to an optimal level, reducing or increasing arousal in relation to "arousal potential" (Berlyne, 1967). In other words, he set his homeostat at a different point than Hull's.

Purely associative theories of motivation

At the time that W. K. Estes (1958) developed his theory of motivation, his important work had been in what was called statistical learning theory. Because it was a probabilistic theory of association, it could not easily accommodate a generalized drivelike motivation concept. (This is a limitation of most mathematical models of learning, for example, the very influential Rescorla–Wagner [1972] model. Even in an extension of the Rescorla–Wagner model to include frustration-theoretic concepts [Daly & Daly, 1982], there is no explicit nonassociative variable.) When Estes decided to put motivation into his theory in the form of stimulus sampling, he adopted a theory based entirely on cue function: As we have seen, the activating or arousal function is difficult for mathematical learning theory, particularly for a stimulus-sampling theory like Estes's in which the stimulus is represented as a population of stimulus elements, which in the manner of the associationistic theory of Guthrie are hooked and unhooked from responses, reinforcement and nonreinforcement having equal but opposite effects on response strength.

Figure 2.5 is a summary of Estes's theory of drive in stimulus sampling terms, and it postulates the existence of four kinds of stimuli: deprivation

Motivational and associative mechanisms 23

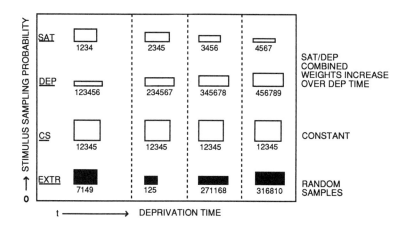

Figure 2.5. Stimulus sampling theory of motivation. Area of rectangle represents *weight* of stimulus set; horizontal dimension represents *number* of stimulus elements; vertical dimension represents *average sampling probability* of elements. (From Estes, 1958. Reprinted from 1958 Nebraska Symposium on Motivation, by permission of University of Nebraska Press; © 1958 by the University of Nebraska Press.) See text for details.

stimuli (DEP), satiation stimuli (SAT), conditioned stimuli (CS), and extraneous stimuli (EXTR). In other words, what is proposed is a definition of the drive concept as four populations of stimulus elements, three in addition to the CS elements that were proposed in the original statistical learning, stimulus sampling theory. Apart from the CS elements, which remain constant over deprivation time, the other three samples of elements (each element represented by a numeral on the base of the rectangle) change over deprivation time. In the case of satiation and deprivation stimuli, the change is systematic. In the case of other extraneous stimulus elements, it is random. The area of each rectangle represents the weight of the set of stimuli; the horizontal dimension represents the number of stimulus elements in the set; and the vertical dimension represents the average sampling probabilities. In this model are represented the kinds of mathematical necessities that are required to express motivation in stimulus-sampling terms. The motivational part of the model is the conceptualization of satiation and deprivation stimuli whose combined weights increase over deprivation time because of changes in stimulus sampling as the weight of deprivation stimuli increases and the weight of satiation stimuli decreases. The increase in the combined weights of the deprivation and satiation stimuli over time allows for the relatively greater importance of deprivation in motivation. The conditioned stimuli, the external stimuli that control learning and responding, remain constant, and corresponding to Hull's concept of behavioral oscillation ($_sO_R$), there is a set of extraneous stimuli that fluctuate randomly.

24   *Frustration theory*

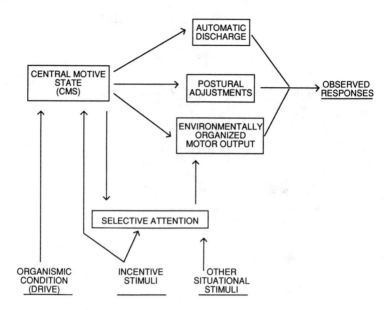

Figure 2.6. Theory of the central motive state. (From Bindra, 1969. Reprinted by permission of the New York Academy of Science.) See text for details.

What are the implications of such a theory? As in the case of a similar position advanced by Bolles (1958), they are, in a sense, negative: Generalized drive strength is ruled out. There is no representation of arousal in learned performance. Learned performance involves only stimuli; these stimuli are sampled according to certain rules, and it is the combination of the strengths of these sets of stimuli that are sampled that determines the strength of a response at any given time. This is an example of a complex motivation theory without a nonassociative component, of a motivation theory that is purely associative. It is a way of handling the motivation concept without departing from a purely probabilistic approach.

Bindra (1969) proposed a motivational theory of what he called the *central motive state* (CMS). The fundamental difference between this theory and other theories of motivation is its emphasis on attentional factors. Figure 2.6 is a block diagram of the theory. The CMS, as Bindra conceives it, is contributed to by organismic, or drive, conditions and by incentive stimuli, that is, by expectancies of reinforcing events. Bindra's theory is actually a theory of behavior, not of motivation: The CMS is the motivational mechanism, and selective attention is the associative mechanism. Bindra's theory could almost have been included in Table 2.1 as an instance of the two-systems approach: Like other theories of learned behavior, response strength is a joint function of nonassociative and associative variables, but in this case the language is a little different. Incentives are *learned* motivational conditions. They are expectancies of the goal, and they appear to have both associative and nonassociative properties. In this respect, there appears to be a fairly direct correspondence between Bindra's theory and the Hull–Spence theory of incentives.

The CMS is, in Tolman's terms, a first-level intervening variable, as is selective attention. (The independent variables are at the bottom of the figure.) The CMS appears to have both nonassociative (D-like) and associative (S_D-like) components. (But as we shall see in the next section, they are more like Hull's incentive-motivational, nonassociative and associative components, K and s_G.) Selective attention is purely associative. The second-level intervening variables are autonomic discharge, postural adjustments, and environmentally organized motor output. These intervening variables are like excitatory potential (E) in Hull's system, those that are linked to the observed response. This kind of intervening-variable theory incorporates features of both Hull's (1943) and Tolman's (1938) approaches: It divides the determinants of behavioral output into its motivational and associative precursors. In one respect the theory is more like Tolman's than like Hull's, in that the associative factor is called selective attention, rather than simply habit (H). Selective attention has a kind of surplus meaning that is closer to Tolman's cognition, or sign-gestalt-expectation, than to the mathematically defined H.

The block diagram that describes Bindra's theory introduces the concept of incentive. It does not define the concept but treats it as a kind of independent variable. Up to now, we have been dealing with the cue and drive functions of motivation. It is time now to consider and define the third function of motivation, the incentive function. In order to do this we will need to introduce and consider the more general topic of acquired or learned motivation.

Primary and acquired motivation

It is quite common in motivational writings to make a distinction between primary (unlearned) and acquired (learned) motivational conditions (e.g., Brown, 1961). Primary motivational conditions are those that arise directly out of some departure from physiological homeostasis – for example, from an antecedent food or water deprivation, a hormonal imbalance, or noxious events such as painful shock or the absence of an expected reward. Acquired motivations depend on learning, specifically on classical or Pavlovian conditioning. There are two kinds of these learned forms. Both therefore have the status of conditioned responses (CRs). In one of these cases the unconditioned responses (UCRs) are primary motivational states such as hunger and primary frustration; in the other, the UCRs are goal events such as rewards and frustrative nonrewards, and the CRs are goal anticipations or expectancies. The latter kinds of motivation are positive and negative incentives, which Seward (1951) has referred to as *tertiary*, to distinguish them from the *primary* and *secondary* motivations.

Let us now be clear about the difference between the secondary and tertiary (or incentive) motivations from simple examples in both appetitive and aversive systems. In the case of appetitive systems, secondary motivation involves, as the unconditioned stimulus (UCS), the deprivation of food and, as the UCR, hunger. The CR is then conditioned hunger, the environmental arousal of hunger or appetite by an originally indifferent stimulus (the CS) in the absence of a basic physiological hunger. In appetitive tertiary or incentive motivation, the UCS is a positive goal event (e.g., food or water), the UCR is the consummatory response of eating or drinking, and the conditioning is of the consummatory response to some indifferent CS: A conditioned form of the *consummatory response* defines this kind of motivation. This anticipatory reward is in the class of constructs that Hull (1931) generically called r_G–s_G, the *fractional anticipatory goal response*, r_G being an incentive-motivational conditioned response formed on the basis of Pavlovian or classical conditioning, and s_G being the interoceptive stimulus feedback from r_G. When we deal with incentive motivations, the UCR is now designated R_G, and the UCS is S_G. Because r_G is the generic term for the anticipatory goal response, I have used the symbols r_R and s_R to designate, specifically, anticipatory reward and its feedback stimulation.

In the case of aversive systems, secondary and tertiary (incentive) motivations are more difficult to untangle. It is clear that if the ongoing instrumental response is driven by aversive motivation – for example, pain based on shock or primary frustration – external cues present *during that response* can serve as CSs to elicit their secondary form, conditioned fear or conditioned frustration. In these cases, the conditioned expectations of shock offset or removal of frustration can be regarded as cases of tertiary motivation, the positive incentive of anticipatory "relief."

In the case of fear, an early experiment (Amsel, 1950a) is instructive. Rats were trained in a runway to escape from shock into a safe goal box. Subsequently, they were not shocked in the runway, and escape was now from a conditioned form of the shock (fear or anxiety), a secondary motivational condition. In these two cases the tertiary (incentive) motivation, anticipated relief from shock or from fear, could be applied to removal of either the primary or the secondary source of motivation. If, however, the animal was shocked at the goal at the termination of what had been an appetitive instrumental response, a new negative incentive would be added; anticipatory fear. According to such an analysis, then, conditioned fear based on traumatic pain can be regarded as a case of either secondary or tertiary (incentive) motivation, depending on whether the CS eliciting the fear is present when the animal is shocked during the occurrence of a

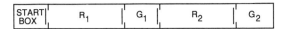

Figure 2.7. The double-runway apparatus, introduced by Amsel and Roussel (1952), includes a start box, a first runway (R_1), a first goal box (G_1), a second runway (R_2), and a second goal box (G_2). R_1 was 3.5 ft long and R_2 was 10 ft long. The start box and goal boxes were 1 ft long. The rationale for its use in studying frustration is given in the text.

response or at the goal when the response has been completed. Because, as we have seen, the generic r_G–s_G must be reduced to its specific cases, the three kinds of incentive motivation were designated r_R–s_R for anticipatory reward, r_P–s_P for anticipatory pain (or fear), and r_F–s_F for anticipatory frustration (Amsel, 1958a). To these, Mowrer (1960) added a fourth, which he called *relief*.

The double-runway apparatus (Figure 2.7), which has been used extensively in frustration research (see Chapter 3), provides a particularly good way to conceptualize the differences between the various forms of motivation in the case of frustration, our major concern in this book. This apparatus is two runways in series, the goal box of Runway 1 becoming the start box of Runway 2. The sequence is: start box – Runway 1 – Goal box 1 – Runway 2 – Goal box 2. In the typical experiment on the *frustration effect*, the animal is run in both runways and rewarded with food in Goal box 1 (G_1) and in Goal box 2 (G_2). In Runway 1, then, the motivation is at first primary and appetitive (e.g., hunger), but after a series of rewards and nonrewards in G_1, when the response can be said to be motivated by competing incentives based on both reward and nonreward (r_R and r_F), the response in Runway 1 slows and becomes more variable. In Runway 2 the motivation is also at first primary and appetitive, but following nonreward in G_1, this response is motivated not only by primary hunger and anticipatory reward (r_R), but also by the carried-over nonspecific generalized drive strength (D) from primary frustration that results from nonreward in G_1, by the secondary frustration that is the conditioned form of this primary frustration, and by the incentive of relief from frustration by the food that is always in G_2. As we shall see in Chapter 3, despite this potential motivational complexity, the double-runway apparatus and its variants have been used primarily to conceptualize and measure the contribution of nonspecific drive strength (D) from primary frustration.

To return to our discussion of the distinction between the cue and drive properties of motivation in Hullian terms, these conditioned motivational responses, whether they are secondary or tertiary, are like need states. They produce characteristic stimuli (s_D, s_G) and contribute to generalized (D-like) arousal. They are, in a phrase used by Brown (1961), "sources of motivation." The difference is only that in these secondary and tertiary cases the sources of motivation are learned. Nonetheless, like sources of primary motivation, the outputs of these learned motivational conditions have nonassociative properties as well as associative properties. According to what might be called the "Yale–Iowa motivation theory," since its major proponents were Hull, Spence, N. E. Miller, and J. S. Brown, each motivational source contributes to the generalized undifferentiated pool of drive strength, and each contributes a characteristic drive stimulus, which defines that particular state. To emphasize a point made earlier, in the two associative cases, the difference is only in the nature of the UCS in the conditioning paradigm. For example, if the UCS is defined by appetitive deprivation (e.g., hunger), the CR is the evocation of a conditioned form of hunger by a CS; if the UCS is a goal event (e.g., food), the CR is an incentive, conceptualized as an anticipatory goal response (r_G). Acting as needs, both also contribute to generalized drive or arousal, which in the first case Hull called D and in the second, K. In the second case, the directive stimulus (s_G), the feedback stimulus from r_G, operates like any drive stimulus – except that it directs the organism toward the goal.

Motivation theory: generalized and selective aspects

Hull's (1943) original formula for response strength, in its various relationships to excitatory potential (E), made it a multiplicative function of habit strength (H), drive (D), incentive (K), and a number of other factors [$R = f(E) = f(H \times D \times K \times \cdots)$]. This multiplicative formula implies that if any item in the right-hand side of the equation is zero, E and therefore R are zero. To emphasize the functional identity of D and K, which we have been emphasizing, Spence (1956) changed the formula so that E was a function of H multiplied by D plus K.

It is important for present purposes to have developed (with some redundancy) the idea that sources of motivation, including conditioned sources, contribute not only nonassociatively, but associatively as well. Spence's identification of the K factor as equivalent to D and as emerging out of classical conditioning, was a prominent example of a nonassociative mechanism derived from incentive, and there has been some tendency by neobehaviorists to treat incentives partly in terms of the energizing or drive function. This prompted Bolles (1972), quite erroneously, to argue that the associative function of incentives had been neglected. What Bolles himself overlooked, however, was that the *original* statement of incentive function by Hull (1931) proposed and identified *only its associative function*. He also overlooked many other treatments that prominently featured this function of incentives (Brown, 1961; Kendler, 1946; Logan, 1960; Mowrer, 1960; Seward, 1951). I would add that almost all of my own work in frustration theory (e.g., Amsel, 1949, 1958a, 1962, 1967) stressed the associative function of incentives. Whereas Rescorla and Solomon (1967)

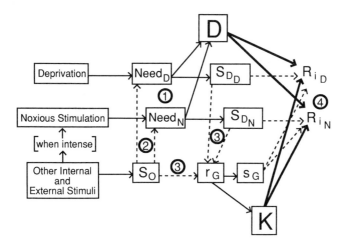

Figure 2.8. Schematic representation of the nonassociative and associative aspects of the Hull–Spence motivation theory. (1) *Primary motivation:* Needs D and N arise from a departure from homeostasis or from noxious stimulation, respectively. (2) *Secondary motivation:* Needs arise when other stimuli (S_O) are paired with primary need states; a conditioned form of the need state is elicited by S_O. (3) *Tertiary motivation:* r_G is produced when S_O, S_{DD}, or S_{DN} is paired with a goal event; a conditioned form of the goal event is elicited by S_O, S_{DD}, or S_{DN}. (4) *Instrumental responses:* R_{iD} and R_{iN} are affected associatively (S_{DD}, S_{DN}, and S_G) and nonassociatively (D, K). Heavy arrows represent drive function (nonassociative); dashed arrows, cue function (associative); light arrows, direct consequence.

provided an excellent treatment of the associative function of incentive, they were not the first to emphasize this feature, as Bolles appears to indicate.

In the final portion of this chapter, we will work toward a conceptualization of the motivational system in both its nonassociative and associative aspects. Such a theory can be portrayed diagrammatically, as in Figure 2.8. It is a conditioning model of what can be referred to as "behavioral dynamics," or even behavioral symptoms, and it incorporates all of the conceptual elements required for the treatment of frustration-theoretic concepts in the chapters that follow. The diagram is actually an extension of a kind used by G. A. Kimble (1961) and is a more extended version of the Hull–Spence view of cue, drive, and incentive function.

In review, the steps are as follows: Deprivation and noxious stimulation produce conditions of need. Each of these contributes to the nonspecific, generalized drive strength (D), and each also has connected with it a characteristic drive stimulus (S_D). Other external and internal stimuli (S_O), when intense, are noxious events, which also become conditions of need. These S_Os, when they are associated by conditioning to needs arising from deprivation or noxious events, produce specific secondary needs, which also contribute to D and S_D. When associated by conditioning to the goal event, these S_Os also produce r_G-like responses that, again, contribute to generalized arousal through the K factor and have associated with them characteristic stimuli, generically s_G. Further, the drive stimuli (S_D) themselves can enter into associative relationships with motivational responses. That is, in an animal that is hungry and running for food, the stimulus of hunger can arouse the expectancy of food, as can the external stimuli. Consequently, any incentive system involves in its arousal a kind of compound stimulation consisting not only of the manipulable external stimuli, but also of hypothetical internal or drive stimuli.

All of the stimulus and drive factors are determinants of instrumental behavior: The generalized drive factors (D) emerge out of homeostatic factors such as deprivation or intense stimulation, or out of incentive conditions (K), and contribute nonassociatively to behavior; the stimulus factors, of course, contribute in an associative manner.

What has been outlined here and called a theory of motivation is in actuality a theory of behavior with strong motivational components. In this respect it is similar to Bindra's theory in the sense that it, too, is a theory of behavior, in Bindra's case in terms of the central motive state and selective attention. The central motive state, of course, would correspond to both the D and K factors, and selective attention to H, the habit factor.

Facilitating and interfering properties of motivation

A view I have held very strongly for some time (Amsel, 1950b) is that increased motivation should not be regarded as always facilitating behavior. This is a very important idea in the chapters that follow. To say that a rat (or a person) is strongly motivated is not to say that its (his or her) response, the particular criterion response under consideration, will show an increase in intensity or frequency or a reduction in latency.

This idea has been around in several different forms. One form appeared in the study of the role of anxiety in simple conditioning and complex learning (e.g., Spence, 1958; Spence & Taylor, 1951; Taylor & Spence, 1952). This work, based mainly on the concept of anxiety defined by the Manifest Anxiety Scale (MAS, Taylor, 1953) and on experiments in human eye-blink conditioning and human complex learning, demonstrated that if the learning situation is simple, that is, if the criterion response is so dominant in the response hierarchy that no other response can interfere with it, anxiety facilitates performance. In eye-blink conditioning experiments it was shown that subjects from the high end of the MAS conditioned faster, and reached higher asymptotes, than those from the low end. In contrast, if the task was paired-associate or complex serial learning, that same relationship did not hold. In fact, if the learning was difficult, high anxiety actually reduced, or interfered with, performance.

Just preceding this work, and quite independently, were experiments showing that if rats were shocked and put directly into a situation in which all they had to do was drink water, they drank more (Amsel & Maltzman, 1950). However, if thirsty rats were put into the specific situation in which the anxiety or fear had been conditioned, so that they drank in the presence of this fear, they drank less (Amsel, 1950b; Amsel & Cole, 1953). The *conditioned emotional reaction* of Watson and Rayner (1920) and the *conditioned emotional response* of Estes and Skinner (1941) are, of course, earlier and almost parallel examples of this later work, the difference being that for the earlier investigators the criterion response was instrumental, whereas in our experiments it was consummatory.

In this last sense, the conceptualization of increased motivation increasing or decreasing some ongoing criterion response can be diagrammed as in Figure 2.9. It involves a two-process account of the possible effects of fear; but it can obviously also be applied to other motivational systems, like frustration. As we have seen, a two-process account is simply one in which classically conditioned incentive motivation is embedded in what is essentially instrumental trial-and-error learning. Trial-and-error learning, the differential strengthening of one out of a number of possible responses by selective reinforcement, in turn requires a conception such as Hull's (1934) *habit-family hierarchy*. In our terms, fear is a conditioned need state (r_P) based on pain. It contributes to generalized drive strength (D), and like all other hypothetical internal responses, it provides a characteristic feedback stimulus (s_P). Along with the environmental stimuli (S_E) and the CS itself (which, in its role as a part of the stimulus complex for the instrumental response hierarchy, we have labeled S_C), it forms a new compound stimulus whose effect is to rearrange that hierarchical order of response strengths.

We now have a stimulus complex or compound controlling the response hierarchy in which one of the new elements is the feedback stimulus from the conditioned response of fear (s_P) and another is S_C, the CS which elicits that fear. If one ignored the contribution of these new stimuli and focused only on the increased effect on D of the added motivation, the conclusion would have to be that since this increased D activates all of the responses that exist at a particular moment in the hierarchy of responses, the multiplicative effect of this stronger D would serve to increase the differences among these responses: The hierarchy would spread and fan out, but the order of strengths would be unchanged. In this case, the dominant response in the hierarchy would be even more dominant.

However, since there are also the new habit factors associated with S_C and s_P in this new stimulus complex, the order of associative strengths for responses in the hierarchy is unlikely to remain the same. One must there-

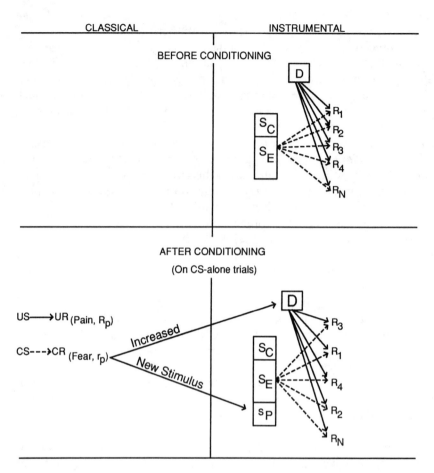

Figure 2.9. Diagrammatic representation of the view that increased motivation may facilitate or interfere with ("inhibit") ongoing instrumental responding. Illustrated here are the possible effects of conditional fear in a two-process account. D represents generalized drive strength; S_C, specific eliciting stimulus before conditioning; S_E, environmental (background) stimulation; s_p, feedback stimulus from the conditioned fear response (r_p) or from the conditioned form of pain (R_p). Vertical arrangements of these stimuli are stimulus compounds. Solid arrows represent direct (unlearned) effects; dashed arrows, habit-strength associations. See text for a detailed account.

fore consider not only the effects of increased D, but also the effects of the changed stimulus complex on the hierarchical arrangement of behaviors. As we have seen, the effect of the addition of fear is often unpredictable, except in terms of the history of the organism. In most learning experiments the effect is to produce what is called "conditioned suppression." An animal is placed in a Skinner box, trained to a fairly high level of responding to a lever – for food, say – and then, on several occasions, even independent of feeding, *in that situation,* is given brief electric shocks. When reintroduced into the situation, the animal does not respond for food; it freezes and/or it does other things – urinates or defecates. In this case, the associative strength of the criterion response, formerly R_1, has weakened, even though the generalized drive level has obviously increased.

An account of this kind is based on the idea that new sources of motivation will facilitate or interfere with previously learned ongoing behaviors, depending, generically, on the new S_D and s_G that are introduced, and depending also upon the existing hierarchical relationship of responses to these new drive and incentive stimuli in that situation. In Chapter 3 and those that follow, this kind of analysis is applied to frustration theory and to the kinds of dispositional learning and memory it explains.

3 Frustration theory: an overview of its experimental basis

At the end of Chapter 1, brief reference was made to our recent work on the ontogeny of a number of reward-schedule effects in infant rats. This developmental work emerged out of many years of research in which the adult rat – and to a lesser extent other animals – served as a laboratory preparation to study dispositional learning, the acquisition of certain emotional-temperamental characteristics that seem to be common to humans and other mammals. One of our particular concerns has been to study the effects of *frustrative nonreward* on arousal, suppression, and persistence in the context of a theory of such effects, frustration theory. This chapter provides an overview of this theory and of the experimental work on which it was based and which it has predicted. It deals with the first two items of strategy presented at the end of Chapter 1 (Table 1.2). The chapters that follow deal in more detail with the specifics of the theory and experimental work and provide material relative to the other four items in the table, which have to do with the order of appearance in ontogeny of the reward-schedule effects and their neurobiological substrate.

Setting the boundary conditions

It is important in defining the boundary conditions for frustration theory to differentiate three kinds of dichotomous outcomes of behavior, because they are often regarded as synonymous. The theory, as presented in this book, is focused on the role of reward and nonreward (or reduced or delayed reward) in learned behavior and its development; it does not address directly the dynamics of success and failure. It addresses the correct–incorrect dichotomy only in the limited sense that reward and nonreward occur differentially in the context of choice behavior, that is, stimulus-selective (discrimination) learning, response-selective (trial-and-error) learning, or a combination of both (conditional discrimination learning).

In common usage, it is not unusual to say that a response is successful or correct if it leads to a reward and that it fails or is incorrect if it leads to nonreward. In the present context, particularly, this overlooks the fact that in the acquisition followed by the extinction of a response, and in random (intermittent or partial) schedules of reinforcement, the same response is sometimes rewarded and sometimes not – that there is no difference in the behavior that precedes reward and nonreward in such schedules. In an early experiment, Grosslight and Child (1947) trained feebleminded adults to pull a sequence of levers for a reward of candy. In 10 training trials, the reward was omitted on 1 or 2 of the trials, and these omissions, or "failures" as they were called, increased persistence in extinction. In present terms, these would not be called failures, but simply nonrewards. Frustration in these cases is defined by the fact that a particular behavior, based on a learned reward expectancy, is followed on occasion by nonreward. Failure and success, in contrast, usually imply differential behaviors, as in solving or not solving a problem, achieving or not achieving a result. The same can be said for responses that are correct and incorrect, except that there has to be an objective definition (or criterion) of "correct" (less so for "success"), apart from whether or not the response is rewarded. Some time ago I defined this difference as follows:

Strictly speaking, the word "incorrect" describing a response implies differential responding and different consequences for alternative responses. The differential response situation – a situation in which a response can be incorrect – has been termed "trial-and-error learning" or "response selective learning." Selective learning implies differential consequences – "correct" (perhaps arbitrarily designated correct) responses are rewarded; "incorrect" or "error" responses are nonrewarded or even punished. At the beginning of such a sequence of correct and incorrect responses occurring in chance arrangement, reward will occur on less than 100 percent of the trials, the frequency of occurrence depending upon the number of alternative responses and the initial probabilities of occurrence of these correct and the incorrect alternatives.

Partial reinforcement implies a single, prepotent response, already dominant in the hierarchy – or, at least, other responses are ignored – which is rewarded not consistently and continuously but intermittently. The important factor is that there is inconsistency in reward occurrence but not in response occurrence. Objectively, the same response is sometimes rewarded, sometimes not. (Amsel, 1960, pp. 508–9)

The concept of frustration in psychological theory

An excellent overall discussion of the scientific treatment of the concept of frustration can be found in a book by Lawson (1965, chap. 2). Lawson points out that a number of psychologists have attempted to derive a theoretical meaning of frustration from its vernacular usage, and then to move from the theoretical meaning to experimental operations. If we add to this assigning to the term an explanatory burden within an existing framework, this sequence is a fair description of the development of frustration theory as it is discussed in this chapter.

Lawson divided theories of frustration into two kinds: "self-contained theories" and theories "integrated with general behavior theory." In the first category he placed Rosenzweig's (1934, 1944) frustration theory, the frustration-aggression hypothesis (Dollard, Doob, Miller, Mowrer, & Sears, 1939), the frustration-regression hypothesis (Barker, Dembo, & Lewin, 1941), and the frustration-fixation hypothesis (Klee, 1944; Maier, 1949, 1956). These theories are called "self-contained" because, according to Lawson, they identified the study of frustration as a topic in its own right – that is to say, they were hypotheses about the defined concept, each with a somewhat different thrust than the others, and none particularly linked to or emergent from a more general theoretical position.

I would add that, in these self-contained theories, frustration was defined in a number of ways (see Zander, 1944), because in most cases the definitions were derived from a variety of reactions to stress, conflict, and thwarting in humans. However, in at least one case, the frustration-fixation hypothesis, the experimental work involved animals as well as humans. In this instance, as we shall see, the concept of fixation in relation to an insoluble problem and the concept of persistence as a result of partial-reinforcement training are not dissimilar. As we will show in the work described in Chapter 5, in which discrimination learning can be made very difficult by certain prediscrimination treatments, frustration theory can be regarded as an explanation for fixated responding, in the context of the insoluble discriminations that were employed by Klee, Maier, and others.

The second category of theories, the "integrated" ones, are identified by Lawson as the Child and Waterhouse (1952) revision of the frustration-regression hypothesis, Brown and Farber's (1951) treatment of frustration as emotion conceptualized as an intervening variable, and my own (Amsel, 1958a) "frustrative nonreward theory." As Lawson points out, and as I have indicated in my own case, the distinguishing characteristics of such theories include their closer alliance to more formal (and more general) behavior theory, the related recognition that these frustration theories involved many independent variables already familiar from theories of learning, and a growing recognition that "there was no unique overt behavior characteristic of frustration situations" (Lawson, 1965, p. 27).

In Hull's formal papers on learning theory in the 1930s, there are references to frustration, both as an experimental operation in his analysis of the goal gradient in maze learning (erecting a physical barrier against a response, Hull, 1934) and in the form of a specific "frustration hypothesis" in his work on the goal gradient in children (Hull, 1938). Hull's hypothesis states that "whenever an excitatory tendency is prevented, for any reason, from evoking its accustomed reaction, a state ensues substantially like the experimental extinction or internal inhibition long known to be characteristic of conditioned reactions" (p. 278, n. 10). Williams and Williams (1943) tested barrier frustration plus the absence of food against frustration by simple absence of food (extinction) in a runway and concluded that the barrier-frustration manipulation produces a larger decrement in speed than simple extinction. While Hull's concept of frustration stressed barriers rather than nonrewards, and while it was never formally incorporated into his own systematic theorizing, it remained, as we shall see, a peripheral interest of his and of those at Yale who later proposed the frustration-aggression hypothesis (Dollard et al., 1939). The eventual realization of what Lawson called a frustration theory that was "integrated with [Hull's] general behavior theory" came in the 1950s with the Brown and Farber theory, and with my own.

This is not a book on frustration *theories*. For a more detailed description and analysis of all of these theories until the mid-1960s, Lawson's discussion is highly recommended. (Another extensive bibliographic treatment of the subject can be found in Lawson & Marx [1958], and briefer examinations of the history of the concept of frustration as it is applied to theories of learning, specifically, can be found in Brown & Farber [1951], Amsel [1958a, 1962], and Mowrer [1960].)[1] There will, however, be references and allusions to some of these alternative conceptualizations of frustration in this chapter and in some that follow, particularly Chapter 6. We will also have occasion to examine explanations of phenomena that are alternatives to those offered by "frustration theory," a term that, in the present volume, designates my own theoretical position on the effects of "frustrative nonreward."

Metatheoretical overview

Both Hull (1952) and Spence (1956) accepted the view that "frustration" (or some other emotional factor) accounts for the incentive-contrast effect (Crespi, 1942; Elliot, 1928), a suppression effect that occurs when magnitude of reward is shifted from large to small, but neither provided a detailed account of how frustration might enter into the structure of behavior theory. Other researchers had also reported signs of emotional upset in animals at the beginning of extinction (e.g., Miller & Stevenson, 1936; Skinner, 1938), but again these observations were never formally incorporated into a theory of learning.

In 1951 I proposed a conceptualization of the role of frustration and, more specifically, anticipatory frustration as a third factor to be added to Hull's (1943) two-factor theory of inhibition, and applied it to the Elliot–Crespi incentive-contrast effect (Hull's two factors were reactive inhibition, I_R, and conditioned inhibition, S^IR). This conceptualization was later extended to cover a number of other reward-schedule phenomena (Amsel, 1951, 1958a, 1962, 1967, 1986; Spence, 1960). In the same year two other theoretical treatments of frustration appeared in the literature. Brown and Farber (1951) published an article in which they, too, advanced a theory

1. An excellent review and analysis of the experimental study of frustration until about 1938 can be found in Rosenzweig, Mowrer, Haslerud, Curtis, and Barker (1938).

of frustration, also in a Hullian framework. Their elegant theory defined frustration as emerging from conflicting response tendencies, as well as from nonreinforcement and thwarting. The frustration to which they referred is what I subsequently called *primary frustration* (Amsel, 1958a), and there was, in the Brown and Farber theory, no reference to its conditioned form. (The theory was appended to a more general conceptualization of emotions as intervening variables, and it is perhaps for this reason that it does not seem to have had the impact on experimental work that one might have expected.) Seward (1951), in an article titled "Experimental Evidence for the Motivating Function of Reward," introduced a concept like primary frustration defined by "the consumption of any quantity of food short of satiation [that] involves some frustration of the unsatisfied need (F_{RG}); the less food the more frustration" (p. 137). He added the following, defining a kind of anticipatory frustration mechanism: "As the animal's set for this quantity of food is generalized to the starting box of a runway, a corresponding degree of frustration (F_{rsG}) will also be aroused there and will act as a negative drive to oppose the positive drive to food" (p. 137). In a footnote, he defined a more explicit anticipatory mechanism as follows: "We could arrive at the same result by assuming that F_{RG} is conditioned to the starting box along with food anticipation, thus becoming sF_{rsG}. Although not indispensable here, this assumption may prove to be a decided advantage in dealing with extinction" (p. 138). Because Seward's version was based on a kind of contrast phenomenon and involved anticipatory frustration, and Brown and Farber's version was based on an analysis of conflict and identified only primary frustration, Seward's version and my own of 1951 were the more similar.

The conceptual form of frustration theory was derived from a portion of neobehavioristic theory of the Hull–Spence variety that combined elements of Pavlovian and Thorndikian conditioning. This approach to theorizing has a clear antecedent in Hull's conceptualization of "goal attractions and directing ideas ... as habit phenomena" and of the *fractional anticipatory goal response* (r_G–s_G; Hull, 1931). Later termed "conditioning model" theory (Lachman, 1960), it bears a decided similarity to the cybernetic concepts of "hope" and the "go mechanism" in the theories of Mowrer (1960) and Miller (1963), respectively.

Conditioning model theory can be regarded as a form of what Lloyd (1986) has called "functional explanation," and it assigns states and processes to various behaviors that are not necessarily tied to particular biological processes, although as in the case of Hull's theorizing, and lately in my own (see Chapters 7 and 8), such processes are sometimes suggested. The virtue of such theorizing, which, as we have seen, was already evident in Hull's (1930, 1931) earliest papers on learning theory – it was later called "two-process theory" (Rescorla & Solomon, 1967) and the "second learning process" (Trapold & Overmier, 1972) – is that the functions (the states and processes) that are postulated have an empirical base; they are defined

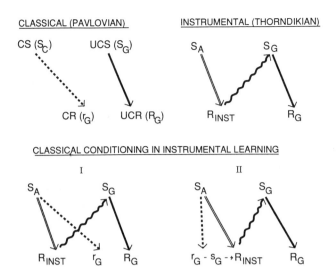

Figure 3.1. Schematic representation of classical and instrumental conditioning and their interactions in two-process accounts of learning. (From Amsel & Ward, 1965. Copyright 1965 by the American Psychological Society; reprinted by permission.) See text for details.

and constrained by the existing laws of classical conditioning. The hypothetical internal states and processes, the mediating machinery, are assigned functions consistent with their empirical, conditioning counterparts (see Amsel, 1962). I see this constraint as a major strength – and this, in part, is what this book is intended to demonstrate in developing the explanation of a family of dispositional learning phenomena.

In such theorizing, the role of goal events (S_G and R_G) is all-important, the major functions of the goal stimulus (S_G) being to serve as the UCS for a goal response (R_G) and to provide for the classical conditioning of "fractions" of the R_G in instrumental learning to cues antedating the goal. A schematic representation of these two kinds of simple learning and their interrelationships is presented in a Hull–Spence type of diagram (Figure 3.1). The top portion of the figure shows, separately, classical (Pavlovian) and instrumental (Thorndikian) conditioning. Interchangeable symbols have been provided for the stimuli and responses of classical conditioning: The unconditioned response is also designated R_G; the unconditioned stimulus, S_G; the conditioned stimulus, S_C; and the conditioned response, r_G. Clearly, then, the schema or paradigm for Pavlovian conditioning applies also to the conditioning of r_G (the anticipatory goal response), except that in this case S_G (the UCS) and R_G (the UCR) occur at the end of the instrumental response – for example, in the goal box of a runway or at the appropriate time at the feeder in an operant chamber. The bottom part of Figure 3.1 shows how classical conditioning, once formed, is involved in instrumental learning and provides the essence of the conditioning, or two-

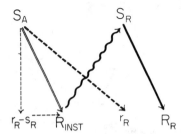 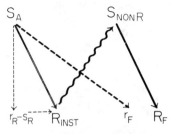

Figure 3.2. Schematic representation of (*left*) the conditioning of anticipatory reward (r_R) and (*right*) the manner in which r_R and its feedback stimulus (s_R) in the absence of reward (S_{nonR}) combine in the evocation of primary frustration (R_F) and in the subsequent conditioning of the response of anticipatory frustration and its feedback stimulation (r_F–s_F). (From Amsel & Ward, 1965. Copyright 1965 by the American Psychological Society; reprinted by permission.) See text for details.

process, model of instrumental behavior with which we will be concerned. Two stages are represented. In Stage I, the CR (in this case, r_G) is formed in the context of instrumental learning; and in Stage II, this r_G, once formed, moves forward in time and, through its feedback stimulation (s_G), becomes part of the mechanism, along with S_A (the cues of the apparatus), for the evocation of the instrumental response (R_{INST}).

The relationships shown in Figure 3.1 are based on a general conception of the primary reinforcing properties of goal events. In an article relating frustration to the partial-reinforcement extinction effect (PREE; Amsel, 1958a), three kinds of primary goal response were conceptualized: reward (R_R), punishment (R_P), and frustration (R_F). To these three, Mowrer (1960) added a fourth, relief. Such a classification of goal responses sets relief and punishment into the same relationship as reward and frustration. At first, I dealt only with the latter two members of this fourfold classification, that is, with reward and frustration in learning. However, others (e.g., Banks, 1966; Martin, 1963) later showed that the kind of reasoning involved in treatments of reward and frustration could also be applied to an analysis of punishment and relief.

Figure 3.2 represents assumptions about the manner in which primary reward (R_R) and frustrative nonreward (R_F) are involved as goal responses in simple instrumental learning. It shows, schematically, the conditioning of anticipatory reward (r_R) and anticipatory frustration (r_F) to the cues of the apparatus or situation (S_A) in the context of instrumental behavior. The left-hand side of Figure 3.2 shows that instrumental responses followed by S_R and R_R occasion the development of r_R (anticipatory reward), which moves forward in the temporal sequence, so that its feedback cues (s_R) become part of the stimulus compound eliciting that instrumental response. When r_R, through its feedback stimulation, s_R, shares control of the instrumental response, the behavior can be said to involve appetitive incentive motivation. The right-hand side of Figure 3.2 shows that when the incentive motivation of reward (r_R) is operating, and the goal event is nonreward (S_{nonR}) instead of S_R, primary frustration (R_F) results. The presence, in sequence, of both anticipatory reward (r_R) and nonreward (S_{nonR}) is, then, the necessary and sufficient condition for the elicitation of primary frustration (R_F). In the presence of r_R, the unconditioned response to the absence of the unconditioned (nonreward) stimulus, S_{nonR}, is R_F, which becomes the UCR for the conditioning of r_F (right-hand side of Figure 3.2) to the cues of S_A. The action of this conditioned response of anticipatory frustration (r_F) is then to move forward in time (backward along the instrumental sequence) to affect the instrumental response through the action of s_F (not shown in Figure 3.2), presumably in a manner antagonistic to that in which r_R–s_R affects the instrumental response.

To recapitulate, in my 1951 paper I expressed the view that Hull's two-factor theory of inhibition – reactive (I_R) and conditioned ($_sI_R$) inhibition – was an incomplete account of extinction because it did not include as a factor in extinction the active aversive properties of frustration. Recall that, in Hull's (1943) theory of inhibition, total inhibitory potential ($\dot{I}_R$) was an additive function of two sources of inhibition: I_R, reactive inhibition, was taken to be a temporally labile source related only to the effort involved in a response, whereas $_sI_R$, conditioned inhibition, was the classically conditioned form of I_R and was, as the subscripts indicate, an associative state and therefore permanent. In the determination of response strength (R) at any given moment, $\dot{I}_R$ was subtracted from $_sE_R$, the excitatory potential, to form $_s\overline{E}_R$, the effective excitatory potential.

It should be pointed out that Rohrer (1949) had suggested earlier that Hull's I_R "[be] conceived as resulting from frustration rather than 'sheer reaction' " (p. 477). This was based on an experiment in which Rohrer simply showed that extinction was more rapid after massed than after spaced trials, a finding that was as compatible with an interpretation in terms of I_R as with frustration; there was no actual evidence for what he called "frustration drive." I proposed that, in addition to I_R and $_sI_R$, a third factor, anticipatory frustration (r_F–s_F), was important for understanding a number of nonreinforcement-related effects. This was followed by the demonstration of the *frustration effect,* taken to be an indicant of the response of *primary frustration* (R_F; Amsel & Roussel, 1952), and this demonstration provided a basis (the UCS and UCR) for the conditioning of *anticipatory frustration* (r_F–s_F). Subsequent developments of frustration theory provided accounts of the PREE and, as we shall see, other paradoxical effects emerging out of the interaction of rewards and nonrewards in a variety of discriminative and nondiscriminative instrumental learning situations.

In informal terms, the basis of the theory of 1951 was that the replacement of a reward by a nonreward, reduced reward, or delayed reward can result in the aversive motivational state, primary frustration. Primary frustration (R_F) occurs when the parameters of reward change so that the existing r_R–s_R is greater than that which the reduced or delayed reward on that trial could support. For example, the strength of r_R–s_R established with a large reward is greater than that with a small reward. Accordingly, the theory requires that R_F occur when a small reward, or in the limiting case, nonreward, is presented after training with a large reward.

Properties of frustration: a recapitulation

With the conditions that produce primary frustration specified, the theory of 1951 outlines the four properties of frustration. The first is unconditioned (primary) frustration (R_F), a hypothetical unconditioned reaction to the frustrating event, which acts nonassociatively to exert a transient motivational (energizing) effect on responses with which it coincides, increasing particularly the intensity with which these responses are performed. That is to say, the immediate consequence of R_F is a short-term increment in generalized, energizing drive or arousal. The second property of frustration, to which some experimental attention was given, is the primary frustration drive stimulus (S_F), a feedback stimulus from R_F that acts, like any other stimulus, to cue, guide, and direct behavior – in other words, that acts associatively (e.g., Amsel & Ward, 1954). It is important to note that this hypothetical stimulus (S_F) is not the result of learning, as are, for example, s_R and s_F, the feedback stimuli from r_R and r_F.

In an analysis of what he calls two-factor theories of frustration, Marx (1956) discussed the evidence for the proposition that frustration contributes an "increment in drive" and provides "unique stimulus–response relations" (p. 125). His conclusion, based mostly on research in which frustration is defined by a "blocking" operation rather than by nonreward, is that the case for a two-factor theory is convincing in certain respects and that caution is required in generalizing from one frustration-inducing operation to another – in this case from the blocking to the frustrative-nonreward operation. With this caveat I of course agree, and in this book there is no attempt to generalize between these two operations; the definition throughout is unequivocally the latter one, and the basic theory advanced here and in all its elaborations depends on this definition.

The third and fourth properties of frustration, and the ones that bear perhaps the heaviest theoretical burdens, are conditioned (anticipatory) frustration and its feedback stimulation (r_F–s_F). These factors refer to the ways in which frustration influences responses that *precede* the frustrating event, and here the theory relies on the logic derived from its Pavlovian antecedents:[2] With repeated occurrences, stimuli (CSs) paired with primary

---

2. A Pavlovian treatment of reaction to discrepancy or "discordance," analogous to the suppressive properties (r_F–s_F) in frustration theory, can be found in the writings of Anokhin (e.g., 1974). In this treatment, an inhibitory stimulus is defined as one that signals the omission of the unconditioned stimulus. In Anokhin's terms, the existence of inhibition

frustration come to evoke a classically conditioned form of R_F, designated r_F. As with r_R, of which it is the aversive counterpart, r_F is initially evoked by stimuli in the region of the goal event, and later in the region of instrumental response, and is assumed to increase in strength as a function of nonrewarded trials, reaching an asymptotic strength appropriate to the strength of R_F.

The frustration effect

As we saw earlier, the frustration theory of persistence emerged out of a series of experiments, beginning in the early 1950s, which seemed to demonstrate that frustrative nonreward (nonreward in the presence of anticipated reward) can be regarded as influencing behavior in several ways. First of all, frustrative nonreward appears to have an invigorating or potentiating (D-like) effect on any behavior that immediately follows it, the so-called frustration effect (FE). This effect can be thought of as an attribute of primary, *unlearned* frustration, a natural reaction of probably all higher animals (certainly reptiles, birds, and mammals among the vertebrates) to nonattainment of expected goals, to thwarting, and to encountering physical or psychological barriers or deterrents in the path of goal attainment. In the first study of the FE (Amsel & Roussel, 1952) we employed an experimental arrangement that subsequently came into fairly general use. The apparatus, two runways in series, consists of a start box, Runway 1, a first goal box (G_1), which becomes the start box of Runway 2, which leads to a second goal box (G_2). This is called a double runway or tandem runway (see Chapter 2 and Figure 2.7). It was designed in accordance with a conceptualization of the way in which failure to attain reward for one response (in Runway 1) affects an immediately subsequent response (in Runway 2). The basic FE result is that, after reward expectancy has been built up in Runway 1, vigor (speed) of approach in Runway 2 to G_2 is greater after nonreward than after reward in G_1.

There was nothing particularly new about this conceptualization. It was, in one sense, like most experimental treatments of natural phenomena, simply a more abstract representation of the mechanisms that operate immediately after an expectation or anticipation is violated. The double

---

depends on the existence of excitation. His analysis stresses the operation of what he calls the "action acceptor," a feedback mechanism that registers discrepancies between motivated responding and the goal. When such a discrepancy is detected between expected and actual goal events (e.g., the presence or absence of food), an aversive motivational state results, and this suppresses appetitive conditioned responding. There is an obvious difference between such thinking in the Pavlovian case and in ours; however, as I pointed out some time ago (Amsel, 1972b) in distinguishing among classical, Pavlovian, and instrumental conditioning, the conditioning procedures in many Pavlovian laboratories do not have the "purity" of (say) eye-blink conditioning procedures at a 0.5-second interstimulus interval, but tend to involve instrumental components. Concepts of discrepancy and frustration are more apt in this case.

runway had an additional feature, however, that had simple theoretical importance: It separated out, and made a clear distinction between, the to-be-frustrated response in the first runway and the frustration-driven response in the second runway. As was indicated in Chapter 2, in terms of the definitions in the theory, the frustrated or to-be-frustrated response in the first runway becomes an indicant of *conditioned (anticipatory) frustration* (r_F) at the same time as the frustration-motivated response in the second runway is an indicant of *primary* or *unconditioned frustration* (R_F). If running in Runway 2 is always rewarded, as it was in most of these experiments, then reward for running in Runway 1 could be manipulated and scheduled so as to provide a basis for estimating and measuring relative degrees of r_F and R_F. As we shall see, other uses have been made of the double-runway arrangement. In one study, for example, the first runway provided a set of discriminanda (say, black or white alleys), while the second runway remained a neutral gray, so that primary frustration was monitored in Runway 2 during the formation of a discrimination in Runway 1 (e.g., Amsel & Ward, 1965; see Chapter 5, this volume). In another study, the response in the first runway was rewarded on a single-alternating schedule of rewards and nonrewards, and primary frustration following nonrewards was monitored in the second runway while a discrimination based on the carried-over cues or memories of reward and nonreward was being formed in the first (Hug, 1970c). In yet another study, the second runway was converted to two arms and goal boxes of a T-maze, and cues attending reward or nonreward in the first-runway stem of the maze (S_{DR} and S_{DF}) were shown to control spatial choice behavior in the arms (Amsel & Ward, 1954).

The basic conceptual features of the double runway have been incorporated into a variety of experimental situations that involve responding in two or more runway segments, or analogous bar-press segments, in series, so that rewarding or not rewarding the responses in earlier segments may have some direct potentiating (D-like) effect on the response in later ones. Scull (1973) provided a selective review of studies of the FE, and the alternative interpretations of the effect, covering the double-runway literature for the first 20 years. Here are a variety of examples (additional, more recent examples of experiments in this time period on frustrative nonreward [FNR], on the FE, and on other subjects covered by frustration theory are provided in the Appendix). In this 20-year period effects similar to the FE were found with monkeys and rats in bar pressing (Davenport, Flaherty, & Dyrud, 1966; Davenport & Thompson, 1965; Hughes & Dachowski, 1973); with pigeons in key pecking (Staddon & Innis, 1969), though the interpretation of this work differed from ours; and in a number of studies with children (see Ryan & Watson, 1968, for an early review). The FE was studied under a number of other conditions. For example, there was a demonstration that the energizing properties of frustrative nonreward (FNR) could be duplicated by mild electric shock in the first goal box (G_1) of the double runway (Bertsch & Leitenberg, 1970). There were demonstrations of central nervous system effects; for example, that the FE results from the omission of electrical stimulation in a rewarding area of the brain (Johnson, Lobdell, & Levy, 1969; Panksepp & Trowill, 1968) and that lesions of the amygdala eliminate the FE (Henke, 1973; Henke & Maxwell, 1973), while lesions of the hippocampus (Swanson & Isaacson, 1967) and of the septum (Mabry & Peeler, 1972) have no effect on it.

In some other experiments, conducted mainly with pigeons (e.g., Azrin, Hutchinson, & Hake, 1966; Davis & Donenfeld, 1967; Gallup, 1965), the response, whose probability and intensity are increased just after reward is withheld, is striking or otherwise attacking another animal. Under these conditions, the intensified response has been regarded as aggressive. In cases such as these, this energizing of behavior following frustration supports the idea that primary frustration leads directly to aggression, the so-called frustration-aggression hypothesis (Dollard, Doob, Miller, Mowrer, & Sears, 1939). As we shall see, another possible analysis of these cases involves the concept of response-family hierarchy (see Chapter 2): Specifically, the generalized activation resulting from primary frustration simply energizes the dominant response in the hierarchy, which is controlled by stimuli arising out of frustration in the context of the existing external situation (see Figure 2.9). Such an interpretation would fit the following results in rats (Lambert & Solomon, 1952) and in humans (Adelman & Rosenbaum, 1954): The closer to the goal that a response is "frustrated" by blocking, the greater is the resistance to extinction of that response. In both of these cases the authors interpreted their results as supporting the idea that the closer a response comes to completion, the greater is the acquired (secondary) reinforcement of that response. Another interpretation is based on the fact that the operations defining secondary reinforcement and primary frustration are identical: the absence of a reward or reinforcer that had previously been present following approach to a goal region (see Amsel, 1968, for eight references to this idea): The closer to the goal a response is blocked, the greater are the energizing effects of primary (and perhaps also secondary) frustration.

The double-runway literature (and to a lesser extent the operant analog of this literature) quite naturally contains a number of experiments performed to test the frustrative nonreward (FNR) hypothesis and to offer alternative interpretations of the FE. As I have already indicated, Scull (1973) provided an extensive review not only of this literature, but of these interpretations for the 20 years following the work of Amsel and Roussel (1952). I will here address myself to a small subset of these.

Perhaps the most extreme examples are two experiments by McCain and McVean (1967) that were taken as evidence against the FNR hypothesis of the FE. In the first, they ran a condition in which zero reinforcement was given in G_1 and found that when, in a second phase, they switched to

50% reinforcement in G_1 there was an "FE" that occurred soon after the shift in reward conditions. Even more dramatically, in a second experiment they showed that after zero reinforcement in G_1 there was an indication of the FE when, in the postshift phase, a single reinforcement was followed by a nonreinforcement. The implication of these experiments, and particularly the second, is that an effect resembling the FE can be shown to occur without any prior extensive buildup of reward expectancy.

A series of experiments by Sgro and others (e.g., Sgro, Glotfelty, & Moore, 1970) examined the effects of delay of reward in G_1 and showed that, unlike nonreward, delay of reward *reduced* second-alley speeds. In a subsequent experiment, Sgro, Showalter, and Cohn (1971) offered a competing response interpretation of the finding that previous exposure to delay in G_1 depresses speeds in Runway 2 after a shift to nonreward in G_1. It is not difficult to accept the fact that a period of delay in G_1 would not lead to the FE in Runway 2. The frustration occurs during the delay, and the reward that follows reduces this frustration and therefore Runway 2 speeds. Shifting from delay to nonreward in G_1 would clearly be expected to produce competing responses, particularly the response of retracing from Runway 2 to G_1, at least in the part of Runway 2 contiguous to G_1. It is for reasons such as these that the use of the double runway was for the most part restricted to the study of nonreward (or zero magnitude of reward) and reduced (nonzero) magnitude of reward in their various forms.

Unlike delay of reward, reduced magnitude of reward and time of nonreward detention in G_1 affect the FE. As to the magnitude experiments, the FE occurs in Runway 2 with reduced reward in G_1, as well as with nonreward (McHose & Ludvigson, 1965; Patten & Myers, 1970), and the size of the FE is directly related to magnitude of reward reduction (Bower, 1962). Successive reductions of reward in a three-runway sequence summate to increase FE (Bower, 1962). Gradual as well as abrupt reductions of reward in a single runway arouse primary frustration (R_F), as indexed in a subsequent phase of hurdle-jump training to escape the lowest magnitude of reward given in the earlier runway phase (Daly, 1974b). Of course, escape from FNR is indicative of its aversive properties, whereas the FE is taken as an indicant of its generalized drive (D) properties. There is also a positive relationship between the initial magnitude from which reward is reduced and the size of the FE, and this has been shown in within-subject (Peckham & Amsel, 1964, 1967) and between-subject (Krippner, Endsely, & Tucker, 1967) designs. The relation between magnitude of the FE and time of detention in the nonrewarding goal box can be expressed as an inverted V-shaped function, the values of detention in one experiment being 2.5, 10.5, and 30 seconds on nonrewarded trials (Horie, 1971).

Several experiments have tested, or at least addressed, the *demotivation hypothesis,* first proposed by Seward, Pereboom, Butler, and Jones (1957). This hypothesis states that the difference in speed of running in the second alley, following R and N in G_1, may be due to a depression in performance following reward rather than to facilitation following nonreward – that the consumption of food in G_1 may produce a temporary state of satiation or demotivation. In a comment on this article, which is seldom mentioned by proponents of the demotivation argument, I argued that our original and subsequent experiments on the FE involved very small amounts of food in G_1, whereas the Seward et al. experiments involved amounts eight times and four times these amounts (Amsel, 1958b). I also cited evidence from experiments that showed that prefeeding amounts of food in the range of our experiments actually facilitated performance, in terms of reduced errors in mazes and increased speeds in runways (Bruce, 1937; Morgan & Fields, 1938), a set of findings that had been summarized by Maltzman (1952) with the term *process need*. The most important support for the FNR interpretation of the FE against the demotivation hypothesis came a couple of years later when Wagner (1959) ran the appropriate control condition in a double-runway FE study, a group that was never rewarded in G_1. This group did not show the FE following nonreinforced trials, showing that it is omission of an expected reward that produces the FE. Mackintosh (1974), who provides an excellent summary of experiments on this issue, points out that the most satisfactory evidence for an interpretation of the FE in terms of FNR comes from a within-subject design (Amsel & Ward, 1965). In this case, a prediction from frustration theory is supported: the gradual appearance and eventual disappearance of the FE related to different stages of discrimination learning (see also Hug, 1970c; Peckham & Amsel, 1967; and Chapter 5, this volume, for details). Demotivation cannot be a factor in these experiments.

Another interesting hypothesis about the FE was that the increase in speed following nonreward in G_1 reflected not simply a nonassociative energizing effect of frustration, but a kind of associative mechanism; in cognitive terms, this amounts to "trying harder" to reach G_2 after nonreward in G_1. This was tested in the context of the concept of correlated reinforcement (Logan, 1960). The idea was that if reward in G_1 was negatively correlated with speed in the first alley (the slower the animal runs, the greater the reward), then performance in the second alley should be different than if reward in G_1 was uncorrelated with speed in the first alley. If the FE is an indication of trying harder, negatively correlated reinforcement in G_1 should cause the animal, when switched to nonrewards in G_1, to decrease speed in the second alley. The hypothesis was not confirmed. The FE was present and of the same magnitude following nonreward in G_1 whether reward had been correlated or uncorrelated with speed in the first alley (Logan, 1968).

Cognitive dissonance theory (Lawrence & Festinger, 1962), which relates behavioral persistence (the PREE) to the effortfulness of behavior, provided another hypothesis of the FE. The reasoning was that since partial reinforcement and effort expended in acquisition have in common the effect of prolonging extinction, the size of the FE should be directly related to effortfulness in Runway 1. In the only relevant experiment of which I am aware, the FE was demonstrated under all conditions of effort, but there was no relation between effort and the magnitude of the effect (Grusec & Bower, 1965). (A more extensive discussion of cognitive dissonance theory is provided in Chapter 6.)

Finally, a test of another hypothesis of the FE was made in an operant analog of the double runway (Berger, 1969) to determine whether the effect would be seen, or be as great, if the reward in G_2 for the frustration-motivated response was qualitatively different from the reward withheld in G_1. The hypothesis in this case was called the "preparatory response interpretation," and it held that the invigoration in the second response segment reflected the fact that the animal had been "prepared" to make the response because of the similar motivation existing in the first segment. An experiment with rats used food and shock avoidance as the rewards and showed that increased bar-press rates following the absence of food or the presence of shock as compared with their opposites occurred even though the "rewards" were different: The FE does not depend on the presence of the same reward in both goal boxes. This result, of course, speaks also to the demotivation hypothesis and to explanations in terms of associative mechanisms such as "trying harder." The Berger experiment is also interesting in relation to interpretations of the *omission effect,* which is taken to be the operant counterpart of the FE. One such interpretation is that the FE results from depressed performance after reinforcement, rather than increased performance after nonreinforcement, in the first link of a two-link response chain. Labeled *temporal inhibition* (Kello, 1972; Staddon & Innis, 1969), this interpretation of the FE is a kind of associative version of the demotivation hypothesis. In the operant case, a "blackout" following the first-link response (the analog of Runway 1) is substituted for nonreward in G_1. But unlike the situation in the Berger experiment, both response links in the Kello and the Staddon and Innis studies are followed by the same appetitive reinforcement. If temporal inhibition is a kind of appetitive demotivation, it should not demotivate and slow down a response motivated by shock avoidance. If, however, temporal inhibition is a kind of work or reactive inhibition (analogous to Hull's I_R), it should be generated whether or not reinforcement occurs, the difference being only that an increment of excitation should offset the inhibition on reinforced trials (Hull, 1943, p. 289).

After belaboring the niceties of interpretations of the double-runway FE and its operant analog, and after considering a small sample of the literally hundreds of experiments performed since 1952 on versions of this procedure (only about a half-dozen in my laboratory), I conclude that the paradigm has had its usefulness but that the energizing or activating properties of primary frustration can be defined quite well without an experimental paradigm involving two responses in series; it can be defined in the context of the absence of a reward for a single well-learned appetitive response.

We have already alluded to the effects of FNR on escape behavior (Daly, 1974b). However, several experiments performed at Tulane University, the home of the double runway, make this point equally clearly. The first of these experiments demonstrated that activity in an open-field apparatus following nonrewarded trials was greater than that following rewarded trials (Gallup & Altomari, 1969). A second experiment used a stabilimeter as the floor of the goal box of a single runway and demonstrated the same effect of absence of expected reward (Dunlap, Hughes, O'Brien, Lewis, & Dachowski, 1971) as did a third (Dunlap, Hughes, Dachowski, & O'Brien, 1974), which, like the original double-runway study (Amsel & Roussel, 1952), also addressed the temporal lability of the effect. The point here is that to demonstrate the generalized activating property of FNR it is not necessary to become involved with the interactive effects of two goal events in tandem arrangement, each preceded by a response; it is necessary only to observe and quantify the fact that one of the consequences of FNR is the unconditioned amplification of the behavior, whatever it is, that is prepotent immediately following the frustrating event.

A related point perhaps worth considering, at least as a historical note, is that the FE was advanced originally as an objective basis for defining primary frustration (R_F), the hypothetical unconditioned response in the conditioning of anticipatory frustration (r_F), an essential component in the interpretation of negative incentive contrast (Amsel, 1951), and the PREE and aspects of discrimination learning (Amsel, 1958a). A means of defining R_F to provide a basis for r_F was the important consideration and not the double-runway apparatus or the FE as a phenomenon. However, as we have seen, this apparatus and its operant analog have had their own usefulness in terms of the behavioral effects they have revealed and the interpretations of these effects.

The frustration drive stimulus

The frustration drive stimulus (S_F), the second characteristic of primary frustration to which I alluded earlier, exercises control over behavior as a characteristic internal cue. It is defined as arising as feedback from the primary frustration reaction (R_F–S_F) and is differentiated from the cues presumed to arise as feedback from conditioned frustration (r_F–s_F) (Amsel, 1951, 1958a). The role of stimuli arising out of primary frustration is to provide cues for choice behavior (Amsel & Prouty, 1959; Amsel & Ward, 1954); for the direction of aggression (e.g., Azrin, Hutchinson, & Hake, 1966; Dollard et al., 1939); for escape responses that remove an animal from frustration in a goal box on nonrewarded trials (Adelman & Maatsch, 1955; Daly, 1969a), from the frustration of reduced rewards (Daly & McCroskery, 1973), or from the negative goal event in discrimination learning (Daly, 1971). While the hypothetical function of the frustration drive stimulus is clear, conclusive evidence for its role in choice behavior (e.g., Amsel & Ward, 1954) has been questioned on the grounds that visible but inaccessible food is a better stimulus for such behavior than no food (Tyler, Marx, & Collier, 1959).

Stimuli arising out of conditioned frustration, in contrast, are signals that alert the animal to an upcoming aversive event and provide some basis for escaping from these cues (e.g., Daly, 1969b, 1974a) or for avoiding such an event (e.g., Amsel & Surridge, 1964). The conceptualization of stimulation of this second sort has been very important in our work because, as the signaler of negative upcoming goal events, it has provided the mechanisms for relating frustration not only to suppressive effects in extinction and discrimination learning, but also to persistent behavior in these forms of learning and to regression (Amsel, 1971a). We shall deal first with the concept of persistence, and later the case will be made that, in terms of our analysis, regression may be another form of persistence.

Theory of the partial-reinforcement extinction effect

Perhaps the most frequently cited portion of the theory is a set of hypotheses that concern the influence of r_F–s_F on an appetitive instrumental response that continues to be performed in its presence (Amsel, 1958a). (A parallel set of hypotheses described the role of primary and conditioned frustration in successive discrimination learning; see Chapter 5.) The prototypical case is the discrete-trial, partial-reinforcement (PRF) experiment in which, for the experimental group, both rewards and nonrewards occur on some percentage of trials for the same instrumental response (the PRF condition), whereas the control or continuous-reinforcement (CRF) group is rewarded on every trial. The basic finding is called the partial-reinforcement extinction effect (PREE), and it describes the fact that the rate of experimental extinction of a response is lower following PRF than following CRF training: Animals (and humans) are more persistent after exposure to a PRF than a CRF schedule of reinforcement. (We will have many occasions to examine such a result in this book.) The four-stage hypothesis of the PREE (Figure 3.3) outlines a sequence of events leading to the development of persistence in the presence of a stimulus (S±) to which the approach response (R_{APP}) is both rewarded and not rewarded from one trial to another on an unpredictable basis (see Amsel, 1958a): (1) The occurrence of rewards early in PRF training results in the conditioning of r_R–s_R to the stimuli (S±) of the situation. (2) Once r_R–s_R is sufficiently strong, nonrewards (NR) evoke primary frustration (R_F) and R_F becomes the UCS for the conditioning of r_F–s_F to the S± cues. (3) As r_F becomes stronger, its feedback stimulation (s_F) evokes responses that compete with the established instrumental response, and the mean amplitude of that response decreases, mainly because of increased response

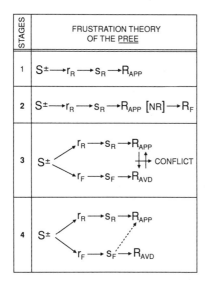

Figure 3.3. Four-stage frustration theory of the PREE. See text for details.

variability. (4) If the subject continues to perform the response to the goal (because reward is found there on some trials), s_F comes to evoke the instrumental response because of the repeated contiguity of s_F and approach to the goal (dashed arrow Stage 4). The mechanism invoked is instrumental counterconditioning – the conditioning of approach to s_F. Other possible outcomes of PRF training, however, are (a) that the animal will remain in conflict (at Stage 3) or (b) that the conflict of Stage 3 will be resolved by the counterconditioning of avoidance to s_R (not shown in Figure 3.3) and that the result of the PRF training will be desistance (faster extinction) instead of persistence.

As we shall see, the assumption that the instrumental response becomes counterconditioned to s_F, the feedback from r_F, has provided some theoretical leverage in the explanation of certain paradoxical effects of reinforcement, some of these clearly beyond the phenomena to which the theory was initially applied – the FE, the PREE, negative contrast, and discrimination learning.

In the case of the PREE, the assumption that anticipatory frustration-related stimuli (s_F) come to evoke the instrumental response during extended PRF training provided a plausible alternative in 1958 to what has been called the Hull–Sheffield (V. F. Sheffield, 1949) hypothesis of the PREE. According to this hypothesis, persistence in approaching the empty goal box in extinction after PRF (relative to CRF) training has the following explanation: Some portion of the stimulation to respond, on any trial following the first trial in an acquisition sequence, is composed of carried-over traces of stimulation feeding back from the consummatory goal response on the previous trial. In a CRF schedule, the argument goes, these carried-over traces are always rewarded-aftereffect cues, whereas in a PRF schedule they are mixed – some are aftereffect cues from reward and some from nonreward. Since, on all extinction trials after the first, all carried-over cues are from nonreward, there is greater generalization decrement (less similarity) between CRF acquisition training and extinction than between PRF training and extinction. Consequently, CRF training results in the more rapid extinction. As we shall see, this explanation of the PREE became the basis for Capaldi's (1966, 1967) more formalized (and more elegant) *sequential hypothesis*.

Between the time of V. F. Sheffield's (1949) article and Capaldi's revitalization of the Hull–Sheffield hypothesis, some experiments suggested that this hypothesis was an insufficient explanation for *all* cases of the PREE, since this effect could be shown to occur under conditions in which direct short-term carryover from the previous trial could not operate as a cue for the next trial. Experiments by Weinstock (1954, 1958) showed that the PREE occurred when trials were spaced as widely apart as 24 hours. (A later experiment showed a very robust PREE at a 72-hour intertrial interval [Rashotte & Surridge, 1969].) Two other experimental tests of Sheffield's interpretation of the PREE (Wilson, Weiss, & Amsel, 1955) also showed – in conditions different from hers, involving dry food and water as rewards instead of the wet mash she used, and spaced trials as well as the massed trials she employed – that in all of these conditions the PREE showed up as a very strong effect.

The frustration theory interpretation of the PREE, as it appeared in 1958, was in part, then, the formalization of the point of view that the Hull–Sheffield interpretation of the PREE could not account for PRF-related learned persistence under all conditions of its occurrence. The frustration theory position was that, for other than the highly massed-trial case, an explanation based on something other than directly carried-over cues was required. The proposal, as we have seen, was that, in extinction following PRF training, the anticipatory-frustration-produced stimulus (s_F) continues to evoke the instrumental response, whereas in extinction after CRF training, responding deteriorates rapidly because frustration occurs for the first time, as does, subsequently, its conditioned form (r_F–s_F): The feedback stimulus (s_F), to which the approach response was not counterconditioned in CRF training, now evokes strong avoidance rather than continued approach. This same kind of explanation of the PREE was provided in the discussions of the results of two experimental studies (Kendler, Pliskoff, D'Amato, & Katz, 1957; Wilson, Weiss, & Amsel, 1955), the difference between the two being that Kendler et al. preferred a neutral designation of the conditioned effect of nonreinforcements in the PRF schedule, while our somewhat earlier characterization of it was in terms of frustration.

Overview of the explanatory domain of the theory

In this and subsequent chapters, greater detail and specific references will be provided for the assertion that our neobehavioristic conditioning model has provided some theoretical integration of a number of phenomena besides the frustration effect, the Elliot (1928) and Crespi (1942) negative incentive-contrast effect, simple discrimination learning, and the PREE. The theory has been extended to account for the early appearance and later disappearance of primary frustration in discrimination learning and the effects of prediscrimination exposure to stimuli on subsequent discrimination learning (Chapter 5). It has addressed the retention and durability of persistence (Chapter 4), and more recently developmental considerations of these problems (Chapter 7). Some of the areas in which explanations in terms of frustration theory have been applied by a number of investigators are the overlearning extinction effect; the overlearning reversal effect in discrimination learning; the phenomenon of subzero extinction (a finding analogous to the depression effect); the paradoxical Haggard (1959) and Goodrich (1959) partial-reinforcement acquisition effect (PRAE), in which partially rewarded animals show higher speeds of running to the goal box in acquisition than continuously rewarded animals; the action of certain drugs like alcohol and sodium amytal in attenuating the PRAE and the PREE; certain other phenomena of contrast, including simultaneous negative contrast and aspects of operant behavioral contrast; the appearance of certain "adjunctive" behaviors, like schedule-induced polydipsia and aggressive behaviors in Skinnerian operant experiments; the role of the limbic system in the FE, the PREE, and certain other phenomena of reinforcement; and the transfer of persistence across situations and motivational-reinforcement conditions, including the effects of prior experience on later behavior and "regression" to earlier successful modes of behavior.

An extension of the theory provided a more general account of persistence, which views the consequences of partial reinforcement as only one instance of a more general case (Amsel, 1972a). The more general theory, and some of the experimental work related to it, are presented in detail in Chapter 4. The theory holds that if an instrumental response is initially disrupted by a class of distracting stimuli, and if habituation eventually occurs to such stimuli, the response will have been counterconditioned to these stimuli with the consequence that it will later be relatively unaffected by these and other distracting stimuli. This theory makes frustration theory a special case of the more general one. As we shall see, this general position bears some similarity to earlier neo-Guthrian accounts of the PREE in terms of habituation (Estes, 1959; Weinstock, 1954). It stems from our own work on transfer of persistence and also from the work on fear–frustration commonality – the idea, with which I do not agree, that as

Gray puts it in its extreme form, fear equals frustration (Gray, 1967; Terris, German, & Enzie, 1969; Wagner, 1966). My own position is that fear and frustration have in common the capacity to disrupt behavior and to contribute to the acquisition of persistence. One of the effects of adopting this more general position has been to move our work in the direction of a more ontogenetic-developmental perspective and subsequently toward very recent work relating the developing behavioral effects to the developing brain (Chapter 8).

But before we go on to consider these later developments, the remainder of this chapter provides a more detailed overview of some of the earlier ones. This presentation is weighted rather heavily on the concept of persistence, which has been so central to this work. However, the Appendix provides references to areas of research that have placed a number of other reward-schedule phenomena under the general explanatory umbrella of frustration/persistence theory; and, in Chapter 7, where we review the developmental aspects of this work, greater attention will be paid to these other phenomena.

Frustration, punishment, and persistence

The term *persistence,* for present purposes, refers to a tendency for organisms to pursue goal-directed activities despite nonreinforcement, punishment, obstacles, or deterrents – in general, in the face of any kind of negative indication. For some years *differential resistance to extinction* has been the classical, learning-theory term that specifically identifies the two levels of persistence that define the PREE. Among the other forms of persistence (see Chapters 4 and 5) are retardation of discrimination (Amsel, 1962), which we have called "resistance to discrimination" (Amsel & Ward, 1965); continued approach in the face of punishment, which has been called "courage" (Banks, 1966, 1973; Brown & Wagner, 1964; Fallon, 1968; Linden & Hallgren, 1973; Miller, 1960); and, as I have suggested, a form of "regression" (Rashotte & Amsel, 1968; Ross, 1964). To the extent that these have been studied, the evidence is that the mechanisms operating are similar to those in resistance to appetitive extinction. In the present usage, then, *persistence* is a general term encompassing all of these cases, and it can be said to involve a single basic factor common to all of them – learning to approach in the face of cues signaling some degree or probability of negative consequence. For persistence to develop there must be some uncertainty of outcome; there must be significant probabilities both that reward will be present and that it will be absent following a response; or there must be significant probabilities that a response will be terminated both by reward and by punishment.

Let us first consider the kinds of behavioral persistence that are related

An overview of its experimental basis

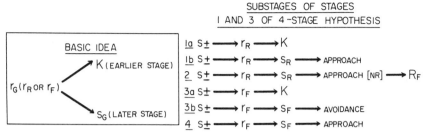

Figure 3.4. Six-stage frustration theory showing substages (1a and 1b, 3a and 3b) of Stages 1 and 3 of the four-stage hypothesis of frustrative effects in PRF acquisition. These substages are derived from the basic idea of earlier and later stages of r_R and r_F effects shown in the box at the left. These substages are necessary to account for the separate energizing and directive effects (in that order) of r_R and r_F. (From Amsel, 1967. Reprinted by permission of Academic Press.)

to approaching a goal despite stimuli that signal frustration or punishment. In these cases, what we have called "active" persistence (e.g., Amsel & Ward, 1965) involves conditioned emotional responses and the counterconditioning of feedback stimuli from these responses from avoidance to continued approach. There have been several alternative conceptualizations of the mechanisms that mediate persistence (Capaldi, 1967; Lawrence & Festinger, 1962; Sutherland, 1966; Sutherland & Mackintosh, 1971 – to mention some of the more influential ones); however (and perhaps in their favor), these have been applied less generally than mine – almost exclusively to the interpretation of the appetitive PREE. (We describe these alternative theories in some detail in Chapter 6.) The frustration theory of the PREE has been extended and generalized, though in its simplest form it represents the process of acquiring persistence as a hypothetical four-stage sequence of events (see Figure 3.3). In the four-stage model, as we have seen, Stage 1 of PRF training describes how the rewarded trials operate to affect the conditioning of r_R; nonreward trials cannot effect any significant amount of frustration until r_R develops in strength. In Stage 2, when r_R is already strong and is a factor in the evocation of the instrumental response, the occurrence of nonreward results in primary frustration (R_F). During this second stage there is also the beginning of a buildup of r_F, conditioned on the basis of R_F as the unconditioned response. In Stage 3, anticipatory frustration evokes avoidance and anticipatory reward evokes approach, and conflict results. Finally, in Stage 4 of PRF acquisition, the anticipatory frustration-produced cues (s_F) come to evoke approach as well as avoidance. This instrumental counterconditioning of continued approach to cues signaling nonreward is the frustration theory mechanism for persistence in appetitive learning under intermittent reinforcement.

A later version of the theory (Figure 3.4) makes specific reference to the nonassociative (K) factor in learning under a PRF schedule. It emphasizes that the function of r_F varies with its strength (Amsel, 1967); that while r_F contributes, at all values of its strength, to the nonspecific energizing aspect of incentive motivation, K, it provides an effective s_F only at strengths above some threshold value. At all of its values, including weak intensities, r_F contributes to generalized drive or arousal relative to its intensity. As r_F becomes stronger, however, a directive feedback (s_F) emerges and provides an aversive stimulus that evokes unconditioned and conditioned responses, antagonistic to the referent instrumental behavior. Obviously, such competing responses evoked by s_F provide a plausible (at least partial) account of extinction. As we shall see, perhaps less obviously they provide an explanatory account of other phenomena as well.

An account similar to the frustration theory explanation of persistence has been given for what Miller (1960) has called "courage" – learning to approach in the face of stimuli signaling punishment (for specific treatments, see Banks, 1966, 1967; Martin, 1963; Wagner, 1966, 1969). The difference between punishment and frustration at a goal is only that frustration emerges out of nonreward in the context of anticipatory reward, whereas direct physical punishment does not depend on an associative history and can be directly introduced into a previously rewarding situation at any time. (I pointed out earlier that r_R, r_F, and r_P stand for specific anticipatory versions of primary reward, frustration, and pain/punishment, respectively, having the characteristics of the more general r_G [anticipatory goal response], and that Mowrer [1960] later made things more symmetric by adding "conditioned relief" [r_{REL}], so that r_{REL} is in relation to r_P as r_R is to r_F.)

In many experiments, we and others have studied the manner in which an animal learns to approach a goal in the face of cues signaling that, instead of being rewarded, the animal may be frustrated or punished. The more experience of this kind an experimental animal acquires, at least up to a point, the more likely it is that the animal will be persistent in this kind of situation. The circumstance under which frustration, fear, or, as we shall see (Chapter 4), disruptive events in general lead to persistence is that in which intermittent reward and nonreward (or intermittent reward and punishment) are applied to the same behavior in approximately the same situation. We can, of course, apply these simple principles to nonexperimental settings. For example, when an infant cries and is lifted out of the crib and petted, the infant will learn to continue crying until picked up. If the infant is picked up only sometimes, this inconsistent treatment will probably produce a persistent crier. That is to say, when the parents decide in desperation to let the infant "cry it out," crying will continue for a long time. If the crying is always rewarded, this will, of course, establish it very strongly as a response; but under these circumstances of continuous reward it will also tend to extinguish very quickly. On the same principle, if the same behavior is sometimes rewarded and sometimes frustrated or punished, the child will tend to persist in the face of frustration or punishment. And it is of course possible, on the basis of a principle of fear–frustration commonality (Brown & Wagner, 1964; Wagner, 1966), that early but inconsistent frustration will result in later persistence in the the face of punishment and vice versa. These and other instances of transfer of persistence will be discussed in the next section and in more detail in Chapter 4. As we shall see, they have important implications for the developmental study of emotion, temperament, and personality, which is to say, for the endpoints of dispositional learning.

Transfer of persistence effects

We have been considering a means of conceptualizing the intermediary factors that control persistent behavior in terms of a classical-conditioning model. The argument goes that, in the PRF experiment, the control by s_F (or s_P) of responding learned in acquisition operates in that same physical situation to mediate persistent behavior when that same response undergoes extinction. But is this pattern of persistent behavior highly specific to the situation in which it was learned? Or is persistence that has been learned in relation to one particular situation likely to show up in some others? To put it another way, do the behavioral effects of persistence transfer from one situation to another? Can a persistent personality be learned?

A very important component of all the chapters that follow is the generality and transfer of the products of dispositional learning from one situation to another. For example, experiments to be discussed in Chapter 4 involve reward-schedule comparisons *within the individual subject,* and those discussed in Chapter 5 involve the facilitation and retardation of discrimination learning. These experiments have shown that there is at least some situational transfer of suppression and persistence effects based on frustration. The procedure in these within-subjects experiments is, for example, to reward an approach response every time it is made in the presence of one stimulus, say in a black alley, and to reward that same response some lesser percentage of the time (say, 50%) when it is made in a white alley. The finding of these particular experiments is that when the response is extinguished in both alleys, there appears to be generalization of the PREE across both, unlike the differential persistence in the CRF and PRF between-subjects conditions (e.g., Amsel, Rashotte, & Mackinnon, 1966; Brown & Logan, 1965). These transfer-of-persistence experiments permit the interpretation that there is a degree of interoceptive stimulus control by s_F in extinction that overrides the external stimulus control of the differential alley color. In frustration theory terms, the mechanism proposed in this example of transfer of persistence from one alley color to the other (or the generalized PREE, to use Brown & Logan's

[1965] term) is *mediated generalization* of the $s_F \to$ approach countercon-ditioning, which operates, specifically, in the transfer of persistence from the PRF alley to the CRF alley. That is to say, as soon in extinction as anticipatory frustration occurs in the CRF alley, the persistence mechanism "switches on" in this alley: The addition of s_F to the stimulus complex brings with it a rearrangement of the habit-family hierarchy, and the response of approaching in the presence of s_F overrides the response of avoidance elicited by the relatively less important external cues.

If, as in our later work (see Chapter 7), the purpose is to study the ontogeny of persistence, a more representative within-subjects experiment is one in which the CRF and PRF training occur separately, in different stimulus contexts, in successive phases of time (at separate ages), rather than being mixed in a single time phase, as is the case in most within-subjects CRF/PRF or discrimination experiments. In the separate-phase paradigm, the experimental sequence is, for example, PRF $\to$ CRF $\to$ extinction (EXT) and the control sequence is CRF $\to$ CRF $\to$ EXT, and we look in the terminal extinction phase for effects of earlier PRF (as compared with CRF) training conducted in different stimulus conditions. The question is this: Can the effects of prior – even early – reward-schedule experience in one situation transfer to the extinction of behavior learned later in a different situation? Such transfer-of-persistence experiments will be reviewed in Chapter 4.

Regression as a transfer-of-persistence phenomenon

A number of early studies dealt experimentally with the relationship of frustrative extinction to the potentiation of behavior and to its variability and aggressiveness (Barker, Dembo, & Lewin, 1941; Miller & Miles, 1935, 1936; Miller & Stevenson, 1936). A portion of our work can be thought of in the context of these early experiments, and some of it, as I have suggested, can be interpreted as showing that frustration, both unlearned and learned, is implicated not only in the behavioral consequences of arousal or invigoration, suppression, aggression, variability, and persistence, but also in regression, the return to an earlier "successful" mode of behavior. But an important feature of, and prediction from, frustration theory is that while aspects of both primary and conditioned frustration intensify behavior (in the form of the FE and the PRAE) and may lead to aggression, it is only conditioned or anticipated frustration that is involved in persistence, whether the persistence takes the form of increased resistance to extinction or regression.

In what sense are we talking about regression? A counterconditioning view of persistence, as we have seen, implies that a connection of some kind is formed in acquisition between an initially disruptive (emotional) mediating event and some ongoing behavior. If this mediational control is powerful (the argument goes), s_F might elicit approach not only in the situation in which the counterconditioned connection was originally formed, as in the usual PRF experiment, but also in other situations in which anticipated frustration comes into play; that is to say, there may be a mediated transfer-of-persistence effect. It has already been suggested that examples of this explanatory mechanism can be applied to within-subjects transfer-of-persistence experiments. (Keller and Schoenfeld [1950] refer to the possibility that regression can be a by-product of extinction, and Mowrer [1940] showed experimentally that rats reverted to an earlier method for reducing shock when a panel-pushing response to reduce shock was extinguished. He regarded this demonstration as a prototype of regression.)

Still, how do we get from transfer of persistence to regression? As we shall see in detail in Chapter 4, persistence acquired under PRF conditions can survive a later block of CRF to affect subsequent extinction. Beyond this, however, we shall see that persistence acquired in one situation can have effects in a different situation, involving different responses and different motivational-reward conditions. Another way of putting it is that there can be regression to a mode of persistent behavior learned in the context of earlier PRF acquisition, and this regression can be said to be mediated by anticipated frustration. In this case, *to persist is to regress*.

Another case of transfer of persistence involves a series of experiments in which idiosyncratic response rituals learned under Logan's (1960) condition of discontinuously negatively correlated reinforcement (DNC) – a condition in which the animal learns to run slow (or take time) in order to be rewarded – emerged when these responses were extinguished, not in the presence of the DNC stimulus, but in a new situation in which animals had received uncorrelated CRF training, that is, in which there was no correlation between running speed and reinforcement. The interesting thing about these experiments is that persisting in extinction in the face of anticipated frustration takes a form of returning to an idiosyncratic response ritual that was learned in a situation different from the one just prior to extinction – responding to cues signaling frustration in a manner learned much earlier in a situation different from the current one. Again, this is a case in which persistence can take the form of regression.

A case can then be made that, in frustration theory terms, a frustration-regression hypothesis is supported if what we mean by frustration is the response-evoking properties of feedback cues (s_F) from *anticipatory* (conditioned) frustration (r_F). A case can also be made for a frustration-aggression hypothesis, if by frustration we mean *primary* (unconditioned) frustration (R_F) and its feedback cues (S_F). The difference, then, would be in the nature of the response-evoking stimulation, emerging in the former case out of the expectancy of frustration and in the latter case out of its direct experience. We will come back to these considerations in Chapter 4.

Summary of the explanatory scope of frustration theory

As I indicated earlier, most of the work described in this overview of frustration theory is heavily weighted on the concept of persistence. I have chosen to follow this plan of presentation because it develops a logical sequence of theoretical-experimental steps. However, as we shall see, the explanatory scope of the theory extends well beyond the empirical particulars that can be organized under the heading of "persistence," and this will become clear in the chapters that follow. (Provided in the Appendix is a partial list of experimentally studied phenomena that have been deduced from, or have at least a partial explanation in, frustration theory.) Other theories have addressed several subsets of these phenomena (see Chapter 6), but I think it fair to say that none of these theories has addressed all or even most of them.

The study of arousal, suppression, persistence, and regression, the end products of dispositional learning, has given rise to the study of these phenomena in transfer experiments (Chapters 4 and 5), and later in more developmental and psychobiological ways (Chapters 7 and 8). An exciting possibility is that when we study persistence in a developmental manner, we may discover that it is possible to learn general dispositions or characteristics of temperament that govern tendencies to persist (or to desist) and that the strength of such tendencies may be related to occasions, perhaps at sensitive stages of development, on which goal-directed behaviors occur in the face of frustration, physical punishment, or other kinds of disruptive events. The chapters that follow examine this possibility.

4 Survival, durability, and transfer of persistence

As will be apparent in this chapter, our work has been greatly influenced by an experiment performed some time ago, independently, by Theios (1962) and Jenkins (1962). Theios, using a runway apparatus and rats as subjects, and Jenkins, working with pigeons in a discrete-trial Skinner box, demonstrated that when one group of subjects acquires a response under CRF and another group under PRF conditions, and then both are exposed to CRF training before an extinction phase, the persistence acquired in PRF acquisition carries through the block of CRF trials, interpolated between acquisition and extinction, so that the PREE survives the common interpolated experience. This was an important finding for what I now call "dispositional learning" because, at the time, it contraindicated hypotheses based on expectancy (Humphreys, 1939b) or discrimination (Bitterman, Fedderson, & Tyler, 1953; Mowrer & Jones, 1945) as necessary for the explanation of the PREE. These two hypotheses are similar in that both interpret the PREE as being a more short-term effect, relating to a detectable change in percentage of reinforcement from the end of acquisition to the beginning of extinction, in the first case in terms of the degree of disconfirmation of an expectancy and in the second case in terms of the discriminability of that change.[3] A related position accounts for the PREE in terms of degree of generalization decrement, from acquisition to extinction, of the sequential effects of carried-over stimulation from reinforced and nonreinforced trials (Capaldi, 1966). A more detailed analysis of these and other interpretations of the PREE will be provided in Chapter 6.

From our point of view, the Theios–Jenkins design was also an experimental model for studying persistence in terms of developmental-learning mechanisms – that is to say, in terms of long-lasting, even permanent, associations formed early in life that can be activated later through internal, mediational mechanisms. (We will see in Chapter 7 how this model applies to developmental work with infant rats.) The design of Theios and Jenkins is featured in experiments that follow.

---

3. A similar result was reported in eye-blink conditioning in humans (Perry & Moore, 1965). The results of this classical-conditioning experiment were explained in terms of a modification of the Humphreys expectancy version of the discrimination hypothesis.

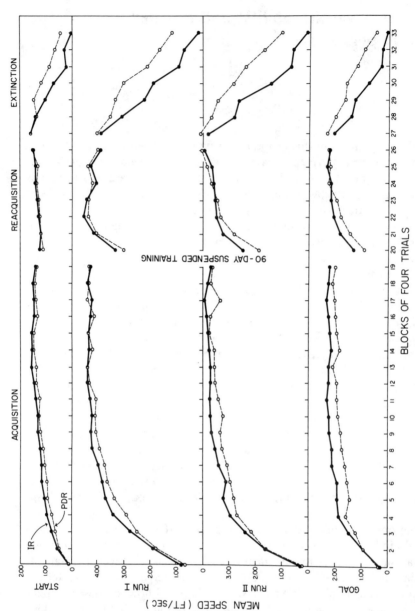

Figure 4.1. Performance in the four alley segments during the acquisition, reacquisition, and extinction phases. IR, Immediate continuous reward (CRF); PDR, partial delay of reward. Reacquisition is under IR (CRF) conditions for both groups. (From Donin, Surridge, & Amsel, 1967. Copyright 1967 by the American Psychological Association; reprinted by permission.)

Derivatives of the Theios–Jenkins experiment

Several published experiments from our laboratory extended the limits of the basic Theios–Jenkins results, in both between- and within-subjects experiments, and I will describe a few of these. They address the concepts of survival and transfer of persistence, and have been influential in our more general thinking about persistence.

Survival of persistence

In an experiment conducted at *one trial per day,* Donin, Surridge, and Amsel (1967) demonstrated that learned persistence acquired under conditions of partial delay of reinforcement (PDR; Crum, Brown, & Bitterman, 1951), in which all trials are reinforced but on half the trials reward is delayed for 30 seconds, carries through a period of 90 days of no training ("vacation"), and then through a block of CRF trials to affect extinction. Figure 4.1 shows the results of this experiment. These results make it even more difficult to hold that the necessary mechanism for instrumental persistence is the direct carryover of a stimulus trace or expectancy of the goal event from one trial to the next, or that the transition from acquisition to extinction is critical. In another three-phase-plus-"vacation" experiment (Rashotte & Surridge, 1969), one trial was run only *every three days.* Again the experimental design was the Theios–Jenkins sequence: A CRF phase was interpolated between acquisition and extinction of a running response, measured over five 1-foot segments of a runway. Long-term retention of learned persistence is dramatic under these extreme conditions, there being virtually no extinction in the PRF condition (Figure 4.2), and, of course, the same considerations about direct stimulus carryover or transition from acquisition to extinction as factors in learned persistence appear to apply.

These experiments extend the earlier findings of Theios and Jenkins on survival of persistence. First, the experiments are conducted with very large intertrial intervals and "vacations," making it even more unlikely, as I have suggested, that any form of simple discrimination of the transition from acquisition to extinction can account for the results. Second, the finding is extended to include not only PRF but also PDR, a condition in which every acquisition trials ends in reward. It seems not unreasonable, then, to think that in the case of training involving widely spaced rewards and nonrewards, persistent behavior may be mediated by some internal process that has the characteristics of a classically conditioned response. (Of course, other conceptions of the mediating mechanism are possible.) These experiments also extend the conditions under which persistence can be formed – but as we shall see, *not necessarily the mechanisms responsible for its formation.*

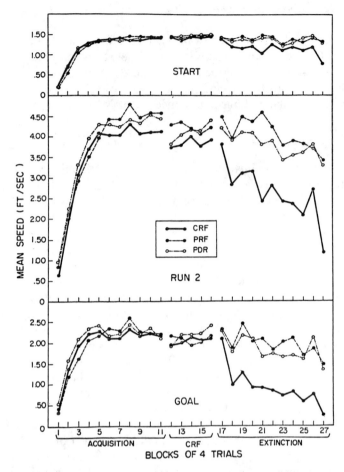

Figure 4.2. Relative persistence (resistance to extinction) after partial reward and partial delay of reward even after interpolation, between acquisition and extinction, of a block of continuously rewarded trials for all groups; and even though the experiment was run at one trial every 3 days. (From Rashotte & Surridge, 1969. Copyright 1969 by the Experimental Psychology Society.)

Durability of persistence

Other experiments, which we have classified as demonstrating "durability" of persistence are conducted in four stages rather than three: acquisition → extinction → reacquisition → extinction. The first acquisition ordinarily includes both CRF and PRF (or PDR) conditions. The question is whether the PREE (or the partial delay of reinforcement extinction effect, PDREE) is still evident in the second extinction (Phase 4) after extinction in Phase 2 and reacquisition under CRF conditions *for both groups* in Phase 3.

Durability of persistence, defined in this way, has been demonstrated in several experiments. The following are examples:

1. An experiment involving CRF and PRF in Phase 1 showing durability of persistence in two strains of mice (DBA and C57B1), the difference between strains being only that DBA mice ran faster in all phases (Wong, Lee, & Novier, 1971)
2. An experiment in rats showing the effects of manipulating the number of PRF trials in Phase 1; durable persistence was shown after 32 PRF trials with multiple-pellet reward, but not after only 4 such trials (Traupmann, Wong, & Amsel, 1971)
3. An experiment showing that the PREE, acquired in Phase 1 under highly spaced (one trial per day) conditions, can be sustained through 3 successive days of massed-trial (10-second intertrial interval) extinction (during which the PREE was evident mainly from a plot of only the first trial of each day), a 6-day "vacation," a spaced (one trial per day) reacquisition under CRF conditions for both groups, and can then show up clearly in an extinction test, also at one trial per day (Amsel, Wong, & Traupmann, 1971)
4. An experiment in which the first phase involved a PRF–CRF comparison for one group and a PDR–CRF comparison for another (Wong, Traupmann, & Brake, 1974)

In the last experiment, relative persistence was just as great after PDR as after PRF in the first extinction (Phase 2), but whereas persistence due to PRF training showed up as durable in Phase 4, it did not after PDR training. This result had also been suggested in an earlier study (Surridge, Mock, & Amsel, 1968).

Transfer of persistence

Let us extend the conditions even further (though still not the mechanisms). We have been dealing so far with persistence that is developed and measured in the same physical situation. However, an instrumental counterconditioning view of persistence, as we have so far employed the term, implies that a connection of some kind is formed in acquisition between an initially disruptive external or internal stimulus and some ongoing behavior. In the case of internal stimulus control, if this mediational control is powerful the counterconditioned response might be evoked not only in the physical situation in which the connection was formed, as in the usual PREE experiment (or the Theios–Jenkins variant of this experiment), but in any situation in which the internal mediating cues come into play. Another experiment (Ross, 1964) is a case in point. The experiment is outlined in Table 4.1. Again, note that the design has in it the essential features of the Theios–Jenkins experiment. However, there are also important addi-

Table 4.1. *Design of the Ross experiment*

	Phase 1: preliminary learning	Phase 2: acquisition running response	Phase 3: extinction running response
Apparatus	(A) Short, black wide box	(B) Long, white narrow runway	B
Motivation	Hunger	Thirst	Thirst
Experimental conditions	*Running* Continuous (RC) Partial (RP) *Jumping* Continuous (JC) Partial (JP) *Climbing* Continuous (CC) Partial (CP)	*Running* Continuous reward	*Running* Continuous nonreward

tional features: First of all, the preliminary learning phase (Phase 1) involves not one but three different responses, each learned under either CRF or PRF conditions, each then continuously reinforced in Phase 2 and extinguished in Phase 3. Running and jumping (across a short gap) in Phase 1 were meant to be compatible with the Phase 2–Phase 3 response, while climbing was seen as incompatible. If under a PRF schedule in Phase 1 subjects learn to make the required response to cues from anticipated frustration, and if these responses are controlled by internal mediating stimuli, they may emerge in Phase 3 extinction in a different physical situation, under different motivational conditions – in fact, in circumstances in which there is *no history of persistence training*.

To summarize the results and their implications, Ross's results, like the earlier work of Theios and of Jenkins, and the later ones from my laboratory, demonstrated that persistence acquired in PRF acquisition survives a block of CRF trials to affect extinction. In addition, however, Ross demonstrated that persistence acquired in one situation can have effects in a different situation, that persistence can transfer to different responses and different motivational conditions from those existing when it was developed. Reactions to anticipated frustration, learned when the subject was hungry and in a black box, emerged when the subject was thirsty and in a long, high, white, narrow runway, even though there had been no persistence training in that runway. It can be said that in this experiment the rat "regresses" in the Phase 3 extinction to the response acquired under PRF conditions in Phase 1 (Amsel, 1971a). Eleven out of 13 of the PRF trained-to-climb animals actually climbed in the long, high, white, narrow runway in extinction, and because climbing was incompatible with running, there was the appearance, in a comparison of CRF- and PRF-trained groups, of a *reversed* PREE – the PRF group "extinguished" faster than the CRF group. These findings, again, can be taken as pointing out the importance of internal relative to external stimulus control in the emergence of PRF-related persistence in extinction. Even more, however, they make it difficult to account (in stimulus–response terms, at any rate; it is always easier to explain a result cognitively, by putting oneself in the rat's place) for the emergence of the Phase 1 responses in extinction without a counterconditioning interpretation. As we shall see in Chapter 6, it is difficult to see how simple passive habituation to the disruptive effects of frustration (Estes, 1959), or "breadth-of-learning" (Sutherland, 1966), or sequential effects (Capaldi, 1966, 1967), or cognitive dissonance and increased attractiveness of the goal (Lawrence & Festinger, 1962) can account for the emergence of the climbing response in Phase 3 extinction, a response never seen in Phase 2 CRF acquisition. Still, whereas this experiment again extends the conditions under which persistence can be demonstrated, it still *does not require an alteration of the basic counterconditioning mechanisms of its formation.*

In an attempt to extend the Ross experiment as an analog of human failure and depression, Nation, Cooney, and Gartrell (1979) conducted an experiment in which male and female college students who experienced either PRF (persistence training) or CRF on either of two instrumental tasks (finger shuttle or button press) as "therapy" to mitigate later failure-induced "depression" were subsequently exposed to protracted failure (extinction). After this initial extinction phase, subjects in both conditions were given CRF training on a common instrumental task (button press) as "therapy," followed by a second extinction test. The results of the experiment were that the increased persistence occasioned by the PRF "therapy" was durable; it survived interpolated periods of extinction and CRF. Furthermore, these effects of persistence training were generalizable; the persistence advantage associated with the PRF "therapy" transferred across topographically different responses (Nation, Cooney, & Gartrell, 1979).

Transfer of idiosyncratic response rituals

A final experimental example of how persistence acquired in dispositional learning can transfer, and seems to depend on counterconditioning, is derivative of the Theios–Jenkins experiments and also of the Ross experiment. This is a within-subject experimental design in which animals acquire idiosyncratic response rituals under PRF-like (disruptive) conditions to one stimulus, while learning to respond normally under CRF conditions to another discriminative stimulus. If our view of how persistence is acquired and transferred is tenable, these response rituals learned in (say) a black stimulus alley should emerge in the white stimulus alley when the response in the white alley is extinguished. This kind of experiment (Amsel & Rashotte, 1969; Rashotte & Amsel, 1968) depends on Logan's (1960) conception of correlated reinforcement, and particularly on the procedure he calls "discontinuously negatively correlated reinforcement" (DNC). A DNC condition in an alley is one in which the rat must take time to be rewarded. This is analogous to differential reinforcement for low rates of responding in a lever box (Skinnerian DRL). In our case the application of the DNC procedure involved the following response-reinforcement contingency: The rat had to take 5 seconds or longer to traverse the alley if it was to find food in the goal box. If it took less time, it found no food. The beauty of this requirement is that although the animal learns after many trials to take time (as we shall see, by establishing a response ritual), the time it takes varies very narrowly around the 5-second cutoff, and as a consequence, reinforcement occurs about 40% of the time on the average, meaning that a PRF schedule for a response ritual obtains, so far at least as the rat is concerned.

In one version of this experiment, again following the Theios–Jenkins sequence, in Phase 1 each rat ran concurrently in two 5-foot alleys, one black and the other white. Under counterbalanced conditions, the DNC condition obtains in (say) the black alley, and in the white alley the rat runs under the uncorrelated CRF condition; that is to say, it learns to run fast. For each animal in the combined DNC–CRF condition, there is a yoked control in a PRF–CRF condition. In such a condition, the yoked subject is rewarded on PRF trials, uncorrelated with speed, only on those trials on which its DNC partner "earns" reward by taking time. Speeds are taken over five 1-foot segments. The data for each trial can then be plotted as speed against segment of the alley and provide a speed "profile" of each animal's performance. Following Phase 1, in the manner of Theios and Jenkins, Phase 2 involved training in which *all* trials for both groups (DNC–CRF and PRF–CRF) were in the CRF runway, were uncorrelated with time taken, and were rewarded continuously. In Phase 3, both groups were extinguished in the CRF runway.

In this and other experiments, we showed that when run under DNC conditions in (say) black and under CRF conditions in white, most animals do, in fact, learn idiosyncratic response rituals to take time in the DNC-stimulus alley and run quite normally in the alley that signals uncorrelated CRF. This can be seen very clearly in Figure 4.3, which shows acquisition response profiles over five 1-foot segments of the alley for four (the odd-numbered) DNC–CRF rats and their (even-numbered) PRF–CRF controls. (Remember that a response profile plots, for each animal, over individual trials, or a small block of trials, its speed over each of the five successive 1-foot segments of the runway.) Each odd-numbered horizontal panel plots the acquisition speed profiles for a single rat in both the DNC and CRF alleys. Note that, in general, CRF profiles can be described as inverted Vs or Us, while the DNC profiles are, by the end of training, generally not of this shape and are quite idiosyncratic. Furthermore – and

Survival, durability, and transfer of persistence

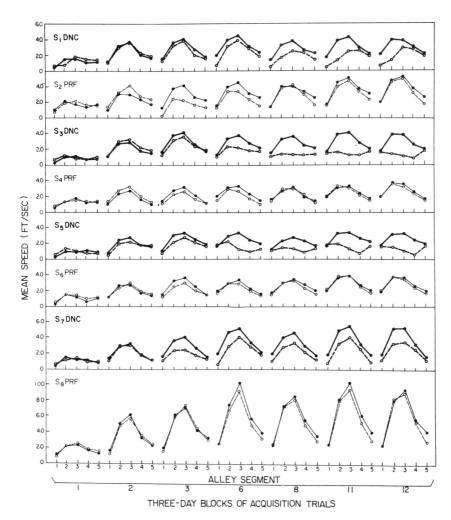

Figure 4.3. Acquisition-response profiles over five 1-foot segments for eight rats, four running in discriminable alleys under DNC or CRF conditions and their four yoked PRF or CRF controls. Each odd-numbered panel presents response profiles of a single animal under DNC and CRF conditions; even-numbered panels are response profiles for single animals under PRF and CRF conditions. Solid lines, CRF; dashed lines, DNC.

the reader cannot, of course, see this in the graph – the behavior is *highly ritualized*. (In more old-fashioned terms, these profiles can be described as idiosyncratic "goal gradients.") Different rats learn the slow-response rituals in very different ways under the DNC condition; that is, different rats take time differently across the segments of the runway. Again, the profiles do not adequately describe the qualitatively different rituals (in some cases almost contortions) these animals go through to take time. (It is almost like watching a child fidget while waiting to be released from the dinner table to go out and play – or, in a more up-to-date example, to watch television.)

We also showed, when this experiment was run in three successive phases, that, following Phase 2, an interpolated period in which only CRF trials were run in *both* stimulus alleys, the DNC pattern does, in fact, emerge in the extinction of responding to the uncorrelated CRF stimulus in Phase 3. The DNC pattern emerges most clearly when, as in the present case, responding to the CRF stimulus undergoes extinction while responding to the DNC stimulus does not (see Rashotte & Amsel, 1968). Figure 4.4 shows the results from two pairs of yoked animals (Subjects 1 and 2, and Subjects 3 and 4 from Figure 4.3), which, in extinction, were run in only the CRF-stimulus alley. (Remember that in the yoked condition in the first phase of the experiment, for every animal running DNC–CRF and *earning* reward in the DNC alley, a mate in a PRF–CRF condition is rewarded on the corresponding trial in the PRF alley, i.e., in the same position in the sequence of trials as its DNC mate.)

The remarkable thing about these data is that a *response ritual* emerges in extinction to a stimulus (in a situation) that never evoked that ritual before. The only possible explanation is that this idiosyncratic pattern of behavior emerges in extinction because of some mechanism in extinction that calls forth an association that was formed in acquisition in reaction to a different external stimulus. The mechanism I have favored is mediated (secondary) stimulus generalization. This is a kind of generalization in which a more recent stimulus (B) elicits an external response (a ritual in our example) that was seen earlier to stimulus A, by virtue of the fact that both A and B have elicited the same mediator; in our explanation, r_F–s_F. I regard this series of experiments as a powerful demonstration of how the mediation of anticipatory frustration and its counterconditioning to ongoing behavior control persistence effects. Obviously, my definition of persistence is broad enough to include the *emergence of a response in extinction that is different from the response the animal learned in acquisition to the same stimulus*. This response to cues signaling frustration, punishment, or other disruptive events may be one that was learned earlier in life in a situation not at all similar to the present one. All of this adds another dimension to the conditions under which persistence can be formed and demonstrated, and even to the nature and idiosyncrasy of the persisting response. Still, the mechanisms I have so far proposed are the same in all cases. They all depend on the same chain of events: primary followed by a conditioned (anticipatory) frustration and the counterconditioning of the persisting response to feedback cues from anticipatory frustration.

To say that learned persistence may be related not only to disruption by frustration but also to any other sort of disruption is to broaden its definition and therefore the theory that explains it. We shall now consider the concept of general persistence and the evidence for it.

Survival, durability, and transfer of persistence

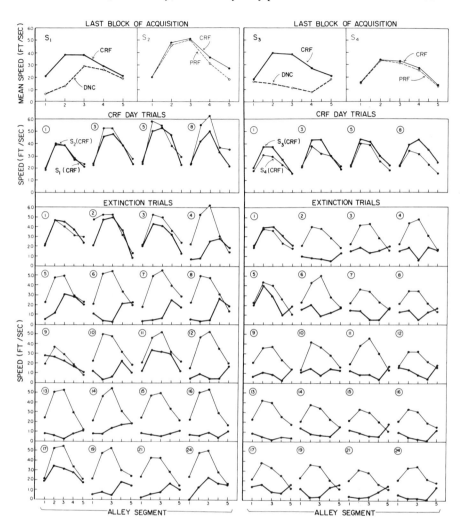

Figure 4.4. Profiles for two DNC–CRF subjects (S_1 and S_3) and their yoked PRF–CRF controls (S_2 and S_4). Each of the top panels shows profiles for one subject on the last block of acquisition trials. (The expanded scale on the abscissa in these panels, relative to the others, accounts for the apparent lack of peakedness of the terminal acquisition profiles.) The second row of panels shows profiles in the CRF runway on Trials 1, 3, 5, and 8 of the CRF day for a yoked pair of subjects. The remaining panels show profiles for these subjects on 24 individual extinction trials. Note that the DNC profile of running speed across segments of the runway reappears in the extinction performance of the DNC subjects. (From Rashotte & Amsel, 1968. Copyright 1968 by the American Psychological Association; reprinted by permission.)

A general theory of persistence

On the basis of experimental work from our own laboratory and elsewhere, I was led to entertain the possibility that when we study persistence in the microcosm of the PRF experiment, we may be dealing with a special case of the variables that affect persistence more generally (Amsel, 1972a). To understand the thrust of this more general theory, we must briefly review its background.

For several years we had been working with the theory of instrumental reward learning detailed in Chapter 3, in which persistence in achieving a goal is taken to reflect learning to respond to feedback cues from anticipated frustration (Amsel, 1958a, 1962, 1967). The basis for this theory was a series of experiments beginning in the early 1950s, carried out by many people in many laboratories, which showed that frustrative nonreward influences behavior in at least three ways: (a) through nonspecific emotional-motivational effects from primary frustration (R_F) that result in sharp increases in the vigor of ongoing behavior; (b) through characteristic feedback stimuli from primary frustration (S_F) that come to be associated with and control the direction of behavior; and (c) through the elicitation of behavior by feedback stimuli (s_F) from the hypothetical responses of conditioned anticipatory frustration (r_F).

The four-stage frustration hypothesis of the PREE, outlined in Chapter 3 (see Figure 3.3), arose from the findings of these early experiments. It conceptualizes how a disruptive process based on nonreward emerges in PRF acquisition and produces conflict in Stage 3, and how the disruption recedes in Stage 4 with the concomitant strengthening of the connection between s_F and R_{APP}.

By the late 1960s there seemed to be some evidence that the classical-conditioning mechanisms in frustration theory could be taken to apply somewhat more generally than I had initially supposed. As we have seen, the counterconditioning mechanism proposed by this theory could be said to operate in the generalized PREE in the transfer of persistence effects across stimuli and across motivational and reinforcement conditions, in delayed as well as in partial reinforcement, and in the learning of response rituals that then emerge in the extinction of responses in stimulus contexts that never before elicited them.

Another way to describe the "neutralization" of the effects of frustrative nonreward in extinction through PRF procedures is to say that there is "habituation" of competing responses to stimuli from nonreward (e.g., Estes, 1959) or from anticipated frustration (Mowrer, 1960) during PRF acquisition, and that this habituation accounts for the "frustration tolerance" of rats (and humans) in extinction. The argument is that in extinction there is less disruption of responding after PRF than after CRF training because the competing responses to stimuli from nonreinforcement have habituated during PRF acquisition. While it had never seemed unreasonable to hold the view that habituation of responses in acquisition affects frustration tolerance in extinction, it seemed to me that "passive" habituation or simple frustration tolerance could not account for the transferred persistence effects in the Ross experiment or for the transfer of ritualized responses in the DNC experiments. A more "active" counterconditioning mechanism seemed to be required. Consequently, I was led to combine the ideas of habituation and counterconditioning into a more general theory of persistence: I turned the argument around to say that a kind of active process is involved in behavioral habituation – that any instance of behavioral habituation to an initially disruptive event involves some degree of counterconditioning, which, in turn, leads to increased persistence in the face of later disruptive events.

This more general theory of persistence (Amsel, 1972a) subsumes the frustration interpretation of the PREE as a special case. It is more general in the sense that persistence is held to result from the counterconditioning to ongoing behavior of stimuli other than the specific frustration-produced stimuli called for in the theory of the PREE. According to this more general view, persistence develops in responding whenever an organism learns to approach, or to maintain a response, actively or passively, in the face of any kind of stimulus that arouses external or internal competing-disruptive responses. The anticipation of frustrative nonreward (and of other negative events) is assumed to be only one such disruptive response. In terms of a stimulus–response schema, the general theory that has just been proposed is diagrammed, in its most simple form, in the right-hand portion of Figure 4.5. (The left-hand portion is the corresponding outline of frustration theory; see Figure 3.3.)

According to this schema, when a disruptive stimulus (S_X) is introduced into an ongoing situation (S_O), an interfering response (R_X) at first competes with the ongoing response (R_O). However, on successive intrusions of S_X into the situation, instrumental counterconditioning of R_O to S_X occurs, and the original ongoing response (R_O) becomes more and more dominant to R_X. In the process, the $S_O \rightarrow R_O$ association becomes more resistant to disruption by stimuli of the class S_X.

It was not possible in 1972 – and it is not now – to give an exhaustive list of members of the class S_X; however, inclusion in such a class of stimuli would, I think, be easy to agree upon, and in general terms, S_Xs can be defined as stimuli to whose disruptive effects organisms habituate. It was clear that the s_F of our theory of the PREE could be regarded as simply *one of these kinds of stimuli* – one that emerges out of a special chain of circumstances in PRF acquisition and elicits responses that disrupt ongoing (usually approach) behavior. Finally, these responses are instrumentally

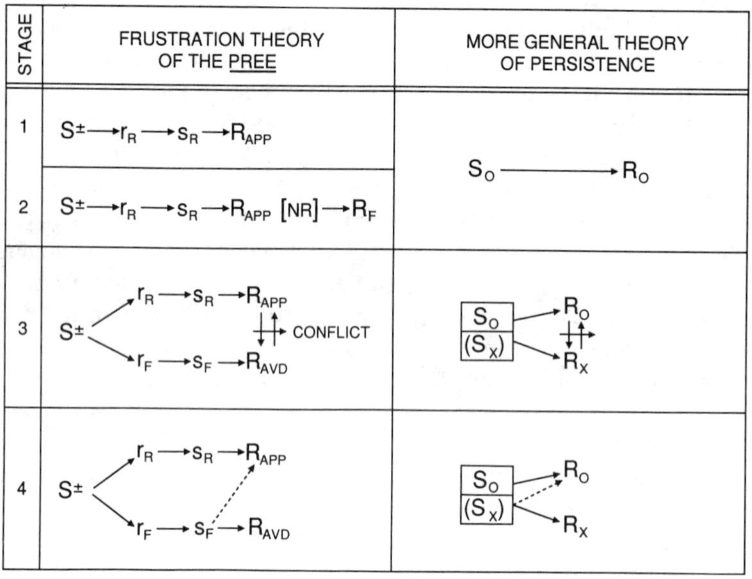

STAGE	FRUSTRATION THEORY OF THE PREE	MORE GENERAL THEORY OF PERSISTENCE
1	$S\pm \rightarrow r_R \rightarrow s_R \rightarrow R_{APP}$	$S_O \rightarrow R_O$
2	$S\pm \rightarrow r_R \rightarrow s_R \rightarrow R_{APP}\ [NR] \rightarrow R_F$	
3	$S\pm \begin{matrix} \nearrow r_R \rightarrow s_R \rightarrow R_{APP} \\ \\ \searrow r_F \rightarrow s_F \rightarrow R_{AVD} \end{matrix}$ CONFLICT	$\begin{matrix} S_O \\ (S_X) \end{matrix} \begin{matrix} \rightarrow R_O \\ \updownarrow \\ \rightarrow R_X \end{matrix}$
4	$S\pm \begin{matrix} \nearrow r_R \rightarrow s_R \rightarrow R_{APP} \\ \\ \searrow r_F \rightarrow s_F \rightarrow R_{AVD} \end{matrix}$	$\begin{matrix} S_O \\ (S_X) \end{matrix} \begin{matrix} \rightarrow R_O \\ \\ \rightarrow R_X \end{matrix}$

Figure 4.5. Frustration theory of the PREE (see Chapter 3) and a more general theory of persistence. See text for details.

counterconditioned to those disruptive stimuli (s_F), increasing their resistance to disruption by s_F (and perhaps also by other members of the class of stimuli S_X).

Another way of putting this more general case is that some amount and form of persistence develop whenever an organism's behavior habituates to a disruptive stimulus (S_X) *because behavioral habituation involves counterconditioning.* The S_X of the diagram may be regarded as either an external stimulus or a mediating stimulus arising out of an emotional response. Such a mediational interpretation of the operation of S_X would make the general case even more like the special frustration theory of persistence. But in the more general case, the extra link in the explanatory chain is not always strictly required, since in the general case the disruption of behavior need not emerge out of a discrepancy, such as that between reward expectation and nonreward, but may be imposed on an ongoing activity directly as a result of external stimulation. The S_Xs in our present analysis can be identified with Pavlov's "external-inhibiting" stimuli, which elicit competitive behaviors that interfere with the active conditioned response. With repeated presentation, the response (R_X, in our terms) to these stimuli also habituates, and the animal returns to normal responding to the CS.

We will later review the evidence that supports a general view of persistence, which, it must be said, is not the same as within-subject transfer of persistence. Nonetheless, the experimental findings on transfer of persistence do have implications for the development of persistence, and in hindsight, it was a consideration of developmental factors that led to the speculation about a more general theory of persistence.

A brief recapitulation of the argument

The experimental data presented in this chapter make it seem plausible that a counterconditioning mechanism *based on frustration* operates with some degree of generality: that it is involved in delayed as well as partial reinforcement, that it operates in the transfer of persistence effects across stimuli and across motivational and reinforcement conditions; that it operates in transferring response rituals learned under DNC conditions to the extinction of responses to stimuli that have never before elicited those rituals, and that it operates in resistance to discrimination and in the generalized PREE. We have shown, then, that the counterconditioning premise in frustration theory can be taken to apply somewhat more generally than we had initially supposed. All of this does not prove that frustration theory is subsumable under a general theory of persistence. But it is certainly not incompatible with such a position.

Now let us go back to the schema presented as Figure 4.5 and try to see how behavioral habituation may be involved in a more generalized version of persistence theory. The four-stage frustration theory of the PREE says that goal-approach response is counterconditioned in Stage 4 to stimuli from anticipatory frustration. It is important to state again that another way to describe these hypothetical factors controlling the PREE is to say that the effects of nonreinforcement are neutralized in PRF acquisition because there is habituation of competing responses to stimuli from anticipated frustration or to other stimuli (Estes, 1959; Weinstock, 1958). We have seen that such a mechanism can be said to account for the frustration tolerance of partially reinforced subjects in extinction – for the fact that the disruption of responding is less (resistance to extinction is greater) after PRF than after CRF acquisition (or that, as Miller [1960] says, in the case of shock at the goal, animals can be trained to resist the stresses of pain and fear). While, as I have indicated, it is not unreasonable to hold the view that response habituation in PRF acquisition affects resistance to extinction, the reasoning of the general theory is incompatible with a simple habituation concept or a simple frustration-tolerance concept or, for that matter, a simple sequential hypothesis of nonreward effects in PRF acquisition, which holds that the immediate carried-over effects or memories of nonreward on Trial N of acquisition come to be associated with approach to the goal on the basis of reward on Trial $N + 1$ (Capaldi, 1967; V. F. Sheffield, 1949). Such explanations cannot account for the kinds of experimental phenomena we have been reviewing. To reiterate, a more active counterconditioning mechanism seems to be required, and the argument can be reversed to say that *a process of active counterconditioning is involved in behavioral habituation,* that instances of behavioral habituation to initially disruptive events involve degrees of instrumental counterconditioning, and that this counterconditioning leads to greater persistence, sooner or later, in the face of these and perhaps even other disruptive events. The general theory brings together the domains of habituation and persistence, provides an active mechanism for behavioral habituation, and makes frustration theory a special case of a more general – and simpler – rule.

"Active" accounts of behavioral habituation

There have been other attempts to account for behavioral habituation in active terms. Stein (1966) proposed a theory of habituation of the arousal reaction, based on classical conditioning, which holds, as does the present view, that habituation is an active process and "not due to simple fatiguing of sensory or effector elements" (p. 352). According to Stein's theory, there is a kind of process in which stimuli that normally evoke excitation or arousal are counterconditioned to inhibition. His diagram of this process is shown in Figure 4.6. The basic distinction for Stein seems to be between novel and familiar stimuli. A novel stimulus elicits an excitatory (E) orienting response, arousal, and signs of alerting. The novel stimulus then becomes a familiar stimulus when, through a kind of opponent mechanism, an inhibitory state (I) is directly activated and is then strengthened through classical conditioning. This strengthened and longer-lasting inhibitory state counteracts arousal and results in habituation. My own later, but independent view[4] was that such orienting-investigatory reactions are instrumental reactions of the R_X sort, which interfere with ongoing behavior, R_O, and that the active process of habituation is the instrumental counterconditioning of R_O to S_X. A mechanism such as Stein proposed has, perhaps, to operate first to decrease the excitatory strength of S_X to evoke R_X, particularly if the response to S_X is strong.[5] Classical counterconditioning is clearly the simpler nervous system mechanism of the two, but it does not, by itself, complete the explanatory job under consideration – the account of instrumental persistence.

A more recent attempt to explain habituation in terms of an active process is Wagner's (1981) *standard operating procedure* (SOP) model, which Wagner characterizes as having some similarity to the *opponent-*

---

4. This position was first proposed in an invited address, entitled "Behavioral Habituation and a General Theory of Persistence," at the meeting of the Canadian Psychological Association in Calgary, Alberta, in June 1968.
5. The same point could be made about weak and strong s_F intensities in relation to the ease with which they can be said to elicit competing behaviors and, conversely, to be counterconditioned (see Traupmann, Amsel, & Wong, 1973).

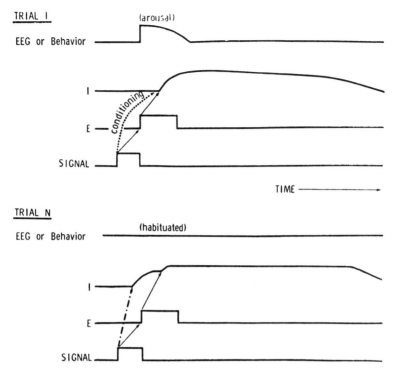

Figure 4.6. Diagram of conditioning model of habituation. *Trial 1:* Signal directly activates excitatory brain mechanism, E, which in turn directly activates inhibitory mechanism, I. Arousal response is elicited while excitatory activity predominates and declines in strength as inhibitory activity grows; furthermore, because inhibition outlasts excitation, the threshold of the arousal response will be elevated for a short time. Since signal onset just precedes activation of inhibitory mechanism, conditioning occurs. *Trial N:* Conditioned activation of inhibitory mechanism to onset of signal overrides direct activation of excitatory mechanism; hence, arousal response fails to occur and is said to be habituated. Further inhibitory conditioning also takes place on this trial despite absence of arousal response, producing below-zero habituation. (Figure and caption from Stein, 1966. Copyright 1966 by the American Psychological Association; reprinted by permission.)

process (O-P) *theory* of Solomon and Corbit (1974). A feature of Stein's and Wagner's theories of habituation, unlike my own, is that initially an opponent reaction emerges *directly* out of an excitatory process. In classical-conditioning terms, Stein's and Wagner's opponent responses are direct reactions to an excitatory response elicited by a novel CS. The O-P theory of Solomon and Corbit, to the extent that it is applied to processes of addiction and tolerance, can be regarded as a theory of the temporal dynamics of the UCS: In this case, the process and state aroused by (say) a drug decrease in strength on successive occasions, because the opponent process (and state) that is a direct consequence of this arousal comes in earlier and stronger, decreasing the reaction to the drug – resulting in

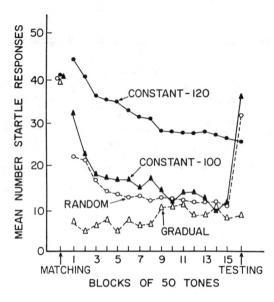

Figure 4.7. Habituation of startle reaction to a tone under a number of conditions (see text). Groups are matched on the basis of initial startle reactions and tested in a final block of 120-dB tones. (From Davis & Wagner, 1969. Copyright 1969 by the American Psychological Association; reprinted by permission.)

habituation or tolerance. In both Stein's and Wagner's models of habituation, the opponent that is activated is inhibitory, but this inhibitory process gains in strength through conditioning, with successive presentation of the CS. Here, then, are three theories, each explaining habituation in "active" terms.

Relevant to the present discussion is work by Davis and Wagner (1969) which showed that an arbitrarily defined startle reaction, measured in a stabilimeter, occurs relatively infrequently in response to a 120-dB tone if that intense tone has been preceded by tones of gradually increasing intensity, from 83 to 118 dB in 2.5-dB steps. If the tone remains at 100 dB throughout habituation training, or ranges from 83 to 118 dB but in a random manner, startle reactions to a 120-dB tone are relatively frequent; that is, there is less habituation to such a tone. Davis and Wagner's basic finding is shown in Figure 4.7. The authors argue that Stein's classical-conditioning model of habituation suggests a kind of stimulus specificity in habituation "such that habituation to a test stimulus should be best provided by exposure to that exact same stimulus" (p. 491). Of course, their own findings support a somewhat different view: that optimal habituation to the 120-dB test stimulus is achieved in the gradually increasing case, that exposures to a constant 120-dB tone result in some habituation, but that the level of responding to that tone is much higher on the test trial than it is in the gradual (83- to 118-db) case.

The general theory that integrates habituation, counterconditioning, and persistence can be related to the Davis and Wagner finding: In all of the PRF studies that refer to the counterconditioning of $s_F \rightarrow$ approach as the mechanism of persistence, the strength of r_F–s_F, which is the learned event that disrupts ongoing behavior, must be thought of as increasing gradually during the acquisition phase. What this means is that all of the various increasing intensities of s_F that must occur in PRF acquisition are counterconditioned to the approach response, so that persistence is maintained in extinction to all of these s_F intensities. One of our experiments (Traupmann, Amsel, & Wong, 1973) addressed a special case of this kind. It showed that if PRF training was preceded by a large number of CRF trials, persistence in a subsequent extinction phase was relatively greater at the beginning and less at the end of extinction than if PRF training was preceded by a small number of CRF trials. The greater the number of preliminary CRF trials, it was argued, the greater the r_R and the stronger the subsequent r_F–s_F conditioned in the PRF training. The reversal of relative persistence early and late in extinction was then attributed to the similarity between the specific s_F intensity to which the approach response had been counterconditioned in PRF training and the specific s_F intensities present at the beginning and end of extinction training.

It is interesting to speculate that in the case of newborn rats, humans, and other not very precocial animals, sensory capacities increase gradually over the first days and weeks of life, so that the strength or intensity of any repetitive, disruptive, external (or internal) event would increase gradually, providing the optimal conditions for habituation (counterconditioning). It is possible, then, that certain precocial animals, like guinea pigs, whose sensory capacities are more fully developed at birth, may be generally less persistent as adults than other less precocial (altricial) mammals – a fascinating idea that is difficult in principle to test precisely. We do have data (see Chapter 7) to show that the guinea pig, unlike the infant rat, shows a full-blown PREE shortly after birth (Dailey, Lindner, & Amsel, 1983). However, this finding does not address the relative persistence of these animals as adults.

Transfer of persistence and the more general theory

It is important now to separate two questions posed by a general approach to a theory of learned persistence: (a) Is instrumental counterconditioning of ongoing approach behavior to disruptive stimuli generally involved as a mechanism in the dispositional learning of persistence? (b) Is there transfer from one persistence system to another? In other words, is there a disposition to persist, a general pool or trait of persistence? Trying to answer the first question necessarily involves us in the second, although to say that a single kind of mechanism operates in most or all instances of learned persistence does not require that persistence involve a single, unitary system: It does not require that persistence acquired in any set of disruptive circumstances necessarily transfer to any other. (We have already seen evidence for this kind of transfer in the case of disruption by frustration.) By the same token, demonstrating transfer of persistence from one set of conditions to another does not mean, necessarily, that the mechanism of persistence is the same in both systems, although our view of this aspect of life would be simpler if this generality did in fact hold. In the final part of this chapter, we will consider briefly an outline of some of the kinds of work that would be helpful in answering questions about transfer of persistence, not just from one situation or motivational condition or response to another, but also *from one set of disruptive mechanisms to another.*

There are eight categories of experiments in which persistence and transfer of persistence have been studied. (Much of this work was detailed earlier in this chapter.) Each category is meant to represent some degree of difference between the experimental condition under which persistence is acquired and the conditions under which persistence is later tested or evaluated.

1. The most obvious, simplest, and certainly most frequently recorded case is the one in which the training is done under PRF conditions, or under conditions of uncertainty with respect to variable magnitude of reward (VMR), partial delay of reward (PDR), or other features of a goal event, and testing is based on extinction of responding in the same situation. In these instances the persistence derives from the PRF, VMR, PDR, or even the DNC condition, even though, as in the experiments we have examined so far (e.g., Donin, Surridge, & Amsel, 1967; Rashotte & Amsel, 1968; Rashotte & Surridge, 1969; Ross, 1964), a CRF phase may precede extinction. Another example, not involving extinction in the final phase, is transfer of persistence between PRF and PDR training and testing under continuous delay of reward (Shanab, 1971). In all of these cases persistence is acquired and tested within the same appetitive motivational-reward condition, but there may be difference in the nature of the nonrewarding, less rewarding, or delay-of-rewarding goal events or procedures.

2. Perhaps a broader transfer of persistence is across different appetitive conditions: from hunger to thirst, as in the Ross (1964) experiment (see also Mellgren, Hoffman, Nation, Williams, & Wrather, 1979). Here, although the disruptive effect occurs in PRF training under hunger in Phase 1, the transfer is through CRF under thirst in Phase 2 to EXT under thirst in Phase 3. However, the transfer of persistence in this case still involves the general properties of frustrative nonreward in the disruption of behavior.

3. Here, the transfer of persistence from acquisition to extinction is still across appetitive conditions; the animal is running for food throughout the experiment as in the first case. However, in this case, the disruptive factor

is related not to nonreward (or reduced or delayed reward or blocking of reward), but to the introduction of disruptive agents that are external to the reward itself (Banks, 1967; Brown & Wagner, 1964; Fallon, 1971; Ratliff & Clayton, 1969). There is in these cases transfer of persistence that is dependent on goal events that appear to involve overlapping systems, frustration, and punishment. Brown and Wagner have shown that when animals are trained to approach food and must take shock in the goal box to get it, they are more resistant to extinction, and conversely, that animals trained under conditions of intermittent reward and nonreward are more resistant to pain and fear. Miller (1960) has shown that training animals to take increasing strengths of shock with food in a goal box increases their resistance to pain and fear, a condition he calls "courage." Some of these experiments, and particularly Fallon's, also demonstrate that the persistence effects derived from PRF training can be enhanced by adding brief shocks on the nonrewarded trials in acquisition. However, an experiment by Scull (1971) in my laboratory failed to reveal a difference between the effects of PRF and CRF training on suppression caused by fear of shock, the idea being that PRF training should reduce such suppression.

We have experiments that demonstrate this same kind of transfer effect in the case in which the disruptive event in acquisition was a loud tone (rather than anticipatory responses based on frustrative nonreward or shock). The tone in these experiments either accompanied the retraction of a lever and the delivery of food reward, or was introduced early in the response chain, in both cases under discrete-trial fixed-ratio (FR–21) conditions (Amsel, Glazer, Lakey, McCuller, & Wong, 1973). Following the Davis and Wagner (1969) experiment, we introduced the tone either gradually, starting from a very low intensity, or at the greatest intensity from the outset. In three experiments, the tone increased persistence in a subsequent no-tone extinction phase. It did not appear to matter whether the tone was introduced at full strength from the outset or was increased gradually. In this respect, the results were not in agreement with those of Davis and Wagner. (In another experiment, rats with hippocampal lesions were also run with appropriate operated controls. The presence of tone increased resistance to extinction in the controls, but seemed to have no effect on the lesioned rats. This kind of work is reviewed in Chapter 8.)

4. This case is a version of the third. It involved adulterating or otherwise decreasing the attractiveness of reward for one group and not for another. In an early (Thorndikian rather than Pavlovian) version of what came to be known as the "US devaluation" experiment (Holland & Rescorla, 1975), food pellets were the rewards in a runway and, in a PRF-like condition, were adulterated with quinine on half the trials for one group but not for another (Wong, Scull, & Amsel, 1970). The idea was that the "partial-quinine" group should be disrupted in their approach and hence be more resistant to extinction. Observations of the rats eating the quinine-

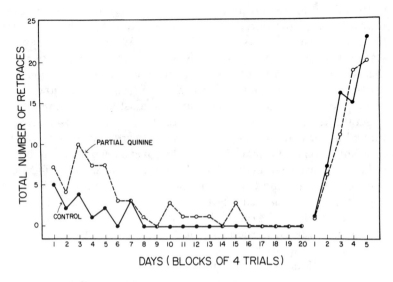

Figure 4.8. Retrace data showing prolonged aversive effects of quinine and their eventual decline in acquisition, with no apparent effect on persistence in extinction. (From Wong, Scull, & Amsel, 1970. Reprinted by permission of the Psychonomic Society, Inc.)

adulterated pellets indicated that the quinine was indeed aversive – that it resulted in a large number of retraces in the alley, relative to those of controls early in training (Figure 4.8). However, there was no resulting increase in persistence: Both groups showed rapid extinction. Hindsight tells us that this may have been because we did not introduce the adulteration in gradually increasing amounts or that the aversiveness was too strong to be counterconditioned in the number of trials run. Another possibility, however, is that the taste system operates in a different way than other sensory systems in rodents (e.g., Garcia & Koelling, 1966).

In an interesting variant of transfer of the persistence experiment, we have shown that the rat's avoidance of a taste conditioned to be aversive, and the reduction in the rat's consumption of the aversive-flavored solution itself, can be attenuated by giving it prior runway training in which the taste to be made aversive is given as reward on a PRF schedule. To state it another way, the rat can be immunized against the avoidance, in Phase 3 extinction, of a taste CS (Saccharin) made aversive with LiCl as the UCS in Phase 2, by giving it independent runway training in Phase 1 in which saccharin-taste reward is given inconsistently – on a PRF schedule (Chen & Amsel, 1980c; see Figure 4.9). The fact that a learned taste aversion, such as one formed when taste and LiCl-induced illness are paired in a Pavlovian arrangement, can be reduced by first rewarding an instrumental response with that taste on a PRF schedule has an obvious practical im-

Survival, durability, and transfer of persistence

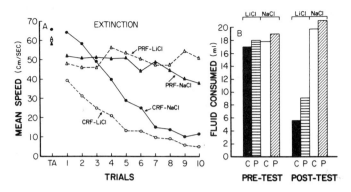

Figure 4.9. (A) Speed of approach to the goal box during terminal acquisition in Phase 1 and in Phase 3 extinction. Note that whereas the CRF-LiCl group showed immediate suppression of approach relative to the CRF-NaCl control, the PRF-LiCl group was as resistant to extinction as its (PRF-NaCl) control. (B) Whereas there were no differences in fluid intake among groups during the pretest (conditioning) phase, both poisoned (LiCl) groups showed suppressed intake. However, even in this measure, the PRF-LiCl group (P) showed a significantly smaller suppressive effect on fluid intake than its CRF control (C). (From Chen & Amsel, 1980. Copyright 1980 by the AAAS.)

plication for food aversions related to radiation treatment or chemotherapy.

5. Transfer of persistence can occur across very different experimental situations – for example, building up persistence by increasing the ratio of lever presses to reinforcements in an operant situation and testing in extinction after the acquisition of a runway response (e.g., Eisenberger, Carlson, Guile, & Shapiro, 1979; McCuller, Wong, & Amsel, 1976). In these experiments, transfer of persistence was investigated across different definitions of reward schedules and different response topographies. In the McCuller et al. experiments, rats were given operant training to press a lever on fixed ratios (FR) of 10, 40, 80, or 120 trials per reinforcement, followed by 12 rewarded trials and 32 extinction trials in a runway at *one trial per day*. Resistance to extinction in the runway was systematically and positively related to terminal ratio requirements of the previous operant bar-press training: The larger the ratio of responses to reinforcements in the box, the greater was the resistance to extinction after CRF training in the runway (Figure 4.10). Two other experiments had involved transfer in the other direction, from partial reinforcement of running to the extinction of a bar-press response. In one case (Wenrich, Eckman, Moore, & Houston, 1967), the result was positive; in the other, performed in our laboratory (Rashotte, 1971), it was negative. The final example I will provide in this category involves only reinforced behaviors (Eisenberger, Terborg, & Carlson, 1979). After bar pressing for food on a variable-interval schedule, rats earned food in a runway for various degrees of effort, then bar-pressed

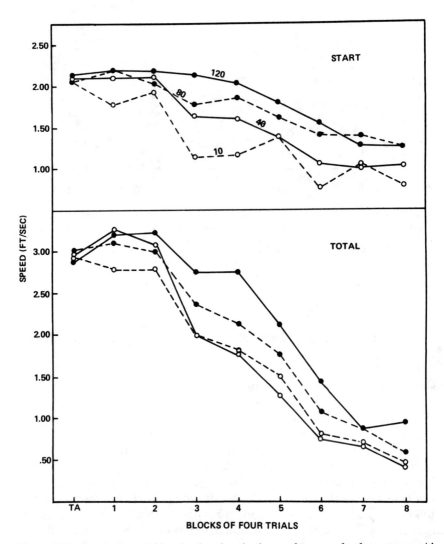

Figure 4.10. Terminal acquisition level and extinction performance for four groups with various terminal ratios of bar presses to reward. The start measure and overall (total) alley measures are shown. (From McCuller, Wong, & Amsel, 1976. Reprinted by permission of the Psychonomic Society.)

again for food. Bar pressing in Phase 3 was directly related to the amount of effort required in Phase 2. This was shown not to be due to greater conditioned general activity. Two explanations were offered: (a) that effort becomes a generalized component of instrumental behavior and (b) that a high level of effort increases the habituation of frustration-produced disruptive responses. Another series of experiments, published by the same laboratory in the same year, led to the same conclusion (Eisenberger, Carlson, & Frank, 1979).

Another experiment (Wong & Amsel, 1976) tested the animals from the McCuller et al. experiment after they had been given a 2-month vacation. Following the vacation, the rats were given 8 days of FR10 bar-press training, 4 trials per day, followed by 9 days of FR10 bar-press extinction at 4 trials per day. Then in the last phase (Phase 5 of both experiments taken together) all rats received 12 rewarded trials in a runway with a 300-mg-pellet reward, followed by 32 extinction trials, all at *1 trial per day*. The data from the FR10 acquisition and extinction are shown in Figure 4.11. Clearly, the *only* differential treatment was the level of FR training before the 2-month vacation, and this affected the extinction rate despite the intervention of the runway training (2 months earlier) and the immediately preceding reacquisition under FR10 conditions. In the final runway acquisition and extinction (not shown), differential persistence in extinction, resulting from the original, differential FR ratio of responses to reinforcement, was still present significantly in the runway start measure. This is a case of very long survival of persistence learned months earlier across a number of intervening experiences common to all experimental groups.

6. In a series of experiments, transfer of persistence was over different situations and different "appetitive" conditions, but not across hunger and thirst as in Case 2. The transfer in this case was from disruption of imprinting by electric shock to disruption of food approach by anticipatory frustration. The imprinting was to a flickering light and a pulsating tone (James, 1959). The subjects, of course, were birds – domestic chicks. Following imprinting, imprinting-shock, and control treatments, persistence was tested under normal nonreward-extinction conditions following both CRF and PRF training to approach food in a runway (Amsel, Wong, & Scull, 1971). If transfer of persistence occurs in this case, it can be taken as an instance of generalized persistence, because the disruptive treatment – imprinting to a flickering light and a pulsating tone, alone or in the context of shock obstruction – can be seen as a version of a habituation treatment involving counterconditioning of approach to originally disruptive stimulation. (It has been argued that there may be a disruptive effect from fear in this kind of imprinting itself.) It can also be thought of as a case of fear–frustration commonality with different "appetitive" rewards. There was some evidence that the imprinting procedure by itself increased persistence in extinction following CRF acquisition; however, the main effect was in the imprinting-shock condition, which increased resistance to extinction following both CRF and PRF appetitive acquisition.

7. This case is an apparent instance of transmotivational training that results in persistent approach to an aversive event without concurrent appetitive reinforcement. The first application of this technique was called "coerced-approach" training (Wong, 1971a, b). The question is: Can an

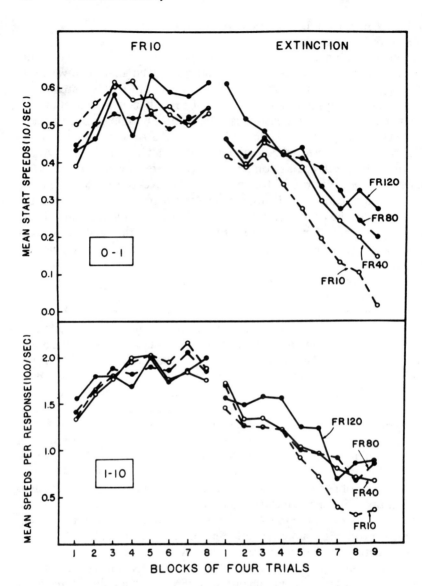

Figure 4.11. Response speed data during FR10 reacquisition and FR10 extinction. 0–1, Latency to first bar-press response; 1–10, speed per bar-press response over 10 responses in FR10. (From Wong & Amsel, 1976. Reprinted by permission of the Psychonomic Society.)

animal be trained to persist in approaching, even though its approach gets it nothing but aversive stimulation? And if so, will the persistence acquired in this manner transfer to the extinction of an appetitively motivated response? Wong conducted two kinds of experiment. The first involved the use of a hollow Plexiglas tube, which could be placed in an inclined or horizontal position. At the end of the tube was an area in which a brief shock could be delivered. In the horizontal position, the animal exhibited passive avoidance of the shock area. When the tube was inclined at about 45 degrees and the rat was placed at the top, its avoidance of the shock area was eventually overcome by a combination of gravity and the smooth interior surface, and it eventually moved into the shock area and received a shock. After several such experiences, the rat approached the shock area even when the tube was returned to the horizontal position. The second kind of coerced-approach experiment involved forcing rats into a goal box, where they got shocked, by punishing them for making competing responses to goal entry: staying in the runway. Wong demonstrated in these experiments that by whatever method the rats were trained to consistently approach an area in which they received a shock, these animals were subsequently more resistant to extinction following training involving food reward.

In the first of another subclass of such experiments (Nation, Wrather, Mellgren, & Spivey, 1980), rats were given PRF or CRF training in a straight alley in either a shock-escape or an appetitive paradigm, after which they received CRF training under motivational conditions opposite to those in Phase 1. In Phase 3, responses were extinguished according to the motivational conditions experienced in Phase 2. The results were that PRF training in the Phase 1 increased resistance to extinction in Phase 3. In a second experiment, this transmotivational PREE was shown to survive interpolated experiences with extinction, a 1-week "vacation" period, and CRF reacquisition. A third experiment examined the influence of intramodal versus intermodal nonreinforcement–reinforcement sequences on the transmotivational PREE. In this experiment, aversive nonreinforcement (failure to reinforce an escape response with shock reduction) was in some cases followed by appetitive reinforcement, and appetitive nonreinforcement was in some cases followed by aversive nonreinforcement. Both of these nonreinforcement–reinforcement combinations resulted in increased persistence. The data from all three experiments are taken to represent cases of generalized persistence.

8. The final case involves the demonstration of transferred effects from ordinary habituation training to appetitive resistance to extinction and vice versa. This is different from the other cases in that the "disruptive" stimulation is not introduced in the context of appetitive learning or even, as in Case 7, in relation to the same instrumental response. It is introduced in the separate context of a simple habituation procedure. (Experiments such as these may bring us closer to an answer to the question, Do habituation training and PRF training lie anywhere on the same dimension?)

This kind of transfer between the effects of habituation and appetitive persistence was demonstrated in three experiments (Chen & Amsel, 1977). In the first, a number of unsignaled shocks were given in Phase 1 "off the baseline." Phase 2 was appetitive runway acquisition, under either CRF or PRF conditions, and Phase 3 was extinction in the same runway. In the second experiment, the shock treatment came in Phase 2 between CRF or PRF acquisition in Phase 1 and extinction in Phase 3. In the third experiment, shocks in Phase 2 intervened between appetitive CRF acquisition in Phase 1 and extinction involving shock as well as nonreward in Phase 3. The shock durations in these experiments were increased in 1-second increments per day from 1 second on Day 1 to 5 seconds on Days 5 and 6. The main finding was that, compared with the results with unshocked controls, shock facilitated acquisition in Experiment 1 and led to increased resistance to extinction and/or punishment in all experiments. In Experiment 1, the effect of shock on appetitive extinction was seen mainly in the CRF group; in Experiment 2, the effect was to increase persistence in both the CRF and PRF groups; and in Experiment 3, shock treatment produced stronger resistance even to punished extinction. These transfer-of-persistence results can be taken as support for an extreme version of the general theory. (However, as we shall see in Chapter 8, Gray and his associates, on the basis of experiments on direct elicitation of the hippocampal theta rhythm, have suggested a nonassociative interpretation of these results.)

The generalized PREE: another transfer-of-persistence phenomenon

In an experiment to be reviewed in Chapter 5, we demonstrated the partial reinforcement acquisition effect (PRAE) in a within-subjects discrimination experiment; response to a stimulus associated with PRF ($S_1 \pm$) was greater than to a stimulus associated with CRF ($S_2 +$) in the start and run segments of a runway but was less in the goal segment. This caused us to ask whether the PREE is demonstrable within subjects. The question is: Will the differential acquisition performance demonstrated in relation to $S_1 \pm$ and $S_2 +$ (the within-subjects PRAE) be followed by differential extinction performance to these stimuli? Will the within-subjects result and between-subjects result be the same?

In abstract terms, the psychological justification for the between-subjects experiment is that it is a model of how separate but similar organisms are affected by different patterns of reinforcement for the same response. In contrast, the within-subjects experiment is a model for the development within the same organism of different systems or processes relative to different environmental events with associated reinforcement contingencies. (Normal successive discrimination learning is the anchoring case.)

A less abstract example of the between-subjects and within-subjects cases comes from the complex of stimulus–response-reinforcement relationships that exists in "the family." Setting aside for the moment complex interactions between hereditary and environmental factors determining personality, and assuming that the family we are considering is composed of mother, father, and two children (why not identical twins?), all of the elements for our comparison are present. Let father be $Stimulus_1$ and mother $Stimulus_2$, and let the children be $Subject_1$ and $Subject_2$. Each child is then a "subject" in both (a) within- and (b) between-subjects experiments to the extent (a) that each, separately, may be on a different discrete-trials schedule of reinforcement in relation to the two parents for the same behavior and (b) the two children may be on different schedules of reinforcement in relation to each parent for the same kind of behavior. The complexity of the relationships that are possible, even in this next to simplest of family situations, will be apparent, and the kinds of questions that this situation raises will be obvious. In the within-subjects case, can the same child learn different patterns of vigor-persistence relationships to the two parents, who serve both as differential stimuli and reinforcing agents? Assuming that this is so, to what extent will persistence learned in relation to the inconsistent reinforcing tactics of one parent transfer to the other parent, who has been more consistent? The between-subjects questions apply to the relationships between each of the parents and both children, or between the much more complex relationships that hold between both parents and both children. Obviously, the between-subjects experiments run in the laboratory, using animals as subjects, address very simple questions compared with those that might be asked about the family relationship; however, the basic between-subjects question is still the extent to which differences in reinforcement can produce two different organisms, one relatively more vigorous or more persistent or both than the other. We have discussed at some length the results of between-subjects experiments. Now we return to a consideration of the transfer-of-persistence phenomenon that has been the subject matter of this chapter and ask whether within-subjects discrimination training to $S_1 \pm$ and $S_2 +$ will result, within the same organism, in differential resistance to extinction to the two discriminative stimuli.

This question was being asked at about the same time and independently in at least three laboratories (Amsel, Rashotte, & MacKinnon, 1966; Brown & Logan, 1965; Pavlik & Carlton, 1965). The answer in two cases (Amsel et al., 1966; Brown & Logan, 1965) was that extinction to the two stimuli was nondifferential, a phenomenon Brown and Logan termed the "generalized PREE." In experiments by Pavlik and others (Pavlik & Carlton, 1965; Pavlik, Carlton, & Hughes, 1965; Pavlik, Carlton, & Manto, 1965) the answer was that the direction of the PREE in the within-subjects case was either normal (as in the between-subjects case) or reversed – slower or faster extinction, respectively, following $S_1 \pm$ than $S_2 +$. There were, however, these differences in method to explain the discrepant results: In the studies conducted by Amsel et al. and by Brown and Logan, the extinction effects were measured in a runway under conditions of separate and discrete trials, whereas two of the studies by Pavlik et al. examined the effects in a free-operant lever-box apparatus, the third in a runway. Of the two lever-box studies, one (Pavlik & Carlton, 1965) showed the reversed PREE, the other (Pavlik, Carlton, & Manto, 1965) a conventional PREE. The first result was obtained when one lever was employed and times of exposure to S_1 and S_2 were equalized; the second, when numbers of responses and of reinforcements to S_1 and S_2 (rather than times of exposure) were equalized. In the runway experiment (Pavlik, Carlton, & Hughes, 1965), a *reversed* PREE was shown in the goal measure, but rates of extinction to S_1 and S_2 were the same in the start measure. It therefore seems fair to say that on the basis of the runway data on which most of the present theorizing is based, the within- and between-subjects extinction effects are different, the difference being that, unlike the between-subjects case, the generalized PREE occurs in the within-subjects case. Furthermore, when the finding takes the form of the generalized PREE (Amsel et al., 1966; Brown & Logan, 1965), the curves of extinction performance to S_1 and S_2 are, in both cases, of the PRF and not the CRF form: Extinction is very slow, and the extinction curves are positively accelerated, suggesting generalized persistence (see, e.g., Figure 4.12).

Why is there no within-subjects PREE? Or, more accurately, why is there a generalized PREE – a PRF-like persistence response pattern in extinction in relation to both the partially and continuously rewarded discriminanda? Frustration theory provides two plausible mechanisms to answer this question; however, they are not mutually exclusive. The first is that *primary stimulus generalization* of r_F–s_F to S_2+ from $S_1\pm$ occurs in acquisition, so that there is counterconditioning in the acquisition of $s_F \rightarrow$ approach in both stimulus contexts. The second mechanism, which I believe is the more powerful, is *mediated stimulus generalization in extinction*. According to this explanation, the within-subjects case permits mediated generalization of the persistence effect ($s_F \rightarrow$ approach) from S_1 to S_2 in extinction; that is, as soon as r_F is elicited in extinction to S_2, which had been the CRF-related stimulus, the internal feedback stimulus (s_F), to which approach has already been counterconditioned in acquisition in $S_1\pm$, becomes a cue for the persistence mechanism in S_2, despite the fact there had been no counterconditioning in relation to S_2+ the external stimulus that had been associated with CRF in acquisition. (Of course, the between-subjects case does not permit such mediated generalization, and therefore transfer of persistence is not possible.)

If we think back to the results of experiments cited earlier in this chapter, the powerful mediating effect proposed for s_F appears to be quite plausible. Recall that Ross (1964) demonstrated in a within-subjects experiment that a specific response (*climbing*) trained under PRF conditions in stimulus

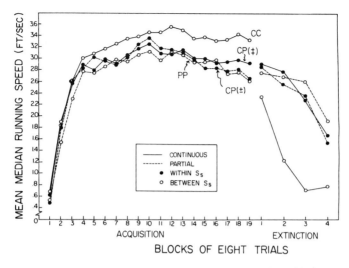

Figure 4.12. Acquisition and extinction comparisons of within-subjects ($CP^+_\pm$ and $CP\pm$) and between-subjects (CC and PP) conditions. The within-subjects group is run an equal number of times to *both* $S_2\pm$, a stimulus associated with a partially rewarded response, and S_{2+}^+, a stimulus associated with a continuously rewarded one. The two between-subjects groups are run to *either* CRF or PRF in both the S_1 and S_2 alleys, to control for possible alley color effects. Extinction is always between subjects: Half of the within-subjects group is extinguished to S_1, the other half to S_2. Note the very large PREE in the between-subjects comparison and the PRF-like extinction to both stimuli (the generalized PREE) following within-subjects training. (Adapted from Amsel, Rashotte, and MacKinnon, 1966. Copyright 1966 by the American Psychological Association; adapted by permission.)

context A, a response never before seen in stimulus context B, emerged in context B during the extinction of a *running* response in that context. And in another within-subjects experiment Rashotte and Amsel (1968) showed that idiosyncratic response rituals, learned in a first phase of training under the "pressure" of a discontinuously negatively correlated (DNC) response schedule of reinforcement, emerged in the third phase in the extinction of a response, learned in a second phase under CRF conditions in an entirely different situation, one in which the ritual had never been observed to occur. In neither of these cases can an adequate explanation be mounted purely in terms of external stimulus control; both seem to require an account in terms of at least a heavy component of mediated generalization, of transfer of responding to anticipatory frustration-produced stimuli.

General persistence and learned helplessness

There have been attempts in the literature to reconcile the apparent discordance between the Chen and Amsel kind of result (and the theorizing on which it is based) and the literature and theorizing, also derived from animal research, on the phenomenon of *learned helplessness* (see Levis, 1976; Maier & Seligman, 1976). This phenomenon, from which the theory gets its name, can be described as follows: After an animal has been exposed to unavoidable and inescapable electric shocks, it appears to be less able to learn to make an instrumental response to escape or avoid those shocks than yoked controls that have not been so exposed; that is to say, some kind of interference with such learning has been created in the helpless animals. The interpretation of the helplessness-interference effects in terms of general persistence theory (Boyd & Levis, 1980; Levis, 1976, 1980; Nation & Boyagian, 1981) has implications that are potentially important for relevant aspects of the study of human personality and psychotherapy. For example, Nation and Boyagian showed that exposure to recurring noncontingent, inescapable shocks of 3 seconds' duration (the average of the Chen & Amsel [1977] shock durations) had the effect of increasing resistance to appetitive extinction, whereas exposure to shocks of 15 seconds' duration reduced persistence (resistance to extinction). Moreover, they showed, as had others in mice (e.g., Anisman, de-Catanzaro, & Remington, 1978), that the pattern of performance at the termination of shock determined the nature of the "persistent" behavior: In the 3-second condition, the behavior was active movement (persistent approach), and after 15-second shocks it was persistent passive avoidance. According to these investigators, the apparently conflicting results from the positions of general persistence and helplessness could be linked to the shock-duration parameter. Nation and Boyagian gave animals no shock, brief (3-second) shock, or long (15-second) shock before appetitive acquisition and extinction. They showed that the 3-second shock increased resistance to extinction and the 15-second shock slowed acquisition and hastened extinction. The implication they drew is that responding in acquisition is counterconditioned to the fear from the brief shocks but not the long shocks, and their conclusion was that general persistence theory can account for helplessness in the 15-second-shock case and persistence in the 3-second case.

Summary

I would summarize this chapter by making two points: The first is that the PREE experiment, and particularly the variant of it I have called the Theios–Jenkins experiment, provides a basis for studying persistence in a developmental way. It is a prototype design and model for the manner in which differential early experiences lay down in the brain differential memories or dispositions to persist, which survive a long period of nondifferential experience to have their effects at some later time in life. My second point is that when we study persistence in this way, we may discover an identifiable dimension of behavior, representing a more or less general disposition in the individual, whose strength is determined not only by experience with frustrative nonreward and punishment, but also by opportunities, perhaps at some critical stages early in life, for habituation to disruptive stimulation.[6] I have argued that the common mechanism in these cases is instrumental counterconditioning (as we have seen, classical counterconditioning may also be involved [Stein, 1966]; see also the discussion of Wilton, 1967, in Chapter 6, this volume), but this counterconditioning may involve disruption not only from anticipatory frustration but from other sources as well, including the proactive effects of simple habituation to novel and/or noxious stimulation.

In Chapter 7 we will examine such proactive effects in developmental experiments, and in Chapter 8 we will review experiments in which the initial persistence-building treatment involves direct stimulation of the septal region of the brain that drives the hippocampal theta rhythm (e.g., Glazer, 1974a; Gray, 1970; Williams, Gray, Snape, & Holt, 1989; Holt & Gray, 1983a). But first we must consider in detail how the mechanisms of persistence and its transfer apply to another case of noncontinuous reinforcement – discrimination learning.

6. For an early review and analysis of research on the measurement of persistence in human adults, see Ryans (1939). The measurement procedures reviewed included such test items as resistance to shock, to distraction, to pain, and to startle stimuli (see Hartshorne, May, & Maller, 1929). Ryans concludes with the statement that "the existence of a general trait of persistence, which permeates all behavior of the organism, has not been established, though evidence both for and against such an assumption has been revealed" (p. 737).

5 Discrimination learning and predisrimination effects

In this chapter, we will review the application of frustration theory to the explanation of another case of reward–nonreward intermittency – discrimination learning. We will consider the role of frustrative factors of arousal and suppression in the formation of discriminations, and the mediating action of anticipatory frustration and counterconditioning in the retardation and facilitation of discrimination learning. The discrimination learning we will consider involves separate experiences with or exposures to the discriminanda, which fits the case of dispositional learning. We should therefore briefly review the history of the involvement of frustration in such "go/no-go" discriminations.

Frustrative factors in discrimination learning

The Hullian analysis of instrumental discrimination learning (Hull, 1950, 1952), which has been and continues to be influential, did not include an active, frustrative role for nonreinforcement. In broad outline, while Hull did maintain that the primary process involved was differential reinforcement, he held that its major function was (a) to neutralize the effect of the background or context stimuli that occur in the presence of both discriminanda, (b) to increase the power of the positive stimulus to evoke the (approach) response, and (c) to decrease the power of the negative stimulus to evoke that response. However, in Hull's theorizing the negative stimulus did not elicit avoidance of nonreward; rather, responses elicited by the S− permitted accumulation of an inhibitory state, reactive inhibition (I_R), which, not offset by the growth of excitatory strength and particularly under massed-trial conditions, led to the extinction of responding to S−. Hull also assumed that the dissipation of I_R satisfied the criterion of drive reduction and led to the development of a negative excitatory potential, conditioned inhibition (S^IR), which accounted for the more permanent extinctive effects in discrimination learning.

In my earliest published paper on frustration theory (Amsel, 1958a), and later, I questioned the development of differential inhibition in discrimination learning based solely on the effort of responding: "There is nothing in the Hullian system which suggests that I_R is in any way related to reinforcement or nonreinforcement; consequently, if S^IR is to be employed in the explanation of discrimination learning, one would have to hold that it develops to both S+ and S−" (Amsel, 1958a, p. 110). I made the further point that the important differential factors in discrimination learning are the positive and negative goal events that determine positive and negative *excitatory* tendencies, and that a factor such as I_R, in and of itself, must affect both tendencies equally and cannot be regarded as differential.

This position on discrimination learning was in part a return to an earlier theoretical treatment (Spence, 1936, 1937), which held that the two major principles of discrimination learning were *reinforcement* and *inhibition*. According to Spence, the principle of reinforcement "assumes that if a reaction is followed by a reward the excitatory tendencies of the immediate stimulus components are reinforced or strengthened by a certain increment, 'I'" (Spence 1936, p. 430); and the principle of inhibition or frustration states "that when a reaction is not rewarded... the excitatory tendencies of the active stimulus components are weakened by a certain decrement, 'D.' It assumes that this weakening is due to an active, negative process, inhibition, which, adding itself in algebraic fashion to the positive excitatory tendencies, results in lowered strength values" (Spence, 1936, p. 430). Spence also made the point that "the weakening effect on an S–R connection of failure of reward... varies directly with the strength of the response, being greater for strong ones than for weak ones" (p. 431). My own analysis of discrimination learning specified that r_R–s_R and r_F–s_F are the mechanisms behind the positive and negative tendencies of Spence. The importance of nonreward as an active factor in discrimination learning had also been suggested in the interpretations of several experiments (e.g., Fitzwater, 1952; Grice & Goldman, 1955; Grove & Eninger, 1952; Shoemaker, 1953).

One test of the role of frustrative nonreward in discrimination learning can be made in the double runway. If this hypothesis has merit, it should be possible to show that the frustration effect (FE), measured in Runway 2 of the double runway, bears certain relationships to various stages of discrimination learning measured in Runway 1. More specifically, evidence of nonreward-related frustration measured in Runway 2 should precede any evidence that subjects are running faster to S+ and slower to S− in Runway 1, and at some point *following* the appearance of discriminative behavior in Runway 1 a reduction in the magnitude of the FE, measured in Runway 2, should become apparent.

The first implication, that in a successive discrimination involving changes in strength of both approach and avoidance to positive and negative stimuli the onset of discrimination will be evident only after nonreward becomes frustrating, is based on the assumption that only under these conditions of instrumental learning does S− elicit the avoidance responses necessary for discrimination learning. The second implication, as we have seen, derives from the assumption that learning a discrimination involves the separate

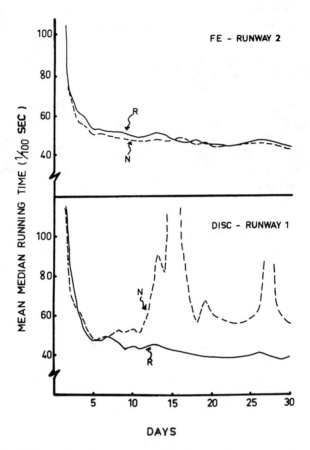

Figure 5.1. The relation of magnitude of FE (top panel) to stages of the discrimination (bottom panel) in the double-runway apparatus. R, Reward; N, nonreward. (From Amsel, 1962.)

elicitation of r_R and r_F by the positive and negative stimuli, respectively. If frustration is dependent on r_R–s_R, the FE should diminish once a discrimination is formed and S– no longer elicits strong r_R but elicits r_F. The relevant experiment (Amsel, 1962; Amsel & Ward, 1965), which involved comparisons between indices of the FE derived from the periods immediately before and after clear-cut evidence of discrimination learning, provided confirmation of these hypotheses (Figure 5.1).

Discrimination learning as a form of partial-reinforcement training

We saw in Chapter 3 that in terms of a conditioning model such as has been outlined, partial-reinforcement (PRF) acquisition and "go/no-go"

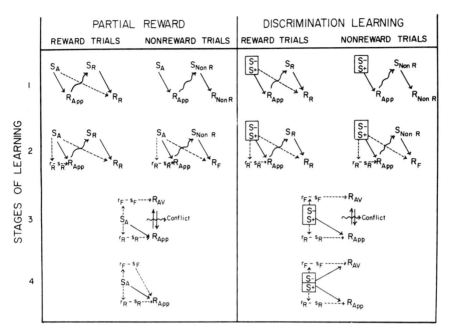

Figure 5.2. Diagrammed sequence of hypotheses relating frustrative nonreward to stages of partial reward and discrimination learning. (Adapted from Amsel, 1958a.)

discrimination learning (DL) can be shown to involve similar processes. To quote from my original paper:

> Partial reinforcement and discrimination learning procedures are highly similar; in fact, they are almost identical if we compare partial-reinforcement training to the early stages of discrimination training with separate (successive) rather than joint (simultaneous) presentation of stimuli. In both, at the outset, S is rewarded on some occasions and not on others for the same instrumental response. The difference is that, in partial-reinforcement experiments, E is training S to make the same response on every trial, whereas in discrimination learning different stimulation is involved when S is rewarded and not rewarded and S comes ultimately to respond (or not respond) selectively; but only, as a rule, after S has learned to respond non-selectively on the basis of partial reinforcement. (Amsel, 1958a, p. 108)

The theory with which we have been working divides DL, like PRF acquisition, into four stages. Each of these stages is assumed to involve processes quite different from the others, especially in regard to the operation of the conditioned anticipatory response factors depicted in Figure 5.2. Now let us work toward a more complete statement of the role of these factors in DL and in transfer from prediscrimination experiences to DL.

It is obvious that, in the cases of both PRF and DL, the predictions that can be made depend entirely on what response the feedback stimuli (s_F)

from r_F evoke. At the level of Stage 3, they are assumed *in both cases* to evoke avoidance responses and should hasten extinction. At Stage 4 of PRF training, they evoke the criterion approach response, or responses compatible with the criterion response, and should retard extinction. In the case of DL at Stage 4, these same stimuli, along with differential external cues, will evoke avoidance. Anticipatory-frustration-produced stimuli can therefore provide the mechanism for persistence, or the rapid abandonment of responses, depending on the context in which they occur, and this, as we shall see, is the case also for transfer effects from prediscrimination experiences to subsequent DL.

In more general terms, this analysis says that persistence and avoidance (desistance) in instrumental learning are, in part at least, frustration phenomena – that the tendency to continue responding in the face of negative indications is acquired under PRF conditions when cues from anticipatory frustration (s_F) become connected to continued approach, and that the tendency to stop responding in DL involves the elicitation of avoidance by s_F. In terms of such a conceptualization of dispositional learning, the PREE does not have to mean that subjects trained on a PRF schedule suffer from reduced discrimination of acquisition from extinction; it may reflect learning to approach rather than to avoid in the presence of cues signaling nonreward. And in these terms, DL cannot depend simply on the signaling properties of external $S+$ and $S-$; it must also depend on the mediating action of positive and negative expectancies. The central question becomes, Does the theoretical complication of adding these hypothetical mediators buy us any increased predictive or explanatory power? Is the extra theoretical baggage worth carrying?

Two specific experimental implications of the foregoing analysis became obvious in the early days of the work: (a) At Stage 3 of PRF (or DL) acquisition (the "conflict" stage), behavior should be more variable than it was at Stage 2 or will be in Stage 4. (b) If PRF acquisition does not reach Stage 4, there should be no PREE, that is to say, if training is discontinued when, under PRF (or DL) conditions, acquisition behavior is still variable, then such acquisition should be followed by decreased rather than by increased resistance to extinction. Both of these deductions received some support from preliminary data (Amsel, 1958a); they have been subjected to further direct and indirect tests in the context of the theory as it is applied to the discrimination and prediscrimination effects to be discussed in this chapter.

A further experimental implication, as we have seen, depends on the following assumption: The level of persistence in extinction following PRF acquisition should be directly related to the level of frustration in acquisition, given that Stage 4 has been reached. Presumably, one way to reduce frustration in PRF acquisition is to arrange for the learning to occur under conditions of reduced excitement – for example, through the use of drugs that act selectively on conditioned aversiveness, such as frustration. There is some indirect evidence that drugs can mitigate the effects of anticipatory frustration (see Appendix). In early experiments, sodium amytal and alcohol were shown to reduce the intensity of anticipatory frustration in extinction (Barry, Wagner, & Miller, 1962). Sodium amytal likewise seems to exercise an effect on PRF acquisition, reversing the usual finding of a PRAE – faster asymptotic speeds, in earlier segments of the response chain, to 50% than to 100% reinforcement (Wagner, 1963a). Here, also, the results are interpreted as supporting the assumption "that anticipatory frustration is particularly susceptible to depressant action of sodium amytal" (p. 474). Such a decrease in the intensity of anticipatory frustration by drug administration must, according to the theory, affect not only the PRAE and PREE, but discriminative behavior as well (see Appendix). In these cases the presence or absence of a tranquilizing or sedative drug determined whether the discrimination that was formed was "emotional" or "phlegmatic" (Amsel & Ward, 1965; see also, later in this chapter, Terrace's [1963a] work on "errorless discrimination").

In DL, Stages 1 to 3 are conceptualized as being the same as in PRF acquisition. The schema of Figure 5.2 indicates that it is only at Stage 4 that DL involves mechanisms that are different from those in PRF acquisition. At the beginning of DL, involving separate and successive presentations of S+ and S−, these two discriminanda are not differentiated. Consequently, as far as response evocation is concerned, they are the same stimulus. They operate in the same manner as S_A on the PRF side until Stage 4, when S+ begins to evoke anticipatory reward (r_R) and S− anticipatory frustration (r_F).

In short, the difference between the two situations depicted in Figure 5.2 is that, unlike discrimination learning, PRF learning affords no basis for differential responding to stimuli. In DL a response originally nondifferential with respect to two physically different stimuli ultimately becomes differential in relation to the two stimuli, on the basis of differential reward and nonreward to S+ and S−. The suggestion has been that in discrimination learning of this sort one of the factors involved is differential evocation by S− and S+ of r_F and r_R, respectively, and that since these processes also affect PRF acquisition, implications of some importance follow.

The latter part of this chapter is composed of material adapted from previous publications (Amsel, 1962, 1971b; Amsel, Rashotte, & MacKinnon, 1966; Amsel & Ward, 1965), which dealt with the effects of frustration on discrimination and on the generalized PREE, and with other implications of the discrimination analysis. For example, we will see that under some circumstances failure over some prolonged period of training to respond differentially to positive and negative discriminanda may not mean that these stimuli are in fact equivalent perceptually, but that previous experience with these stimuli has led the subject to approach in the presence of the negative discriminandum as well as in the presence of the positive one. But first, we must draw a distinction between the results of discrete-trial and operant studies of discrimination learning. We deal here not only with responding to absolutely positive and absolutely negative stimuli, but also with differential percentages and magnitudes of reinforcement in relation to the discriminanda.

Arousal and suppression: positive behavioral contrast and generalization of inhibition

A reliable, though transient, finding of experiments on what operant researchers call *local contrast* is that rates of responding to the stimulus associated with less frequent reinforcement and to the stimulus associated with more frequent reinforcement diverge, the rate to the positive stimulus being greater than in an appropriate control condition (see original experiment by Reynolds, 1961).[7] Another equally reliable finding is that, in a discrete-trial runway procedure, this kind of *behavioral contrast* seen in the operant work does not occur; what is seen is the opposite kind of effect, a convergence of rates (or, in this case, speeds) based on *generalization of inhibition*. Some years ago, I reviewed the history of some of the work on behavioral contrast in operant conditioning (Amsel, 1971b), particularly as it relates to comparable work in Pavlovian positive induction, as follows:

Behavioral contrast... refers to an increase in the strength of responding to the more positive component in a multiple schedule (S+) that accompanies a decrease in the rate of responding and/or of reinforcement to the more negative component (S−) during discrimination training. There is this difference between behavioral contrast, as Skinnerians seem to view it, and positive induction as it is studied in the Pavlovian arrangements: in Pavlov's work the inductive effect was seen to be the direct consequence of an immediately preceding inhibitory event. Positive induction referred to an increase in excitatory strength in the presence of the irradiated cortical inhibition left over from the just-terminated CS−. Apart from statements made by Terrace (1966a, p. 321) and by Nevin and Shettleworth (1966), who regard transient, as opposed to sustained, contrast as having something to do with the just-preceding stimulus, behavioral contrast does not seem to be viewed primarily in terms of the just-terminated stimulus, even though the great preponderance of such studies are conducted with single alternation of S− and S+, so that every positive trial-segment is preceded by a negative one. Contrast seems rather to be regarded as relatively independent of the immediately preceding event,

7. Brief recent reviews of operant positive behavioral contrast and its various explanations can be found in Schwartz and Gamzu (1977) and Staddon (1983). In neither of these is an explanation in terms of potentiation by short-term effects of nonreinforcement offered, an explanation that I have favored (e.g., Scull, Davies, & Amsel, 1970).

to be related to "the fact that the organism has been confronted with a differential reinforcement procedure [Terrace, 1966a, p. 321]," and it has been looked at in three ways: (a) as an increase in responding to S+ *because of a decrease in responding to S−* (Terrace, 1968); (b) as an increase in responding to S+ *because of decreased reinforcement density to S−* (Reynolds, 1961); and (c) as an increase in responding to S+ *because of increased preference for S+ over S−* (Bloomfield, 1967b; Premack, 1969).

Pavlovian positive induction and Skinnerian behavioral contrast are similar in that both refer to an elevating effect on responding to a relatively positive stimulus brought about by some characteristic or feature of a relatively negative stimulus or situation. Pavlov refers specifically to differential conditioning and to the irradiation of excitation and inhibition when he discusses induction. The statements about behavioral contrast are always made in the context of discrimination training, and often in relation to peak shift (Terrace, 1966a, 1968), a phenomenon which does seem, unmistakably, a characteristic of discrimination learning. Interestingly enough, to make the coincidence still more striking, there is evidence from operant-discrimination experiments (Terrace, 1966a) that behavioral-contrast effects, like Pavlovian induction effects and double-runway frustration effects, also diminish after a discrimination has been formed. (pp. 219–20)

The paper then went on to pose the following questions:

Is discrimination training a necessary or only a sufficient condition for demonstrating positive induction and behavioral contrast? And if it turns out to be only a sufficient condition, is it sufficient in all instances of discrimination training or only in certain operant-discrimination procedures and in Pavlovian differentiation experiments which can be thought to be analogous to those procedures? (p. 220)

One approach that was proposed was to examine the effects in DL of S− on S+, with successive and discrete-trial presentations of stimuli, at intertrial intervals (ITIs) long enough that the immediate potentiating effect of nonreinforcement, possible in Pavlovian and Skinnerian experiments, could be ruled out. The question was: With discrete trials and long ITIs, would we find the same effects as in the positive-induction or behavioral-contrast experiments? There had been a number of discrete-trial experiments (reviewed by Black, 1968) on shifts in magnitude of reward and on contrast effects in differential-conditioning experiments in which magnitude, delay, and percentage of reward were manipulated. In these studies, there was virtually no suggestion of positive contrast effects, but there was clear evidence for negative contrast effects and generalization of inhibition, which Black explained in terms of a simple Spence (1936) model, $\bar{E} = E - I$.

It also turned out that these kinds of experiments had been performed in our laboratory quite unwittingly, embedded in studies designed for quite other reasons. In the three experiments I shall describe, the apparatus was a pair of runways, one white and the other black, each of which could be aligned with a common gray start box unit. Response-time measures were taken over three 1-foot segments (start, run, and goal) and were converted to speeds (feet per second). The rats performed under hunger conditions defined by 22- to 23-hour food deprivation, *with an ITI of about 15 to 20 minutes*. Half the trials were run in the white alley, the other half in the black, in a quasi-random order.

In one experiment (Henderson, 1966) five groups were run under the conditions shown across the top of Figure 5.3. These groups differed only with respect to the percentage of reinforcement given for response in the S_2 alley, which varied from 100% to zero. Magnitude of reward was 500 mg of food whenever given. The two extreme groups $S_1 100 - S_2 100$ and $S_1 100 - S_2 0$ provided the conditions for the contrast comparison described earlier. The intermediate groups provided comparisons with intermediate percentage-reward conditions to S_2. Clearly, in none of the three measures was there evidence of a positive induction kind of behavioral contrast, but quite the reverse, a negative induction effect: As the level of reinforcement to S_2 decreased, the level of responding decreases not only to S_2 but also to S_1. The evidence in this discrete-trial experiment is for generalization of the effects of nonreinforcement (of $r_F - s_F$ or "inhibition") in DL.

An experiment by MacKinnon (1967) was highly comparable to Henderson's, except that magnitude rather than percentage of reward was varied in S_2 as a between-groups parameter, so that the numbers across the top of Figure 5.4 represents milligrams of food reward (magnitude of reinforcement) rather than percentage of reinforcement. Obviously, then, the extreme groups ($S_1 500 - S_2 500$ and $S_1 500 - S_2 0$) of MacKinnon's experiment corresponds exactly to Henderson's and compromise the same basic discrete-trial contrast-effect experiment. There was again no evidence of an inductive type of behavioral-contrast effect: Response to $S_1 500$ was weaker when zero reward was given to S_2 than when 500-mg reward was given to S_2. Again, generalization of the effects of reduced reinforcement is a plausible explanation for the data.

Note, however, that in these experiments there is a peculiar kind of "positive induction effect": Whereas the horizontal comparison of response to S_1 reveals a progressively increasing inhibitory effect of lower and lower percentage or magnitude of reinforcement to S_2, there are clear indications, particularly in the second and third columns from the left, of *greater* vigor of responding associated with the *lesser* percentage or magnitude of reinforcement. This within-subjects effect is similar to an effect seen previously only between groups (Goodrich, 1959; Haggard, 1959). It is called the partial-reinforcement acquisition effect (PRAE), and it names the fact that, late in acquisition of a running response, speeds are greater under PRF than CRF conditions in early segments of the alley, but are reversed close to the goal.

Finally, lest the runway type of "contrast" finding be regarded as exclusively a between-subjects design phenomenon, we can provide some data that are of the within-subjects variety and therefore much closer in form

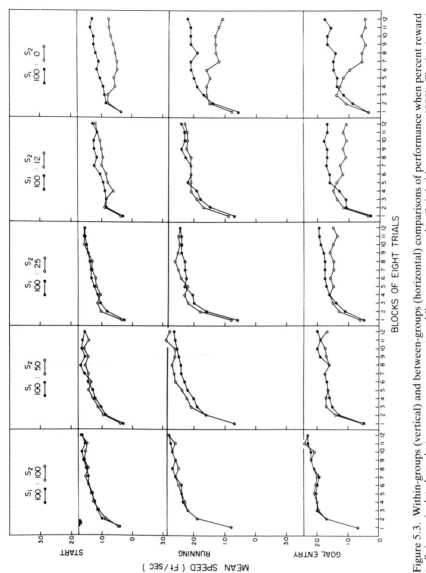

Figure 5.3. Within-groups (vertical) and between-groups (horizontal) comparisons of performance when percent reward to S_2 is manipulated as a between-groups parameter while percent reward to S_1 is held constant at 100%. The horizontal lines are drawn at the level of terminal performance of the $S_1$100–$S_2$100 group to facilitate comparison across groups of performance in the S_1 runway. (Adapted from Henderson, 1966.)

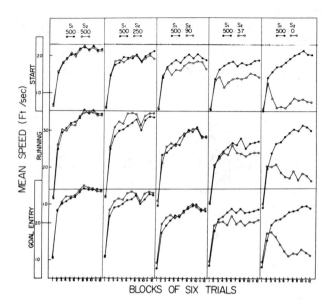

Figure 5.4. Between-groups (horizontal) and within-groups (vertical) comparison of performance when magnitude of reward to S_2 is varied between groups and that to S_1 is held constant at 500 mg. Horizontal lines facilitate between-groups comparison, as in Figure 5.3. (Adapted from MacKinnon, 1967. Copyright 1967 by the American Psychological Association; reprinted by permission.)

to the usual operant, behavioral-contrast design. In a condition that was part of an experiment by Galbraith (1971), the sequence of three experimental phases is as follows: (a) $S_1 + S_2 +$, (b) $S_1 + S_2 -$, (c) $S_1 + S_2 +$. Again, the ITI was about 20 minutes. Clearly, a positive-induction or behavioral-contrast view of this experiment would lead to the expectation that the level of responding to S_1 in Phase 2 would be higher than in Phase 1 and that in Phase 3 responding to S_1 would return to its level in Phase 1. The data from these conditions (Figure 5.5) are clearly not in accord with the first of these expectations, but they are in accord with the second: The shift from S_2+ to S_2- in Phase 2 *lowers* the level of responding to S_1+ in Phase 2; and in Phase 3, when S_2 is positive again, responding to S_1 returns to its Phase 1 level.

Other experiments (Leonard, Weimer, & Albin, 1968; Senf & Miller, 1967) demonstrated effects analogous to Pavlovian positive induction in runways. In these cases the induction was demonstrated in terms of retarding extinction to an $S+$ by alternating presentations of $S+$ and $S-$ in the extinction of responding to $S+$. In both studies, the successful demonstrations of induction in extinction involved an ITI of the order of 10 seconds. A specific test for inductive effects during discrimination training by Leonard et al. showed no such effect when the ITI was 3 to 5 minutes. Another experiment (Ison & Krane, 1969) provided results con-

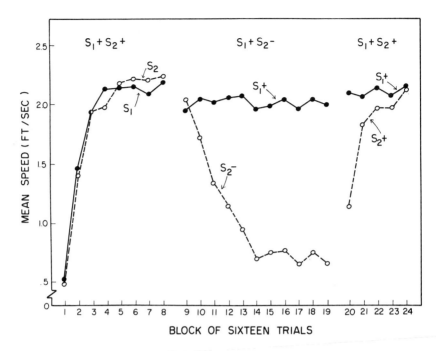

Figure 5.5. Speed of running in an S_1+S_2- discrimination preceded and followed by continuous reward to both stimuli. This is a discrete-trial runway analog of an experiment on behavioral contrast. (From Amsel, 1971; adapted from Galbraith, 1969, by permission of the author.)

trary to positive induction in runway acquisition or extinction training at a 5- to 15-minute ITI, and some evidence of induction in extinction only when the ITI was 30 seconds. These studies suggested that the presence of a negative (nonreinforced) stimulus, as compared with simply a nonreinforcement, increased the excitatory strength of an immediately following S+, but only during extinction when the ITI was very short.

At about the time all these experiments were going on, Terrace summed up what he regarded as a contradiction between the data of behavioral contrast and Spence's theory of discrimination learning as follows: "Behavioral contrast derives its name from the fact that the rates of responding to S+ and S− diverge. According to [Spence's] classical generalization theory, these rates would converge" (1966b, p. 1677). He had made the same point in connection with a more inclusive treatment of behavioral contrast, peak shift, and what he called "errorless discrimination" (Terrace, 1966a). (Errorless discrimination is a kind of DL in which the behavior of the pigeon [in this case] is shaped in such a way that the discriminative stimuli differ almost imperceptibly at the start of training, and their difference is gradually increased so that the discrimination is formed with few if any nonreinforcements – "errors.") The point I made sometime later

(Amsel, 1971b) was that behavioral-contrast effects in operant or other highly massed-trial conditions and the generalized inhibitory effects from discrete, spaced-trial conditions can live side by side and be regarded, respectively, as indicants of the primary excitatory and secondary inhibitory effects of frustrative nonreward. (Investigators of operant DL have not been and are not now fond of explanations in terms of the activating properties of frustration.) I pointed out that under the DL conditions of our discrete-trial runway experiments, which resemble more closely the ones on which Spence's original theorizing was based than Terrace's conditions, response "rates" do indeed seem to *converge*.

The establishment of generalization of inhibition in discrete-trial, appetitive DL experiments, and our interpretation in terms of the mechanism of anticipatory-frustration-produced suppression of responding, set the stage for the discussion that follows which concerns the effects on the rate of subsequent DL of prediscrimination experience with the discriminanda.

Dispositional effects of prediscrimination exposure to discriminanda

Some years ago, I proposed an extension of the frustration theory to handle a variety of circumstances in which learning to discriminate between two environmental cues is preceded by partially or continuously rewarded exposure to one or both of the eventual discriminanda (Amsel, 1962). These preexposures to the discriminanda were termed *prediscrimination experiences,* and they could take a number of forms. In a discrimination involving two cues, S_1 and S_2, it is possible, for example, to have exposed the subject before DL to either or to both of these cues under a number of possible reward–nonreward conditions. If the preexposure is to a single discriminandum, it may then become the $S+$ or $S-$ in the subsequent discrimination phase; if preexposure is to both of the eventual discriminanda, the reward–nonreward relationships between the preexposures and the discrimination itself, as we shall see, will be more complex. The nature of the prediscrimination experience will presumably affect the rate of subsequent DL involving those same two stimuli. If, for example, the prediscrimination exposure is to the to-be-negative stimulus (S_2-) in the discrimination (S_1+S_2-), and if the prediscrimination-phase response to S_2 is partially reinforced ($S_2\pm$), persistence has been learned to $S_2\pm$ if training has proceeded to Stage 4, but not if it has gone only to Stage 3 (see Figure 5.2). Such prediscrimination training should retard the discrimination. If, however, the preexposure to S_2 is continuously rewarded (S_2+), the S_1+S_2- discrimination should develop more quickly, and the greater the number of such S_2+ preexposures, the more rapid that discrimination should be.[8]

---

8. The mechanisms involved in this latter case are not necessarily the same as those later described by Logan (1966) in what he refers to as "transfer of discrimination." In his case,

In the present case, an increasing or decreasing rate of DL, relative to an appropriate control condition, results from the transfer to the DL of mechanisms acquired in the earlier prediscrimination treatment. These mechanisms depend entirely on the CRF or PRF schedules that are involved in the preexposure of the eventual discriminanda. It is plausible that these dispositional mechanisms are the same as those transferred from acquisition to extinction in the case of the PREE, and that they embrace much of what is ordinarily meant by *persistence* and *desistance*. In the case of the PREE, persistence refers to a learned tendency to continue to respond in the face of negative (nonreward) indications. The prediscrimination treatment results in another kind of persistence: learned retardation of DL evidenced by relative failure to respond differentially to two (or more) stimuli, one signaling reward and the other(s) nonreward. If we call the first case of persistence *resistance to extinction,* we can call the second *resistance to discrimination.* I have argued that the main factor in the latter kind of persistence is that the S−, which must evoke avoidance through the mediating mechanism of r_F-s_F, at first elicits approach as a consequence of the inconsistency of reward in relation to its prediscrimination exposure.

A variety of experimental approaches can be employed to study the retarding or facilitating effects on a discrimination of prediscrimination experiences. The importance of this kind of abstract analysis is that it can identify purely internal mechanisms that control long-term dispositions that affect differential responding to discriminably different external stimuli. In short, this kind of analysis provides a basis for understanding why such built-in internal dispositions can override differential external stimulus control.

Discrimination learning and number of prediscrimination rewards

As we saw in Chapter 3, experiments on the double-runway frustration effect (FE; Figure 2.7) were, by the mid-1960s, numerous and went well beyond the simple demonstration of that effect (see Appendix). As they relate to our present concerns only, some of the results were that following a prolonged period of continuous reward, frustrative nonreward produced a relatively immediate FE (Amsel & Roussel, 1952; Wagner, 1959); that the FE developed gradually when the Runway 1 response was partially rewarded from the outset in G_1 (Amsel & Hancock, 1957; Wagner, 1959); that continuous nonreward in G_1 after the FE had already been established resulted in the disappearance of the effect (McHose, 1963); and that the FE was greater

initial training with (a) neither of the eventual discriminanda but two more widely spaced on the continuum, (b) the positive stimulus and a more divergent negative one, or (c) the negative stimulus and a more divergent positive one, facilitated the subsequent discrimination to varying degrees.

when conditions for eliciting r_R were, by definition, better – greater length of Runway 1 (Amsel, Ernhart, & Galbrecht, 1961) and greater similarity of Runway 1 to G_1 (Amsel & Hancock, 1957).

These early results, along with the discovery of a relationship between strength of DL and the FE (Amsel, 1962), led to the prediction that if the initial strengths of the positive tendencies to two discriminanda are equal, rate of DL should be a positive function of the strength of reward expectancy to the to-be-negative one: The stronger the evocation of r_R by the previously positive, but now negative stimulus at the outset of discrimination training, the faster should be the rate of DL. This was the finding that led to my interest in a theory of the effects of prediscrimination exposures to the discriminanda.

A forerunner of this theory of prediscrimination treatments was an experiment by Shoemaker (1953) showing that the effectiveness of a nonreward was directly related to the number of prior rewards. It investigated the relative effectiveness of reward and nonreward trials in a black–white discrimination. In Phase 1, groups of rats were subjected to three levels of prediscrimination training in which they received 4, 12, or 24 rewarded runs to both of the discriminanda. In Phase 2 of the experiment, varying numbers of rewards and nonrewards were given in relation to the black and white stimulus alleys (separately). In Phase 3 of the experiment, there were choice tests, with both alleys available. The result was that the final discrimination performance was better the larger the number of reward or nonreward trials in Phase 2. However, the effect of a given number of nonreward trials on later discrimination was greater than the effect of that same number of reward trials. Also, the effect of nonreward trials on later discrimination performance was directly related to the number of preliminary reward trials (4, 12, or 24) to both of the eventual discriminanda.

In later work (Amsel & Ward, 1965, Exp. 2), we examined the relationship between the strength of primary frustration (R_F), observed independently in Runway 2 in the form of the FE, and DL, indicated by changes in running speed to S_1+ and S_2- in Runway 1, after varying numbers of continuous rewards to both stimuli (S_1+S_2+) in a preliminary phase of the experiment. The expectation was that the larger the number of prediscrimination reward trials, the earlier would be the appearance of the FE in Runway 2 and of differential responding to S_1 and S_2 in Runway 1 when S_2 was made negative. Our results showed a direct relationship between number of prediscrimination rewards to the two discriminanda and rate of DL. In this regard they were in accord with the earlier result of Shoemaker (1953).

Predictions from the theory of prediscrimination effects

On the basis of these findings, a series of predictions were developed about more complicated cases in which DL is preceded not only by CRF, but

Table 5.1. *Kinds of prediscrimination experience in relation to an $S_1 + S_2-$ discrimination*

Stimulus	Experience
One prediscrimination stimulus	
S_1+	Continuous reward, positive stimulus
S_2+	Continuous reward, negative stimulus
$S_1\pm$	Partial reward, positive stimulus
$S_2\pm$	Partial reward, negative stimulus
Two prediscrimination stimuli	
S_1+S_2+	Continuous reward, both stimuli
$S_1\pm S_2\pm$	Partial reward, both stimuli
S_1-S_2+	Discrimination reversal

also by PRF in relation to one or both of a pair of eventual discriminanda (Amsel, 1962). Again, DL was conceptualized in terms of approach and avoidance, and was measured in terms of amplitude (running time) changes to the positive and negative discriminanda. Table 5.1 presents several kinds of prediscrimination conditions in relation to an $S_1 + S_2-$ discrimination. The effects of these prediscrimination conditions were rationalized in the original theoretical analysis; others were added in a later version (Amsel & Ward, 1965). For present purposes we restrict the discussion to those shown in the table.

Figure 5.6 is a schema containing three factors presumed to be important in this kind of analysis. It depicts a class of situations in which only one of the two eventual discriminanda is presented before DL and (a) the prediscrimination stimulus becomes either the positive or the negative stimulus in the eventual discrimination, (b) the response to this stimulus in the prediscrimination phase may be either partially or continuously rewarded, and (c) the number of prediscrimination trials may be relatively large, small, or of some intermediate value. The schema shows the combinations of (a), (b), and (c) in a 12-cell matrix of possible experimental conditions. (The discriminanda are counterbalanced for absolute effects of color, in the case of a black–white discrimination.)

Prediscrimination CRF in relation to one stimulus

We deal first with the simplest case, shown on the right side of Figure 5.6 – prediscrimination CRF exposure to a single stimulus: S_1+ followed by a discrimination in which S_1 remains positive or becomes negative. The predictions here, shown in the top panel of Figure 5.8, are most interesting in relation to those of the preceding section. Briefly, with an increasing number of rewards of the approach response to the prediscrimination stimulus, S_1+, it should evoke increasing r_R, and nonreward in the subsequent

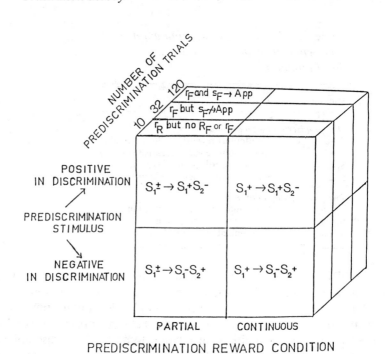

Figure 5.6. Schema showing 12 prediscrimination–discrimination sequences resulting from three attributes of prediscrimination exposure to a single discriminandum (S_1): PRF vs. CRF; positive or negative in discrimination; small, intermediate, or large number of acquisition trials. (Adapted from Amsel, 1962.)

discrimination should be increasingly frustrating. Because of the flat approach gradient, which implies that r_R generalizes strongly from S_1 to S_2, both stimuli should evoke r_R quite strongly in the discrimination, the FE being somewhat greater when S_1 becomes negative in the discrimination than when it stays positive. In either case, then, the flat generalization gradient of approach leads to the prediction that is supported by our own findings and Shoemaker's (1953) results: Increasing the number of prediscrimination CRFs, within the limits of the values employed, produces faster discrimination (see Figure 5.7).

Effects on DL of prediscrimination PRF in relation to one stimulus

The next question is: How should prediscrimination PRF exposure to one of the eventual discriminanda ($S_1 \pm$) affect a subsequent discrimination? The answer in the case of prediscrimination PRF is more complicated; it depends, of course, on the hypothesis that anticipatory-frustration-produced stimuli (s_F) become associated with approach responses in the last stage of PRF training, which means that, at this stage, s_F will become part of a stimulus complex evoking approach and that the earlier (Stage 3) evidence of conflict will disappear.

Of the four stages in PRF acquisition outlined earlier (Figure 5.2), three are theoretically differentiable with respect to the involvement of r_F–s_F: Stage 1, at which r_R has been conditioned, but not yet r_F; Stage 3, at which both r_R and r_F have been conditioned and their stimuli (s_R and s_F) evoke competing (approach and avoidance) response tendencies; and Stage 4, at which both r_R and r_F have been conditioned and both their feedback stimuli, s_R and s_F, evoke approach tendencies. For purposes of this analysis, these stages were taken to correspond to 10, 32, and 120 trials, respectively, under the magnitude-of-reward conditions of the experiments we were then conducting (see Figure 5.6).

A further step in the analysis of prediscrimination effects involved an assumption taken from Miller's (1944, 1948) stimulus–response analysis of conflict and displacement, for which there was some experimental support (see Miller, 1959). The assumption is that the generalization gradient for positive (approach) tendencies is flatter than for negative (avoidance) tendencies, which is to say that the strength of avoidance declines more steeply than the strength of approach as distance from the goal increases. For the present argument, the concept of distance from the goal is replaced by stimulus novelty or similarity between the preexposed stimulus (S_1) and the stimulus that is new in the discrimination (S_2). Combining this assumption with the differentiated stages as they involve r_R and r_F, and identifying $s_R \rightarrow$ approach (which results from r_R) and $s_F \rightarrow$ avoidance (which results from r_F) as the opposing tendencies in Miller's analysis, the three panels of Figure 5.7 represent, from top to bottom, respectively, the state of affairs at the beginning of an $S_1 + S_2$–discrimination following a hypothetical 10, 32, or 120 exposures to the prediscrimination stimulus (S_1). Represented on the left-hand side of Figure 5.7 are the predicted relative magnitudes of approach and avoidance tendencies to the preexposed stimulus (S_1) at the start of the discrimination. Shown on the right are the generalized strengths of these tendencies to the stimulus (S_2) *that is new in the discrimination*. A simplifying assumption, represented in the bottom panel of the figure, is that the strength of the counterconditioned *approach* elicited through r_F–s_F by S_1 and S_2 in the late (120-trial) stage is directly related to the strength of *avoidance* tendencies elicited through r_F–s_F in the middle (32-trial) stage. The assumption is that the avoidance tendency that is conditioned to s_F in the middle stage of training is converted by counterconditioning to an approach tendency in the late stage.

We are now in a position to make some predictions about the experimental conditions in Figure 5.6. These predictions depend crucially on the depth dimension – whether training to the prediscrimination stimulus was

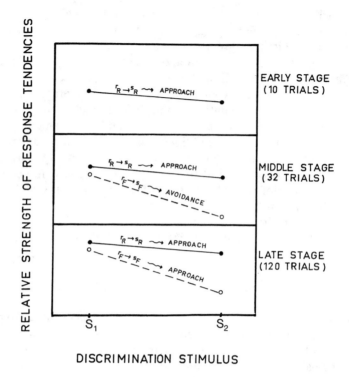

Figure 5.7. Strength of response tendencies to S_1 and S_2 at the beginning of discrimination training following prediscrimination training in which approach to S_1 has been partially rewarded ($S_1\pm$). (Adapted from Amsel & Ward, 1965. Copyright 1965 by the American Psychological Association; reprinted by permission.)

extended to 10, 32, or 120 trials. Outlined for each case of prediscrimination PRF training are predictions as to which discrimination will be learned faster: that in which this prediscrimination stimulus remains positive in the discrimination ($S_1\pm \rightarrow S_1+ S_2-$) or that in which it becomes negative ($S_1\pm \rightarrow S_1- S_2+$).

In the case of 120 prediscrimination PRF trials, the reasoning (see Figure 5.7) is that, at the start of DL, S_1 will elicit r_R and r_F strongly and about equally, but that S_2, which is new in the discrimination, will elicit r_F relatively weakly because of the steeper avoidance gradient. If S_1 becomes negative in the DL after a great many prediscrimination trials, S_1- will elicit strong r_F-s_F and s_F will be associated with continued approach. This will slow DL relative to the condition in which S_1 remains positive in the discrimination. In the latter case, the S_2- stimulus, new in the DL, elicits very weak r_F; consequently, there is little s_F conditioned to approach, there is little to counteract the buildup of avoidance to S_2-, and DL should develop more quickly. The theory predicts the following: After a very large

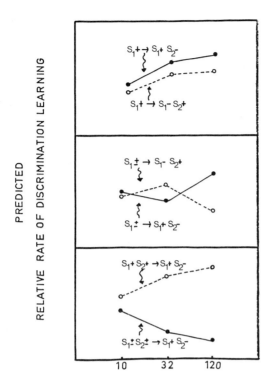

Figure 5.8. Predicted relative rate of discrimination learning following various kinds and lengths of prediscrimination exposure to one or both of the eventual discriminanda. (From Amsel & Ward, 1965. Copyright 1965 by the American Psychological Association; reprinted by permission.)

number of prediscrimination partial rewards of an approach response to S_1, S_1+S_2- will produce faster discrimination than will S_1-S_2+. This prediction and its relations to other predictions for $S_1\pm$ are shown in the middle panel of Figure 5.8.

The most interesting and critical comparison to be made for the purpose of our analysis is between the 120-prediscrimination-trial case and the intermediate, 32-prediscrimination-trial case. In the former, s_F already elicits approach, while in the latter, s_F still elicits avoidance. Consequently, the prediction is reversed from one case to the other. The reasoning for the 32-trial case is that both r_R and r_F are being evoked by S_1 at the start of discrimination, but s_R elicits approach and s_F still elicits avoidance. When, after 32 trials, a switch is made from prediscrimination $S_1\pm$ to the S_1+S_2- discrimination, S_1+ evokes both approach and avoidance tendencies through s_R and s_F, and S_2-, new in the discrimination, evokes mainly generalized approach and little avoidance. When the shift is from $S_1\pm$ to S_1-S_2+, there is a strong tendency for s_F to elicit avoidance in relation to the now negative S_1- (generalizing very little to S_2) and a strong tendency for s_R to elicit approach to the new, and now positive, S_2. This latter situation is much more favorable for discrimination. The prediction is as follows: After an intermediate number of prediscrimination partial rewards of an approach response to S_1, because the response to $S_1\pm$ is still at the conflict stage, S_1+S_2- will produce slower discrimination than will S_1-S_2+. After few enough $S_1\pm$ trials (in our example, 10 trials) to produce some r_R but little or no FE and no r_F, there should be little or no difference between the rates of the S_1+S_2- and S_1-S_2+ discriminations.

Prediscrimination CRF or PRF in relation to both stimuli

The bottom panel of Figure 5.8 summarizes predictions about the rate of DL following CRF or PRF prediscrimination experience with *both* discriminanda, presented separately. It is clear that $S_1+S_2+ \rightarrow S_1+S_2-$ operates essentially like $S_1+ \rightarrow S_1+S_2-$, except that it is not susceptible to generalization effects, since neither S_1 nor S_2 is new in the discrimination.

The prediction concerning $S_1\pm S_2\pm$ as a prediscrimination condition is of a decreasing monotonic relationship between rate of discrimination and number of such prediscrimination trials. The reasoning is simply that PRF of approach responses in relation to both S_1 and S_2 for 32 trials will result in conflicting approach and avoidance response tendencies to both, whereas PRF in relation to both S_1 and S_2 for 120 trials will result in approach responses to s_F in connection with both. While conflict in relation to S_1 and S_2 represents a difficult base on which to build discrimination, persistence in responding in relation to S_1 and S_2 (the PREE in relation to both discriminanda) is an even less readily reversible condition. Hence, the prediction depicted in the bottom panel of Figure 5.8 is that the effect of increasing the number of prediscrimination trials will be reversed from S_1+S_2+ to $S_1\pm S_2\pm$.

Experimental tests of the predictions

The prediscrimination conditions that have been outlined formed the experimental conditions of experiments involving black (S_1) and white (S_2) alleys in a runway as the discriminanda (Amsel & Ward, 1965). Two major sets of comparisons were made involving the seven experimental conditions shown in Table 5.1 (all counterbalanced for color).

Presented first are the results showing the effect of the *number* of pre-

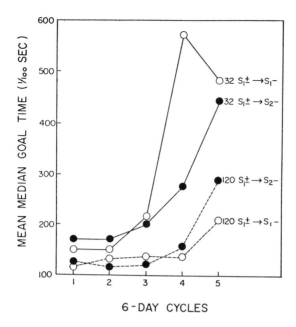

Figure 5.9. Extinction of responding to the negative stimulus (S_2) in the discrimination as a function of the number of prediscrimination trials and whether the prediscrimination stimulus ($S_1\pm$) is positive (S_1+S_2-) or negative (S_1-S_2+) in the discrimination. Shown are levels of responding in extinction to the negative stimulus, which is most affected. (From Amsel & Ward, 1965. Copyright 1965 by the American Psychological Association; reprinted by permission.)

discrimination exposures on discrimination, and second, results that deal with the *kind* of prediscrimination exposure.

Figures 5.9 and 5.10 show discrimination-phase extinction to the negative stimulus (S_1- or S_2-) when it is either the same as the prediscrimination stimulus (S_1) or is new in the discrimination (S_2). Performance to the positive stimuli adds nothing to this result and is not shown.

Figure 5.9 shows the effect of the *number* of prior $S_1\pm$ trials on discrimination when S_1 becomes negative or when the new stimulus (S_2) is negative. First of all, extinction to the negative stimulus in the discrimination is slower following 120 $S_1\pm$ than following 32 $S_1\pm$ trials. The interaction predicted earlier is also evident. After 32 $S_1\pm$ trials, the discrimination (weakening of response to $S-$) is faster for S_1 than for S_2, and the reverse is the case after 120 $S_1\pm$ trials. This interaction becomes even more obvious when the difference between 32 $S_1\pm \rightarrow S_1-$ and 120 $S_1\pm \rightarrow S_1-$ is compared with the difference between 32 $S_1\pm \rightarrow S_2-$ and 120 $S_1\pm \rightarrow S_2-$. The first difference is of about 4 seconds at the highest points of the two curves; the other is about 1.5 seconds.

In Figure 5.10, a comparison is made among the effects of prediscri-

116   *Frustration theory*

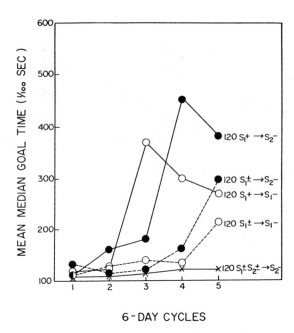

Figure 5.10. Extinction of response to S_1 or S_2 in a discrimination as a function of the nature of the 120-trial prediscrimination exposure, showing fastest discrimination after S_1+ and slowest after $S_1\pm S_2\pm$, with $S_1\pm$ in between. (From Amsel & Ward, 1965. Copyright 1965 by the American Psychological Association; reprinted by permission.)

mination treatments when, in all cases, the number of prediscrimination exposures is 120 trials. It is obvious that the $S_1\pm S_2\pm$ treatment severely retards extinction to the negative stimulus in discrimination (there is none), and that $S_1\pm$ is followed by slower extinction than is S_1+, regardless of whether S_1 or S_2 is negative in the discrimination. This much simply reflects the operation of the PREE, the persistence mechanism, in the discrimination.

A variety of relationships that bear on the details of a frustration theory approach to DL are shown in Figure 5.11. The data are from performance in a double runway, so that the FE in Runway 2 and the prediscriminative and discriminative performance in Runway 1 can be shown side by side. Panels 1 and 5 show, for the prediscrimination phase, the FE measured in Runway 2 and its relation to partially rewarded Runway 1 performance. (In these panels the data from all three prediscrimination conditions are combined.) Panels 2, 3, and 4 show the FE in Runway 2 during discrimination training for each of three discrimination conditions. Note particularly the diminution of the FE in relation to evidence of discrimination in Runway 1. Panels 6, 7, and 8 show the predicted differences in rapidity

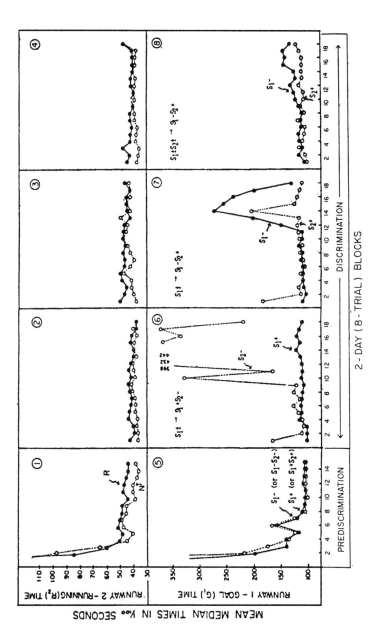

Figure 5.11. The relationships between prediscrimination PRF acquisition (Panel 5) and discrimination performance under the three different prediscrimination–discrimination conditions (Panels 6, 7, and 8), measured in Runway 1, and the corresponding FEs (Panels 1, 2, 3, and 4, respectively), measured in Runway 2. Panel 1 shows development of the FE in Runway 2 as a consequence of reward and nonreward trials during prediscrimination PRF training in Runway 1. In these panels the prediscrimination data from all of the three subsequent discrimination groups are combined. Panels 2, 3, and 4 show the disappearance of the FE in Runway 2 as the discrimination in Runway 1 is formed in each corresponding case. (From Amsel & Ward, 1965. Copyright 1965 by the American Psychological Association; reprinted by permission.)

118   *Frustration theory*

and magnitude of the discriminations among the three groups, all of which were exposed to the same number of PRF prediscrimination trials.

What we have here in one experiment are all of the separate phenomena, and also their interrelationships, that had been predicted by the theory and shown previously in separate experiments. The development of the FE during PRF acquisition had already been demonstrated (Amsel & Hancock, 1957; Roussel, 1952; Wagner, 1959), as had the decreased vigor and increased variability of performance that is seen after 40 trials (see Panel 5, Block 6). This is a highly reliable finding in all groups in this experiment, and such evidence of conflict, in Stage 3 in our analysis (Amsel, 1958a), can be observed in discrete-trials PRF training with relatively large reward magnitude whenever one looks for it. It is not obvious that this kind of prediction can be made by purely associative models (e.g., Capaldi, 1967), or by any kind of cognitive model, or that phenomena such as are represented in Figure 5.11 can be explained by these models, even after the fact (see Chapter 6).

The elimination of the FE attendant upon DL (see earlier data in Figure 5.1) is duplicated in the six right-hand panels of Figure 5.11: Panels 2, 3, and 4 are the FE data (Runway 2) for the conditions shown in Panels 6, 7, and 8, respectively. Recall that the frustration theory interpretation of this phenomenon is that the FE depends on r_R, and that when discrimination is formed in Runway 1, r_R is evoked by $S+$ while r_F is evoked by $S-$. Consequently, nonreward in the first goal box (G_1) following successful discrimination is preceded by r_F – not by r_R – and the FE is reduced. Again, such results appear to be outside the boundaries of purely associative and cognitive models.

Finally, Panels 6, 7, and 8 compare the groups for rapidity and magnitude of discrimination, and they replicate, in Runway 1 of the double runway, the essential relationships in the bottom three curves of Figure 5.10. After 120 prediscrimination trials, extinction is most pronounced when $S_1\pm$ is followed by a discrimination in which the new stimulus is negative; extinction is least pronounced when discrimination follows a prediscrimination condition in which there has been PRF training in relation to both of the discriminanda ($S_1\pm S_2\pm$).

$S_1\pm S_2+$ as a prediscrimination condition and the demonstration of a within-subjects PRAE

One of the problems inherent in exposing subjects to only one of the eventual discriminanda in prediscrimination training is the following: Whenever the switch is from the prediscrimination condition to a discrimination test, there is a clear initial generalization decrement in relation to the stimulus that is new in the discrimination (see the performance on early

Discrimination learning and prediscrimination 119

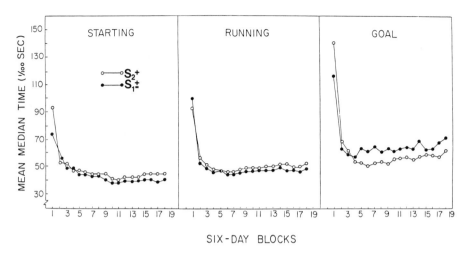

Figure 5.12. The PRAE as a within-subjects effect. Note that the data are in running time, not running speed. (From Amsel, MacKinnon, Rashotte, & Surridge, 1964. Reprinted by permission of the Society of the Experimental Analysis of Behavior.)

trials to S_2 in Panels 6 and 7 compared with Panel 8 in Figure 5.11); and this, to an extent, obscures the differences in rate of discrimination we are evaluating. There is, however, a prediscrimination condition that eliminates this generalization decrement factor. (I introduced this condition at the end of Chapter 4 in discussing the within-subjects generalized PREE.) It involves exposure to both of the eventual discriminanda, while partially reinforcing the approach response to one and continuously reinforcing the response to the other. The general notation of this condition is $S_1 \pm S_2 +$ (Amsel, MacKinnon, Rashotte, & Surridge, 1964).

The first thing that became apparent was that we were working with a new phenomenon of some interest in the *prediscrimination* phase: a within-subjects PRAE, previously shown by others (Goodrich, 1959; Haggard, 1959; Wagner, 1961; Weinstock, 1954) using separate PRF and CRF groups – that is, as a *between-groups* effect. In an $S_1 \pm S_2 +$ condition, start and run speeds were greater (less time was taken) in the presence of the stimulus signaling PRF ($S_1 \pm$) than in the one signaling CRF ($S_2 +$), but goal speeds were lower (more time was taken) in the presence of $S_1 \pm$ than in $S_2 +$. (We discussed the within-subjects PRAE earlier in connection with the later experiments of Henderson [1966] and MacKinnon [1967]; see Figures 5.3 and 5.4.)

These results, shown in Figure 5.12 in terms of time rather than speed, indicate that the mechanisms of PRF versus CRF, as applied to runway acquisition behavior, can exist and operate simultaneously in the single organism. That is to say, the individual organism can behave in a manner appropriate to either a PRF or a CRF schedule, at different times, depending on the difference in a single environmental stimulus. This is an important demonstration that, in the same environmental context, different dispositions can exist in the same individual relative to speed of approach to a goal that has provided consistent or inconsistent reward in relation to specific differential cues. The results are similar to those obtained when one animal is run under PRF and another under CRF conditions in what is otherwise the same experimental situation. The very striking thing about this kind of result, whether for between- or within-subjects conditions, is that, at asymptote, the reversal in relative speeds between the start and run measures, on the one hand, and the goal measure, on the other, occurs in something over a total time of about 2 seconds, a behavioral effect not discernible to the naked eye. An explanation for the superiority in the between-subjects case of PRF over CRF acquisition performance early in the response chain and the reversal near the goal was offered years ago by several investigators (Goodrich, 1959; Haggard, 1959; Wagner, 1961). Wagner's version of the explanation is as follows:

That the partial superiority is not found on those response measures close to the goal may be attributed to less effective conditioning of the approach response to the goal box cues plus s_f. This argument appears quite reasonable in the context of the prior assumptions if it is noted that r_f–s_f may be expected to become conditioned to the approach response when the s_f cues are introduced initially at a weak value, and hence with a negligible tendency to elicit competing responses, and are then increased gradually at the same time that the approach response is being strengthened. This condition obtains much more clearly in the early portions of the alley than in the goal region where, due to the proximity of the primary frustration event, generalized r_f–s_f may be expected to occur earlier and to follow a course of greater intensity, and thus produce responses which compete more effectively with the approach response. (p. 240)

When we went on to test rates of discrimination in this case, the results were the same as with exposure to a single stimulus: Following *many* predistermination trials of $S_1 \pm S_2 +$, the $S_1 + S_2 -$ discrimination is formed earlier than $S_1 - S_2 +$, and following *few* such trials, the $S_1 - S_2 +$ discrimination is formed earlier than $S_1 \pm S_2 +$ (see Amsel & Ward, 1965, Exp. 5, Fig. 7).

Summary

In Chapter 3, I provided an overview of the explanatory domain of frustration theory, emphasizing its accounts of a number and variety of reward-schedule effects. Some of these effects depend simply on the nature of the transition from one level of reward magnitude to another – from a phase of continuously reinforced acquisition to a phase of extinction, or from a phase of large to a phase of small magnitude of reward. Other reward-schedule effects depend on reward intermittency, and here the rewards and nonrewards (or the large and small magnitudes of reward or the immediate and delayed rewards) are intermixed in a single acquisition phase. One such example of intermittency, as we have seen, is what in the instrumental case is called discrimination learning – a kind of reward – nonreward intermittency that normally results in differential responding to a positive and negative discriminandum, unlike the case of PRF training in which there are no differential stimuli signaling reward and nonreward and persistent responding develops.

In this chapter we have examined in a number of ways the role of frustration, both primary and anticipatory, in discrimination learning. We have seen that in the double-runway apparatus, we can study the emergence of these effects side by side in the formation of a discrimination: primary frustration (the FE) in Runway 2 and, in parallel, anticipatory (conditioned) frustration in Runway 1. We have also seen that, as the theory predicts, primary frustration precedes the formation of a discrimination: The FE in Runway 2 is shown until the discrimination in Runway 1 is formed and then virtually disappears. This is unlike the case in which PRF training is given in Runway 1, in which the FE appears to be shown indefinitely following nonreward trials in Runway 1.

The latter part of this chapter is concerned with prediscrimination effects, and the message is that depending on the nature of the prediscrimination exposure to one or both of the eventual discriminanda, discrimination learning can be facilitated or retarded. The proposal is that the mechanisms involved in these effects are the same as those involved in the transfer-of-persistence effects described in Chapter 4. These are the mechanisms of long-term dispositional learning, and they govern tendencies to show differential responding to discriminanda quickly, slowly, or not at all. The primary factors that affect an appetitive discrimination are the extent to which response to the preexposure to one or both of the eventual discriminanda is partially reinforced and, if just one of the discriminanda is involved, whether it becomes the positive or negative cue in the discrimination.

6 Alternatives and additions to frustration theory

In Chapters 3, 4, and 5, we addressed the role of frustrative and, generally, of disruptive factors in the acquisition and extinction of responses in a variety of experimental paradigms. The results of these experiments have been discussed in the context of a more general conceptualization of arousal, suppression, persistence, and regression: We have alluded to "generality" and "transfer" of persistence and, in the case of the theoretical and experimental treatment of the role of prediscrimination experience in subsequent discrimination learning, to "resistance to discrimination" as another form of transfer of persistence. In other versions of a discrimination learning experiment, the within-subjects PRF experiment and the DNC experiments, the "generalized PREE" and the emergence in extinction of ritualized behaviors have also reflected, or at least implied, transfer of persistence. If we now look back to the first chapter and to Table 1.2, we can regard Chapters 3, 4, and 5 as a detailed elaboration of the first two of the six steps in the psychobiological study of related behavior effects. I will now undertake a brief general discussion of the theory. This will focus on the breadth of experimental particulars the theory encompasses and will compare its scope with other theories with which it has some explanatory overlap.

In *A Brief History of Time,* Stephen Hawking (1988) writes the following: "A theory is a good theory if it satisfies two requirements: It must accurately describe a large class of observations on the basis of a model that contains only a few arbitrary elements, and it must make definite predictions about the results of future observations" (p. 9). The first requirement can be restated to say that if a theory's number of "elements" or premises are not an order of magnitude smaller than the particulars or observations for which it accounts, it provides no explanatory advantage over a simple enumeration of the particulars. The second requirement implies that a good theory must lead to new (and, perhaps, nonintuitive) observations that are not members of the class of particulars that gave rise to the theory in the first place. If one accepts Hawking's definition of a good theory, one must also accept that a theory which accounts, say, for only one phenomenon – for example, the FE or the PREE – is not a good theory. Chapter 6 should be read with this assertion in mind.

But before we turn to a discussion of alternative theories that explain one or more of the reward-schedule effects exemplified by the PREE, let us briefly consider a number of well-known theories of association from which *none* of these phenomena can be derived.

Classical association theories

Theories of learning proposing positive monotonic relationships between number of reinforcements and resistance to extinction (e.g., Hull, 1943) are unsuitable conceptualizations of the kinds of dispositional learning that give rise to learned persistence (the PREE) and the related phenomena reported in the preceding three chapters. For example, theories of this kind cannot account for faster extinction following larger rewards in acquisition, the magnitude of reinforcement extinction effect (MREE; e.g., Armus, 1959; Hulse, 1958), or for faster extinction following a greater number of reinforcements, the overtraining extinction effect (OEE; e.g., Ison, 1962; North & Stimmel, 1960). They cannot account for the results of experiments in which number of rewards and resistance to discriminative extinction are inversely related, or in which a positive relationship exists between number of $S_1 \pm$ trials and resistance to discriminative extinction when S_1 becomes negative ($S_1 - S_2 +$) in a subsequent discrimination (Amsel & Ward, 1965).

Statistical association theories of learning (e.g., Estes, 1950, 1959) also are not designed to deal with these phenomena. They are, for example, silent about acquisition functions that require nonmonotonicity; they predict asymptotic performance levels corresponding directly to percentage or probability of reinforcement; and, in general, they do not address "paradoxical" effects. This can also be said of the linear-operator models of learning (e.g., Bush & Mosteller, 1951; Rescorla & Wagner, 1972). This being the case, these models are at best partial theories of the psychological mechanisms that are involved in many cases of acquisition and extinction in nondiscriminative and discriminative learning; they are silent about the many phenomena that are predicted and explained by frustration theory. Of course, these mathematical theories of learning were not designed to deal with such phenomena. To do so they would have to incorporate some means of dealing with phenomena in PRF (compared with CRF) acquisition, such as the increase in variability of performance that occurs at a certain stage (Stage 3, in our analysis) of PRF acquisition (see Panel 5, Figure 5.11) and with the PRAE, the paradoxical within-subjects acquisition effects that vary with separate segments of the response chain. These phenomena and the other particulars of instrumental learning described earlier would appear to require for their integration a theoretical approach that ascribes active properties to nonreinforcement (Amsel, 1958a, 1962, 1967), a characteristic of neither classical Hullian theory nor of the stochastic association theories, to mention two of the earlier, classic approaches. A model (DMOD) combining elements of the Rescorla–Wagner model and frustration theory (Daly & Daly, 1982) does much better in this regard, by adding to the Rescorla–Wagner model associative factors representing conditioned frustration (V_{AV}) and counterconditioning (V_{CC}). However, it still does not account for nonassociative phenomena: It is also a purely associative model.

Alternative theories of the PREE

In 1958 I summarized the existing explanations of the PREE as follows:

Attempts to account for the relative fixation of responses acquired under partial-reinforcement schedules... have been many and varied. Beginning with the "common sense" expectancy interpretation of Humphreys (e.g., 1939b), there followed the "response-unit" explanation offered originally by Skinner (1938) and tested by Mowrer and Jones (1945); the Hull-Sheffield interpretation (Sheffield, 1949) in which traces of previous goal events and stimulus generalization decrement are the important factors; the discrimination or sensory integration hypothesis of Bitterman and Tyler and others (Crum, Brown, & Bitterman, 1951; Tyler, 1956; Tyler, Wortz, & Bitterman, 1953); Weinstock's contiguity hypothesis (1954) which suggests that partial reinforcement provides an opportunity for the habituation of competing responses on the nonreward trials during acquisition; some recent Pavlovian interpretations of Razran (1955), the most convincing of which makes partial reinforcement "a case of the efficacy of repeated post-extinction reconditionings"; a more recent interpretation of Logan, Beier, and Kincaid (1956) in terms of r_G factors; and a still more recent, contiguity-type interpretation of Hulse and Stanley (1956) which suggests that partial reinforcement training increases resistance to extinction because the partially reinforced Ss have acquired during acquisition something to do in the goal box on nonreward trials. (Amsel, 1958a, pp. 112–13)

Cognitive dissonance

By 1962 there was another obvious alternative to frustration theory for conceptualizing the role of nonreinforcement in the acquisition of persistent behavior. In general and nontechnical terms, frustration theory held that persistence is the result of learning to continue to a goal in the face of indications of frustration and uncertainty (Amsel, 1958a, 1962). The new alternative, based on *cognitive dissonance theory,* held that persistence is achieved when organisms "come to love things for which they have suffered" (Festinger, 1961, p. 11). This account proposed that the PREE and persistence are indicants of increased attractiveness ("extra attractions") of a goal region, or of an activity toward a goal, acquired as the result of dissonance created for an animal or person by insufficient reward, in either

Alternatives and additions to frustration theory 125

amount or frequency, or by obstructions placed in the path to a goal. In 1962 I pointed out that

> Festinger's treatment of insufficient rewards is very similar in conceptualization to some earlier work of Olds (1953, 1956). On the basis of experiments with children, Olds concluded that the reward value of a secondary rewarding stimulus (S_r) could be increased – and hence the strength of approach responses leading to S_r could be increased – by delaying the presentation of S_r *after* overlearning. He showed that tokens which could ultimately be exchanged for more primary reward increased in reward value when their presentation was delayed, if these delay trials followed a number of trials in which their presentation was immediate. Olds suggested that such a procedure constitutes "practice at wanting" and that such practice at wanting increases the value of that which is wanted (S_r). Festinger would seem to be taking the same position in his animal research. In one of several experiments dealing with dissonance in rats, he employs a double runway in which a start box is separated from a mid box by an alley, and the mid box is separated from an end box by another alley. Subjects are delayed in the mid box in the manner of Holder, Marx, Holder, and Collier (1957) and Wagner (1959), and the finding is that the subjects delayed in the mid box, before being allowed to run to the end box and find food, will continue to run to the mid box longer on test trials than the subjects always fed in the mid box as well as in the end box. Festinger's (1961, p. 9) interpretation is that the delayed subjects develop dissonance as a result of the delay, and that the "dissonance tends to be reduced by developing some extra preference about something in the situation. The existence of this extra preference leads to the stronger inclination to continue running during extinction trials." (pp. 311–12)

I also argued that

> the S–R treatment of nonreward effects in partial reinforcement (Amsel, 1958a; Kendler, Pliskoff, D'Amato, & Katz, 1957; Spence, 1960; Wilson, Weiss, & Amsel, 1955) does not suggest that nonreward (or delayed reward) increases "wanting" in Olds' terms, or "preference about something in the situation" in Festinger's. This type of neo-Hullian approach is a description of a mechanism operating through anticipatory nonreward which can lead to either rapid extinction of behavior or to increased resistance to extinction, and which can either facilitate discrimination learning or retard it. I will claim that neither increased "wanting" nor "preference" – in fact, no "common sense" cognitive interpretation – can account for all of the effects which nonreward seems to have. (p. 312)

The cognitive account of the active properties of nonreward by Lawrence and Festinger (1962) was not designed to handle the many and diverse experimental observations described in the preceding three chapters, and so far as I am aware, it has not been the subject of experimental research since the early 1960s. It is not unreasonable to hold that for such a theory to be revived and prosper, it would eventually have to encompass all or most of these phenomena. Extra attractions for the activity or the apparatus or the goal events cannot, it would seem, account for the phenomenon of extinction to begin with. Extinction implies continuous nonrewards, and Lawrence and Festinger indicated that zero reward was the optimal condition for building extra attractions in the goal region. This cognitive theory, which had a certain intuitive validity for describing facets of human social behavior, did not – and does not – seem to have the power of a more analytic stimulus–response approach based on conditioning premises. While it accounted for increased resistance to extinction after partial reinforcement, it had, as we have seen, difficulty with ultimate extinction. While it accounted for extinction effects, it would have had great difficulty with the complicated acquisition effects. Why should the rat start and run faster to the "extra attractive" stimuli associated with insufficient reward (as compared with stimuli associated with continuous reward), but enter the "extra attractive" goal area more slowly? How does increased attractiveness account for the increased variability that develops in all PRF conditions, particularly with large rewards? Finally, how does the increased attractiveness of the activity, the goal region, or the goal event account for the findings that an $S_1 \pm S_2 \pm$ prediscrimination condition retards a subsequent $S_1 + S_2 -$ prediscrimination much more drastically than does an $S_1 \pm$ prediscrimination condition? The conditions involve an equal number of nonreward experiences. Why is an $S_1 - S_2 +$ discrimination faster than $S_1 + S_2 -$ after a small number of $S_1 \pm$ prediscrimination trials and slower after a large number of such prediscrimination trials? This would appear to have nothing to do with extra attractiveness.

At the time most of this work was done, there were, outside of the Lawrence and Festinger (1962) treatment of the PREE, no other theories that could be thought to account for such phenomena as the prediscrimination effects and the generalized PREE. There were theories of discrimination reversal, a phenomenon that does, of course, fall within the matrix of experimental procedures that would qualify as prediscrimination treatments (Amsel, 1962, p. 318, table 2). A particularly good early example lay in the continuity versus noncontinuity experiment in which the *presolution* experience is (say) an $S_1 + S_2 -$ discrimination and the solution period involves the reverse discrimination (see Spence, 1940, 1945). Other early work involving prediscrimination treatments took several forms, including Harlow's (e.g., 1949) work on learning sets, Lawrence's (1949, 1952) work on acquired distinctiveness of cues, and studies relating overlearning to discrimination reversal (e.g., Birch, Ison, & Sperling, 1960; Capaldi & Stevenson, 1957; Pubols, 1956; Reid, 1953). One of the characteristics of most of this work was that it involved *simultaneous* rather than *separate* presentation of the discriminanda, a method that does not lend itself to the investigation of the transfer-of-persistence, dispositional-learning phenomena with which we have been most concerned.

Sequential theory

Of the theories of reward-schedule effects since 1962, two stand out particularly: the *sequential theory* of the PREE (Capaldi, 1966, 1967), clearly the more influential of the two, and the *stimulus analyzer theory* (Sutherland, 1966; Sutherland & Mackintosh, 1971).

Capaldi's theory, as we have seen, was in part a formalization of the Hull–Sheffield hypothesis of the PREE. It deals with sequences of reinforcements (R) and nonreinforcements (N) and handles in great detail the consequences on resistance to extinction of the number of N-trials preceding an R-trial. The greater the N-length, according to the theory, the greater the association of the feedback stimulus from nonreward with continued responding and the greater the resistance to extinction of the instrumental response. Capaldi's elegant theory was not designed to account for the transfer-of-discrimination or transfer-of-persistence phenomena with which we have worked. It has no apparent explanation for many of the widely spaced trial phenomena (one trial per day, Donin et al., 1967; one trial every 3 days, Rashotte & Surridge, 1969) in which we have been interested, usually in the context of the Theios–Jenkins experiment, with added "vacations" from the experimental procedure. (In the explanation of the PREE, sequential theory can be said to have difficulty enough with experiments, following Theios [1962] and Jenkins [1962], that merely interpolate a phase of CRF training between PRF acquisition and extinction.) Capaldi's theory has no obvious explanation for the PRAE. Finally, it was not designed to address phenomena that depend for their explanation on the hypothetical response-evoking properties of stimulus feedback from anticipatory frustration (r_F–s_F) – phenomena like the ritualized behaviors in the discontinuously negatively correlated reinforcement (DNC) experiments (see Amsel & Rashotte, 1969; Rashotte & Amsel, 1968; and Chapters 3 and 4, this volume). Capaldi's theory (see especially 1967) has been applied, with great success, to the explanation of patterns and sequences of rewards and nonrewards and sequences of magnitudes of reward, usually with rather short ITIs, although Capaldi's view has been that the duration of the ITI is not a factor in his theory – that memories of R and N carry over relatively long ITIs, even 24 hours, that the N-stimulus or memory remains intact until and unless it is replaced by the R-stimulus from a reinforced trial.

From about 1970 on, Capaldi (e.g., 1971) appeared to regard his sequential theory as a theory of memory, and he moved in the direction of experiments that had less to do with phenomena of extinction and more to do with the "new-look" theories of human memory that influenced investigators of animal learning as well as those of human memory (see Amsel, 1989). In recent years, some of Capaldi's experiments, with rats as subjects, have had to do with serial learning and interitem associations

(Capaldi, Verry, Nawrocki, & Miller, 1984), counting (Capaldi & Miller, 1988a), and anticipation of remote events (Capaldi & Miller, 1988b). It should be obvious, then, that any original tension that existed between the proponents of sequential theory and frustration theory were based on their explanations of the PREE, and that it is now greatly attenuated as the theories have gone in different directions. Each theory has carved out an explanatory domain that overlaps little with the other; still, both can be regarded as partial theories of reward-schedule effects in instrumental learning. Whether, to imitate physicists, a "grand unifying theory" of reward–nonreward effects will integrate these (and perhaps other) partial theories remains to be seen.

Sutherland's stimulus analyzer theory

The analyzer theory of Sutherland and Mackintosh was designed specifically to account for discrimination learning and was later applied to explain the PREE; it has not, to my knowledge, addressed prediscrimination effects except in the case of discrimination reversal. As applied to the PREE (Sutherland, 1966), the theory holds that in PRF (but not CRF) training, because there is no basis in exteroceptive or interoceptive stimulation for predicting which trials will be R and which N, the rat (in this case) switches in a number of (as it turns out) irrelevant stimulus analyzers as a kind of hypothesis-testing procedure to solve this insoluble "discrimination." Then in the extinction phase, following PRF but not CRF training, the rat must, one by one, dissociate or disconnect each of these various analyzers (visual, auditory, olfactory, etc.) from the nonreinforced response, thereby prolonging extinction. It is difficult to see how analyzer theory, which is an attentional theory, could account for the many paradoxical reward-schedule and transfer effects that have been described; to my knowledge, there has been no attempt to do so. These effects, particularly those described in detail in Chapters 4 and 5 and the experimental paradigms out of which they emerge, define mechanisms that can be thought to be important in the developmental and psychobiological aspects of the study of dispositional learning. We will deal with such matters in Chapters 7 and 8.

Wong's behavioral field approach

We have seen that fractionating the measurement of an approach-to-goal response into segments can reveal dynamics of behavior that cannot be observed in the overall response. For example, observations reflecting arousal, suppression, persistence, and regression can all be made in the microcosm of the PRF experiment if several speed measures are taken. In one case, we identified the dynamic relationships among these speeds, in comparisons of CRF and PRF conditions, as the PRAE. This effect involves arousal and suppression: Typically, speeds are higher in PRF than in CRF in the start and run measures, and lower in the goal measure, reflecting, according to frustration theory (Amsel, 1967), the generalized drivelike effects of weak r_F in the early segments and the suppressive effects of stronger r_F–s_F in the goal segment of the runway. A year before the publication of this result by Goodrich (1959) and Haggard (1959), the facilitating effect of reward percentages of less than 100% had been shown in all three measures, at one trial per day by Weinstock (1958); however, even at this long intertrial interval, there was evidence later for the classic PRAE in a four-segment runway (Surridge, Boehnert, & Amsel, 1966).

As we have also seen, another phenomenon that required for its demonstration the fractionation of an approach response in a runway is the ritualized, idiosyncratic behavior that emerges in acquisition under DNC conditions and again, in the same animals, in the extinction of responses learned under normal uncorrelated conditions (e.g., Rashotte & Amsel, 1968). The difference between the DNC case and the case of the PRAE is that, in the former, the animal is required to "take time" (the animal does not simply run slow) in order to find reward, whereas in the latter, more usual case, reward is uncorrelated with time taken (or, in this case, with speed).

In both of these cases, however, what can at first be characterized as a simple appetitive approach response to a goal is shown to be not so simple – either because the speed dynamics of the response change over its course, or because the special reinforcement contingency that obtains causes qualitatively different behaviors to emerge, both within the single subject over the course of the response, where they remain fixed in that subject once acquired, and between subjects, where the qualitative pattern or ritual is different from subject to subject.

The purpose of the foregoing is to introduce Wong's *behavioral field approach* to instrumental learning (Wong, 1977, 1978) and his *stage model of extinction*. Wong's is, at least in significant part, a version of a "try-harder" theory of the PREE, but with this difference: It is an attempt to "incorporate existing findings of extinction-induced behavior into frustration theory (Amsel, 1967)" (Wong, 1978, p. 82). The stage model postulates an orderly, unchanging succession of three qualitatively different behavioral stages in the extinction sequence: habit, trial and error, and resolution. These are what Wong calls "coping strategies." The *habit stage* involves a more or less automatic runoff to completion of the response in the absence of incentive. The second stage of extinction after CRF training, the *trial-and-error stage,* is reminiscent of Sutherland's hypothesis-testing explanation of what goes on in PRF acquisition. Here the animal explores alternatives, and the predominant coping strategy is referred to as "variation aggression." The order of the animal's trial-and-error behavior in this stage must obviously be determined by what Hull called its "habit-family hierarchy," that is to say, by its previous reinforcement history. In this sense, at least, the trial-and-error stage can be regarded as analogous to what happens in our DNC experiments – the emergence of previously learned DNC rituals during the extinction of a second response learned under uncorrelated reinforcement: In both cases the animal's reinforcement history dictates its behavior (given what the situation makes possible) under the stress of extinction. In both cases the emerging responses are anticipatory-frustration-related: In Wong's case, they are characterized as "aggressive"; in the DNC case, they are essentially "superstitious" responses that have allowed the animal to take enough time in the alley to meet the reinforcement requirement. The final stage in Wong's analysis, *resolution*, finds the animal resigned to the negative consequence of its behavior, and "the initial goal is abandoned in favor of subgoals, such as drinking and sand-digging" (Wong, 1978, p. 83). These subgoals are examples of the well-known displacement or adjunctive behaviors of the ethologists and the operant researchers – for example, aggression (Azrin et al., 1966), displacement (McFarland, 1966), adjunctive drinking (Falk, 1971), sand digging (Wong, 1977).

We have been dealing with a description of stages in extinction after CRF training. The stage model has also been applied to the qualitative and quantitative differences in behavior when CRF and PRF acquisition conditions are compared. Here the model distinguishes between "response persistence" and "goal persistence." It predicts for the CRF condition greater persistence in habitual routes and relative absence of competing responses in acquisition, and more response persistence in extinction; and for the PRF condition it predicts more variable behavior – route variation and exploration (much as in DNC) – in acquisition, and greater goal persistence in extinction. The latter, according to the model, is attributable to the trial-and-error coping strategy, a kind of hypothesis testing reminiscent of Sutherland's (1966) "breadth of learning," that the animal acquires under PRF acquisition conditions.

Evidence in support of the stage model as it applies to acquisition and extinction under CRF and PRF conditions is presented in Wong's two papers cited earlier, and it is reasonably good. As described earlier, sequences of qualitatively different behaviors appear over the course of acquisition and extinction, and these sequences differ *across* these phases depending on acquisition conditions: The behaviors described are seen in the course of instrumental-response acquisition (mainly in PRF training) and in the course of extinction (mainly after CRF training).

The idea that the appearance of displacement/adjunctive behaviors in PRF acquisition preempts (or at least attenuates) their appearance in extinction is not a radical departure from observations that have been made in the context of frustration theory, though some of the specific behaviors are different. As we have seen in Chapters 2 and 3, a long-held position, following Hull (1934), is that the addition of frustration-produced stimuli (S_F and s_F) alters the stimulus complex and rearranges responses in the response-family hierarchy. Therefore, one would expect responses to emerge in PRF training – and in extinction following CRF training – to primary frustration and to conditioned frustration and the resultant conflict between approach and avoidance tendencies. Some of these specific frustration-induced behaviors, which can be observed in experiments on the PREE, are urination and defecation (e.g., Amsel, 1958a), retracing (any PREE experiment), vocalization (in chicks; Amsel, Wong, & Scull, 1971), ultrasounding (in infant rats; Amsel, Radek, Graham, & Letz, 1977), and the various ritualized behaviors that emerge in acquisition under DNC conditions (e.g., Rashotte & Amsel, 1968). In FE-type experiments, these behaviors are escape from frustrative nonreward (e.g., Adelman & Maatsch, 1955; Daly, 1969a), aggression (e.g., Gallup, 1965), and increased general activity in the open field (e.g., Gallup & Altomari, 1969) and in a stabilimeter placed in the goal box (Dunlap et al., 1971). Wong's stage model refers to several of these behaviors and makes predictions about their sequence of appearance in acquisition and extinction; such predictions are not made by earlier versions of frustration theory. This, then, is the potentially unique contribution of Wong's behavioral field approach and stage model.

Extensions and revisions of frustration theory

In the late 1960s there appeared two brief "Theoretical Notes" in the *Psychological Review* that can be regarded as extensions or revisions of the original version of frustration theory. One was written by Wilton (1967), who proposed a mechanism for the PREE, separate from but not incompatible with $s_F \rightarrow$ approach; the other was written by Hill (1968).

Wilton's suggested revision

The PREE according to this suggestion is, at least in part, due to differential intratrial stimulus generalization decrement in extinction between animals trained under CRF and PRF conditions. The reasoning, based on frustration theory, goes like this: Under CRF conditions, the strength of approach to a goal depends on the intensity of the antedating r_R–s_R, which is elicited throughout the instrumental chain. Then, in extinction, when primary frustration and its conditioned form, r_F–s_F, are evoked, the presence of s_F changes the stimulus complex that elicits r_R and, by the principle of stimulus generalization decrement, weakens r_R, thereby reducing the tendency to approach the goal. In contrast, PRF training provides the stimulus, s_F, during acquisition, preempting the generalizing effect in extinction. As Wilton points out, such a mechanism is not incompatible with the one proposed in the original theory. The interesting thing about this suggestion, however, is that it puts the explanatory burden for the PREE on the transition from CRF to extinction, whereas the original mechanism, the counterconditioning of approach to s_F, puts the burden on the transition from PRF to extinction. It is worth pointing out that, in fact, most theories of the PREE depend, in one form or another, on the generalization-decrement principle. This was certainly the case in the early theories of the PREE we have reviewed: in the Humphreys expectancy interpretation and in early versions of the discrimination hypothesis. Generalization decrement was, of course, also a central characteristic of the Hull–Sheffield hypothesis, which was later extended and formalized, as we have seen, in Capaldi's sequential theory. It is possible to regard frustration theory (even without Wilton's modification) as an instance of the application of the generalization-decrement principle to the PREE in instrumental learning. The reasoning would be that after CRF training, in which s_F does not occur, its introduction in extinction alters the stimulus complex and results in rapid response decrement for that reason. This would move the burden of the frustration theory explanation from PRF to CRF acquisition, as most other theories (and certainly Wilton's) do. Again, however, this second frustration theory explanation, which does not involve counterconditioning, is not incompatible with the first one, which does. The second explanation can account for many phenomena, but not for some of the most interesting ones – to name just two, the idiosyncratic ritualized behaviors of the DNC experiments, described in Chapter 4 (e.g., Rashotte & Amsel, 1968), and the relation of specific s_F intensity to the appearance of the PREE early or late in extinction (Traupmann, Amsel, & Wong, 1973).

Hill's suggested revision

This Theoretical Note pointed to an interesting paradox in the explanation by frustration theory of PRF effects in acquisition and extinction. As we have seen, the theory attributes relative persistence after PRF training (the PREE) to the counterconditioning of the instrumental response to cues from anticipatory frustration (s_F). As Hill points out, the same mechanism is involved in explaining the PRAE, which is defined by more vigorous terminal responding in early segments of the response sequence and less vigorous responding in the goal segment. The paradox lies in the fact that whereas the PREE occurs in all segments in extinction, counterconditioning does not apparently occur in all segments in acquisition. In a reply to Hill, Hug and Amsel (1969) pointed out that the frustration theory analysis provides mechanisms for dealing with this paradox. One of these is a statement which implies that counterconditioning is not an all-or-none process: "Finally in Stage 4 of partial-reward acquisition...the anticipatory frustration-produced cues (s_F) come to evoke avoidance as well as approach" (Amsel & Ward, 1965, p. 4). The important feature here is that counterconditioning need not involve complete elimination of competing (avoidance) tendencies. There is also the more extensive statement of the theory (Amsel, 1967), in which r_R–s_R and r_F–s_F are assumed to act only in a generalized excitatory capacity when their strengths are weak and to have the additional property of directing behavior when their strengths are greater (see Figure 3.4). Hug and Amsel pointed out that the implication of this assumption is that superior performance under PRF acquisition conditions can result either from weak r_F–s_F or from the approach response being counterconditioned to stronger r_F–s_F. They argued that in the former case the result would be superior PRF over CRF performance in acquisition, but no PREE because of the absence of counterconditioning (a result that had not then and, to my knowledge, has not since been seen, possibly because the typical experiment involves a large number of trials and a large reward magnitude). They concluded that Hill's paradox – inferiority of acquisition performance in the goal measure, but greater persistence in extinction in that measure under the PRF condition – might reflect a degree of counterconditioning insufficient to overcome the stronger $s_F \to$ avoidance in the goal region in acquisition, but sufficient to produce relative persistence in extinction.

Hill's own explanation of the paradox was in terms of the generalization-decrement interpretation offered by Wilton (1967): that in the CRF condition, unlike the PRF condition, s_F is encountered only in extinction. As in Wilton's case, we found no incompatibility between this explanation and the basic theory, but pointed out that, setting aside the acquisition effects, the generalization-decrement principle did not appear to be as plausible an account of such effects as the generalized PREE (Amsel et al., 1966; Brown & Logan, 1965) or the various phenomena involving mediation, such as transfer of responses learned under PRF in acquisition to the extinction of other responses acquired under CRF conditions (e.g., Rashotte & Amsel, 1968; Ross, 1964). I make this point again here for emphasis.

Frustrative and nonfrustrative persistence

If resistance to extinction is the prototype of persistence, the PREE is an indication that persistence is greater following partial than following continuous reward. However, our experiments along with accumulating evidence from other sources suggest that this rule is a *great oversimplification;* that the presence of the PREE and the PREE-like effects in discrimination learning depend on a variety of factors, including *amount* of PRF experience and *amount* or *size* of reinforcement on any given occasion; in fact, that we can make continuously rewarded subjects resistant to extinction (persistent) and partially rewarded subjects nonpersistent. The conjecture, in connection with the latter possibility, is that there are at least two kinds of persistence, "frustrative" and "nonfrustrative," the first of which necessarily involves intermittent reinforcement, the other not.

An example of frustrative (or "excited") persistence is, of course, the ordinary experiment on the PREE in which persistence is built into the partial group by allowing frustration to occur in the acquisition (and extinction) of the response, while persistence is relatively weaker in the continuous group because little or no frustration occurs in acquisition, and the exposure to frustration is only in extinction. (Previously cited experiments have demonstrated that extinction following CRF training is facilitated by a relatively large number of continuous rewards [e.g., North & Stimmel, 1960] and that it is facilitated and retarded after CRF and PRF training, respectively, by relatively large reward magnitudes [Armus, 1959; Hulse, 1958; Wagner, 1961].) Relative nonpersistence might be accomplished in a PRF group by holding frustration down in acquisition (but not in extinction) with a sedative or tranquilizer. The rationale of such a procedure would be that the effect of the drug in acquisition is to suppress the development of frustration during PRF training, thereby preventing the formation of a connection between anticipatory-frustration-produced cues and responding. A necessary feature of the sedative or tranquilizer in this case would be that it have no effect in acquisition on the development of r_R, which would have to be present later in extinction for the occurrence of frustration and rapid elimination of the response (e.g., Capaldi & Sparling, 1971; Gray & Dudderidge, 1971).[9]

Persistence of the "phlegmatic" sort may be built by allowing no excitement in either acqusition or extinction. An example of such nonfrustrative persistence would be acquisition of a continuously rewarded response followed by extinction under the influence of a sedative or tranquilizer. It could also take the form of reappearance of an extinguished response (or response to S−) when a drug is administered following extinction (or discrimination). Persistence, in these cases, would be due to the failure of frustration to operate as an aversive factor.

In this connection, Terrace's (1963b) early work is instructive. He showed that administering a tranquilizer greatly increased the number of "error" responses made by pigeons to the negative discriminandum after perfect discrimination had been learned in the usual way: rewards for responses to S+, nonrewards for responses to S−. However, when discrimination learning was managed in such a way that the pigeon made few if any responses to S− (responses that would be nonrewarded if they occurred), that is to say, when discrimination learning was "errorless," the drug produced no alteration in discriminative performance. Put another

---

9. In some unpublished experiments in the early 1960s, we primary frustration and the PREE with the drug chlorpromazine (Thorazine). After what appeared to be some preliminary positive findings, we spent two years failing to replicate these results, and the work was never published. Hindsight suggests that our failure was preordained because of the drug we chose, which has a generalized suppressing effect on both positive and negative affect. Later experiments employed alcohol, barbiturates, and benzodiazepines (see the Appendix), whose effects are evidently more selective, and many of these experiments yielded the predicted results.

way, it appeared from Terrace's work that if nonrewards were involved in the formation of a discrimination, the aversiveness of S− was a necessary factor in maintaining the discrimination. The removal of this aversiveness by the administration of a drug caused the discrimination to break down dramatically, and the subject returned to responding to S−. If the discrimination had involved little or no aversiveness, as was presumably the case in the "errorless discrimination" procedure (Terrace 1963a, 1963c), a procedure in which S− was introduced very gradually and in such a way that the pigeon never responded to it, then the injection of a drug following training could not attenuate the aversiveness and produce responding to S−.

Persistence in relation to vigor and choice

Recall that when the prediscrimination condition was $S_1 \pm S_2 +$, asymptotic acquisition performance after many trials reflected greater vigor of responding to $S_1 \pm$ than to $S_2 +$. Response to S_1 was also more resistant to extinction than to S_2 in the sense that the $S_1 - S_2 +$ discrimination (response to S_1 extinguished) was retarded relative to the $S_1 + S_2 -$ discrimination (response to S_2 extinguished).

Consider, now, the matter of choice behavior in relation to $S_1 \pm$ and $S_2 +$ after many trials. Given a situation wherein these stimuli are exposed in such a manner as to neutralize all choice-inducing factors other than reactions to the stimuli themselves, which of the two stimuli should be selected ("preferred")? Remember that in this kind of experiment, where start and run speeds are greater to $S_1 \pm$ than to $S_2 +$, there is no opportunity for choice – the rat is faced with a single alley that is either S_1 or S_2. The question is: Does the increased vigor and persistence, presumably related to the connection of s_F to approach in the S_1 alley, imply that the animal will also *choose* S_1 over S_2 when not forced to run in S_1?

At the time we were interested in this problem, we could find only one experiment that came close to addressing it. Using a free- and forced-trials procedure in a T-maze, in an experimental study of reversal following spatial learning, Davenport (1963) varied percentage of reinforcement (100:0, 100:33, 100:67) in original learning and found that most of the rats developed preferences for the 100% side early and showed no tendency to switch to the 33% or 67% side later in training. He recognized that this seemed not to agree with results from runway studies (e.g., Goodrich, 1959; Haggard, 1959; Weinstock, 1954) which give the paradoxical result that asymptotic performance in acquisition is higher under PRF than under CRF conditions – paradoxical if one assumes that response speeds in acquisition should be greater the larger the number of reinforced trials. The Davenport experiment did not provide evidence for a paradoxical asymptotic effect in choice behavior, as our within-subject procedure did for response vigor, and there are two good reasons for this: First, the paradoxical vigor effect does not occur in the goal measure when runway performance is fractionated into start, run, and goal segments. In fact, the goal measure typically shows the reverse effect: CRF generates faster running than PRF in this segment. If behavior at the choice point of a T-maze is more like goal entry in a runway than responding in the other segments, the Davenport result is not unexpected. The more general reason that the two results are not incompatible, however, is that one reflects a drive-dependent *vigor or amplitude* dimension of behavior, and the other, habit-controlled *choice behavior;* that is to say, the higher level of vigor in $S_1\pm$ than in S_2+ reflects the nonassociative or arousing effect of uncertainty at the same time as associative considerations dictate a *preference* for a certain over an uncertain outcome (see Gray & Smith, 1969, for a discussion of this distinction). Indeed, as I pointed out (Amsel, 1967), a relatively complete account of behavior should involve three important dimensions: vigor or intensity, choice, and persistence. A successful, albeit elementary, theory of instrumental behavior encompassing all three of these dimensions would therefore include in its construct language terms connoting not only associative processes (conditioning, counterconditioning, excitation, inhibition, suppression) but, as we saw in Chapter 2, also terms connoting nonassociative processes (drive, arousal).

Summary

A straightfoward summary of this chapter brings me back to a quotation from Stephen Hawkin (1988) with which the chapter began: "A theory is a good theory if it satisfies two requirements: it must accurately describe a large class of observations on the basis of a model that contains only a few arbitrary elements, and it must make definite predictions about the results of future observations" (p. 9). The point here is that frustration theory, starting with a small number of premises, appears to explain a larger number of experimental particulars than do theories exclusively of the PREE – and to have made nonintuitive predictions of phenomena on which the theory was not initially based. Frustration theory is not just a theory of the PREE or of any one phenomenon; its explanatory scope includes phenomena that in nonformal language reflect arousal, suppression, persistence, and regression. We have seen experimental examples from each of these categories of behavior in this and preceding chapters. In Chapter 7, we move on to another level of investigation of these manifestations of learning based on reward-schedule effects: the early development of dispositions relating to such learning.

7 Ontogeny of dispositional learning and the reward-schedule effects

In D. O. Hebb's last book, *Essays on Mind* (1980), there is the following passage:

> The argument [can be made] that the behavioral signs of mind and consciousness are evident only in the mammals, with the possible exception of some of the larger-brained birds; that relatively small-brained mammals like the rat or the hamster may have very small minds (like the penguins of Anatole France's Penguin Island, to whom the Lord gave souls but of a smaller size) – but still minds, whereas fish and reptiles, and most birds, seem to be reflexively programmed and give little evidence of that inner control to which the term mind refers. The best evidence of continuity, in the development from lower to higher mammals, is to be found not only in their intellectual attainments, their capacities for learning and solving problems, but also in their motivations and emotions. Man is evidently the most intelligent animal but also, it seems, the most emotional. (p. 47)

This statement, in Hebb's colorful prose, is an example, in phylogenetic terms, of a kind of thinking that, in its ontogenetic counterpart as well as in levels of functioning in the adult mammal, is seen increasingly in our field (e.g., Amsel & Stanton, 1980; Bitterman, 1960, 1975; Livesey, 1986; Schneirla, 1959; Wickelgren, 1979). My own point (Amsel, 1972b) was (and is) that there is a level of classical conditioning that is purely dispositional, involves implicit memory, and is less dependent on mediation than what is usually called Pavlovian conditioning, and that both levels involve a lesser degree of mediation than instrumental learning. Bitterman allows that learning in fishes and turtles does not depend on learning about reward (learning a mediating reward expectancy), but on the direct reinforcing action of a reward on an associative connection. Livesey's recent view is that the evolution of emotion depends on the evolution of learned expectancies, a position similar to my own about frustration and Bitterman's about "learning *about* reward." Wickelgren observes that there may be a role for Thorndikian and Hullian theorizing in instances in which learning is noncognitive. Amsel and Stanton summarize experiments that raise the possibility that the absence of mediating expectancies (precognitive learning) may characterize learning at very early stages of mammalian ontogeny and even in some kinds of precognitive or noncognitive human learning.

Schneirla, in a statement that could stand as the rallying cry of our recent developmental work, pointed out that simpler animals, both in phylogenesis and ontogenesis, operate on a level of approach and withdrawal, whereas more advanced organisms operate on a level of seeking and avoidance. In Schneirla's terms, seeking and avoidance connote the operation of incentives, expectancies that can be confirmed or disconfirmed. (In our terms, disconfirmation leads to frustration, arousal, suppression, and persistence.) Approach and withdrawal have no connotation of expectancy but of reactions to low and high intensity of stimulation at a more basic associative level. Schneirla's views and the contemporaneous statements of frustration theory (Amsel, 1958a, 1962) were remarkably similar in thrust, though they emerged out of different traditions. Schneirla (1959) wrote, "Psychologists who emphasize disproportions, reversals, and exceptions between stimulus magnitude and response properties are... talking of adult stages at higher psychological levels" (p. 22).

It is interesting, and to a degree ironic, that perhaps the earliest modern treatment of degraded *human* function in learning came from work by Kenneth Spence and his students (Spence, 1966). He described differences between acquisition and extinction effects in human eye-blink conditioning with and without cognitive involvement. He used a "cover story" and a masking procedure so that his subjects would be unaware of the fact that a light-intensity change (CS) was always (or sometimes) followed by an air puff to the eye (US). Under these conditions, extinction – the rate of decline in nonreinforced conditioned responding after both CRF and PRF training – was much slower than under normal cognitive conditions, and under these noncognitive conditions, there was no PREE. Here is the normal adult human apparently operating at a level that I would characterize as unmediated implicit classical conditioning.

Another example of reduced levels of functioning in humans is the difference between normal controls and certain amnesiacs. Korsakoff patients show normal acquisition and retention of eye-blink conditioning, but later cannot recall anything about the stimuli (CS and US) they were exposed to or their contingent relationship (Weiskrantz & Warrington, 1979). In summarizing this and other work on amnesia, Warrington and Weiskrantz (1982) refer to Spence's work and advance the hypothesis that "the amnesic subject can show learning through facilitation by repetition of simple S–R relationships not requiring cognitive mediation" (p. 233). The famous H.M. (who suffered from the amnesic syndrome after bilateral removal of temporal lobe tissue for the treatment of epilepsy) showed good long-term memory for newly learned motor skills but no recollection of having learned the tasks (Corkin, 1968). This, again, is a kind of memory based on learning of which the learner is unaware or "unconscious"; it is in this sense similar to the recently postulated distinction between "implicit" as opposed to "explicit" memory and to the experimental operation of "priming" (Tulving & Schacter, 1990).

Another quite similar distinction is made by Squire (1982) when he divides learning and memory into two basic levels – procedural (an example is simple classical conditioning) and declarative (more complex learning involving reaction to change, such as discrimination reversal). Squire's distinction finds support in the seminal work of R. F. Thompson and his associates (e.g. Thompson, 1983; Thompson et al., 1980), who provided evidence that the hippocampus is involved in, but is not necessary for, classical conditioning (procedural), but does seem to be necessary for more complex (declarative) learning. Tulving (1985) has changed his thinking about his well-known distinction between two kinds of propositional memory, semantic and episodic (Tulving, 1972). He now favors a concept of "triadic memory," in which episodic and semantic memory are no longer simply parallel subsystems of declarative memory; instead, episodic memory is conceived of as "growing out of but remaining embedded in the semantic system" (p. 88), the same arrangement holding between procedural and semantic memory.

Still another distinction of the same sort has been made by Mishkin and others (Mishkin, Malamut, & Bachevalier, 1984; Mishkin & Petri, 1984). They refer to two different "retention systems," a habit system ("knowing how") and a memory system ("knowing that"), only the latter being impaired in most amnesias. In every one of these distinctions (and in many other similar ones), there is the at least tacit recognition of two levels of functioning, one more primary (more primitive) than the other, and therefore less dependent on anticipation, recollection, or expectancy – on confirmation and disconfirmation. One system depends on repetition and incremental associative growth; the other is less incremental and depends more on a single experience or a small number of them. One is regarded as more "Hullian," the other more "Tolmanian."

To repeat, in the neuropsychological study of human memory, but, strangely, not in contemporary animal-based cognitive learning theory, these cognitive and simple associationistic languages have entered into explanatory partnership. All the basic distinctions appear, in one sense or another, to involve at least two levels of functioning, though they go by different names: noncognitive versus cognitive; stimulus–response versus cognitive; procedural versus declarative; procedural versus propositional (semantic and episodic); habit systems versus memory systems. Recall that in Schneirla's (1959) case it was approach–withdrawal versus seeking–avoidance; in Bitterman's (1960), carryover versus reinstatement; in ours, simple classical versus Pavlovian and instrumental conditioning in one version (Amsel, 1972b), and recently, following Thomas (1984), dispositional versus representational.

In recent years, we have examined the ontogeny of several of the reward-schedule effects identified in Chapter 3, which carries out Step 3 of the strategy, outlined early in this book (see Table 1.2), to study these effects for their presence or absence at certain developmental stages and for the order of their first appearance. This examination has strengthened our belief that there are indeed two systems that strengthen associations and

Table 7.1. *Age of first appearance of various appetitive reinforcement-schedule effects*

Effect	Age (days)
Successive acquisition and extinction (Amsel, Burdette, & Letz, 1976)	≤10
Single patterned alternation (PA) (Stanton, Dailey, & Amsel, 1980)	≤11
Partial-reinforcement extinction (PREE) (Letz, Burdette, Gregg, Kittrell, & Amsel, 1978; Chen & Amsel, 1980a,b)	12 to 14
Variable magnitude of reinforcement (VMREE) (Chen, Gross, & Amsel, 1981)	16 to 18
Partial delay of reinforcement (PDREE) (Chen, Gross, & Amsel, 1981)	16 to 18
Partial-reinforcement acquisition (PRAE) (Chen, Gross, Stanton, & Amsel, 1980)	18 to 20
Magnitude of reinforcement extinction (MREE) (Burdette, Brake, Chen, & Amsel, 1976; Chen, Gross, & Amsel, 1981)	20 to 21
Successive negative contrast (SNC) (Stanton & Amsel, 1980; Chen, Gross, & Amsel, 1981)	25 to 26
Slow responding (DNC) (Chen, Gross, Stanton, & Amsel, 1981)	≤63

learned performance – one more "primitive" than the other. The first is the stamping-in action of reinforcers studied by Pavlov, Thorndike, Hull, and other early learning theorists, and such reinforcement systems may be said to be operating in lower phyletic or ontogenetic forms. The second (incentive) system is present perhaps in all mammals, but arguably not very early in life in relatively altricial ones (e.g., the rat and perhaps the human infant). This system involves expectancies related to reward, frustration, punishment, and relief; it is an important aspect of the "mind" to which Hebb alluded and which he linked to motivational and emotional development.

In the context of our interest in what I have called dispositional learning we have been studying, for the first time in a systematic way, the emergence in ontogeny of a number of these well-known reward-schedule effects. Some of these effects have been characterized as "paradoxical," and we have developed a very good working schema relating their first appearance to the chronological (postpartum) age of laboratory rats.

The order of appearance of these effects can be taken to represent increasing levels of functioning. A list of the effects we have studied is given in Table 7.1, and idealized forms of four of these effects are provided in Figure 7.1. Except for the earliest ones, successive acquisition and extinction and patterned alternation, these effects can be said to require

Ontogeny of dispositional learning 141

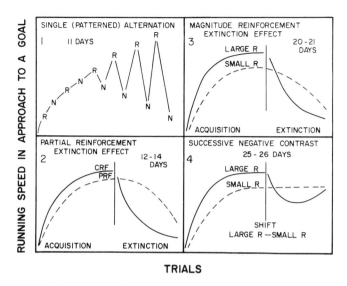

Figure 7.1. Idealized representations of one nonparadoxical (1) and three paradoxical (2, 3, 4) reinforcement-schedule effects and the approximate ages at which they first appear in our experiments. (From Amsel, 1986. Copyright 1986 by the Canadian Psychological Association; reprinted by permission.)

expectancy (reinstatement) and to be "paradoxical" in two senses. The first sense is that of more producing less and less producing more. Examples are that the smaller percentage of reward in acquisition (the PREE), the longer delay of reward (the PDREE), and the more variable magnitude of reward (the VMREE) lead to greater resistance to extinction; the greater reward magnitude in acquisition leads to less resistance to extinction (the MREE); and the greater reward magnitude early in acquisition leads to a lower level of performance when small reward follows large reward (SNC). All of these behavioral effects depend in one way or another on the detection of and reaction to discrepancy, either within a phase or sequence of presentations of reward and nonreward (the PREE, the PDREE, or the VMREE) or between phases, as in the transition from acquisition to extinction (the MREE) or from large to small reward magnitudes (SNC). The prediscrimination effects, including discrimination reversal, discussed in Chapter 5 are more complex examples of phase transitions in which considerations of detection of and reaction to discrepancy also apply.

These effects are also paradoxical in another sense: They are not predicted or explained in any classical or in most modern theories of learning (or in any theory of memory, for that matter) – not in the theories of Thorndike, Guthrie, Tolman, or Hull; not in the mathematical models of Bush and Mosteller (1951), Estes (1950), or Rescorla and Wagner (1972); and not by any of the newer cognitive interpretations. Years ago, Capaldi's (1967) sequential theory and my own (Amsel, 1958a, 1962, 1967) addressed some of these paradoxical effects, as has the recent model (DMOD) of Daly and Daly (1982), which combines the Rescorla–Wagner mathematical form with assumptions taken from frustration theory.[10] The thesis I am here advancing is that the transition from nonparadoxical to paradoxical represents a fundamental transition in levels of functioning.

The two levels of functioning we have identified, nonparadoxical and paradoxical, have some commonality with the bipartite memorial systems I described earlier in this chapter, examples being the habit versus memory and the procedural versus declarative systems. However, further analysis suggests that elements of both these systems operate in both nonparadoxical and paradoxical learning and retention, and that they might better be regarded as being on separate, if not orthogonal, dimensions. In patterned alternation (PA), for example, the declarative element lies in the fact that responding on Trial $N + 1$ depends on a carried-over memory trace from Trial N as in a non-matching-to-sample test of memory; yet PA depends also on multitrial discrimination learning involving these carried-over memories, rather than only on the single-trial non-matching-to-sample event. This many-trial feature of PA learning clearly involves a habit or procedural system, an increase in strength of association based on implicit strengthening through repetition. In contrast, SNC and the MREE seem more weighted on the memory or declarative system, since after the buildup of a particular level of reward expectation (r_R), they depend on a single shift in level of reward and its consequences. The PREE is somewhere in between, but in terms of its age of first occurrence, it is much closer to PA than to the MREE and SNC, emerging as it does just a day or two after PA.

This is in line with a statement by Squire (1987) that "in ontogeny declarative memory develops later than procedural memory" (p. 168), a view previously expressed in the writings of others (Bachevalier & Mishkin, 1984; Mandler, 1984; Nadel & Zola-Morgan, 1984; Schacter & Moscovitch, 1984). This kind of statement about ontogeny can also be made about paradoxical and nonparadoxical functioning and is inherent in an earlier distinction I made among classical, Pavlovian, and instrumental conditioning (Amsel, 1972b). These levels of conditioning, in my terms, represent an increasing order of complexity of associative processing, the simplest level of classical conditioning being clearly procedural and nonparadoxical, the other two involving levels of expectancy (memory) that are formed in classical conditioning but are not involved in its formation.

As we shall see in this chapter and in Chapter 8, results from ontogenetic investigations, from cross-species comparisons, and from investigations involving brain lesions, pharmacological manipulations, electrophysiological

10. In a recent application of this mathematical model to our development results, Daly (1989) has concluded that the order of appearance of the effects listed in Table 7.1 can be accounted for entirely on basis of the growth of anticipatory frustration over the age range represented.

stimulation and recording, and developmental neurology and neuroanatomy can all be taken to support our thesis of transitions in levels of functioning and to contribute to our understanding of dispositional learning in emotional-temperamental development and the factors that facilitate and retard it.

It is not unreasonable, on the basis of our work and the work of others, to believe that frustration, one member of a class of reactions to discrepancy, and particularly anticipated frustration (not just fear based on electric shock, to which I believe a disproportionate amount of attention has been paid in experimental psychology and psychopathology), is a major determinant of emotional development. Evidence has been presented that the persistent (perseverative, idiosyncratic, fixated) behavior that is seen in so many emotional disorders is frustration-related; it is a small step to conclude that the debilitating conflicts that are so common in psychopathology arise at least as much out of the counteracting expectations of hope and anticipated disappointment as they do out of fear and anticipated relief, to use Mowrer's (1960) terminology.

Our recent work has been an attempt to study several related reward-schedule effects, organized into two classes (here termed *nonparadoxical* and *paradoxical*). These phenomena, as we shall see, are known to be influenced in adult experimental animals by drugs that affect, and lesions that destroy, portions of the septohippocampal system of the brain (see Gray, 1982, for an extensive review). These drugs and lesions appear to exert at least part of their action through the reduction of primary and conditioned frustration and their behavioral consequences. Our strategy is based on experimental work that has provided information on the ages at which a number of these frustration-related effects first appear in ontogeny (see Table 7.1). This work can also be regarded as a possible behavioral assay of the effects of such drugs and lesions and of the intrauterine effects of ethanol and other teratogens. In the remainder of this chapter, I will review some of this normative, behavioral, developmental work.

Ontogeny of reward-schedule effects: a review

In attempting to understand the mechanisms that control the development and retention of suppression and persistence in mammals, we conducted experiments to determine at how early an age the manifestation of such tendencies can be seen in the infant laboratory rat. In most of our experiments, the behavior that is measured is the simple approach response, in a straight-alley runway, to a positive (appetitive) reinforcer. The alley, the reinforcer, and the response, "crawling" or "running" ("foraging" in the modern parlance), take different forms depending on the age of the rats. In this section, we begin our review with the oldest animals we worked with and proceed to the youngest.

Young adults, juveniles, and weanlings

As a first approximation to answering questions about age at which persistence can be acquired and retained, we found that the PREE existed in young adult rats whose straight-runway training started at 30 days of age and in weanling rats whose training started at 18 days (Chen & Amsel, 1975). Some age differences were apparent, however. In an immediate extinction test, applied to half the animals in each group, extinction of running after CRF training was very slow and gradual in the rats that had acquired the response as weanlings, while the extinction performance of the young adult rats was like the adult pattern, more abrupt and negatively accelerated. Extinction of running in the alley following PRF training was extremely slow, particularly in the weanling group. Then, in the manner of the Theios–Jenkins experiment (see Chapter 4), following a 45-day vacation period and a phase of CRF reacquisition, the PREE emerged clearly in the animals that had not undergone previous extinction. In those animals (half of each age group) that had been extinguished immediately after original acquisition training, the PREE reemerged in the later test (i.e., it was durable) but was somewhat reduced in absolute size (Figures 7.2 and 7.3). This was our first indication that, even in preweaning rats, persistence could be acquired and retained, at least into young adulthood.

I would point out that this retention of persistence in preweanlings is better than is seen in many tests of retention of fear based on electric shock (Campbell & Coulter, 1976); however, there are at least two major differences between these tests of "memory." First, in the latter test, one shock or a small number of shocks are delivered, whereas in PRF training to induce persistence there are a large number of acquisition trials. Second, the shock test depends on the "recall" of specific cues, a more declarative kind of memory, in Squire's terms, whereas the retention of learned persistence is, in our terms, dispositional memory, is not so cue specific and seems to transfer from one situation to another.

We recognized that these experiments were lacking in certain important respects. First, acknowledged procedural differences at the two ages made direct comparisons about as risky as they are in any other developmental experiment. Second, only two age ranges were investigated, and these were rather grossly defined, making it difficult even to guess where transitions from one pattern to another might occur. In other words, the age ranges in this preliminary experiment were so extended that they must surely have overlapped adjacent stages of development. (As we shall see, we later found that a "stage" of development of this kind might encompass just a single day or two in the rat pup.) In follow-up experiments, we narrowed the age ranges and ran more age groups.

In the first of these experiments (Burdette, Brake, Chen, & Amsel, 1976), there were four age groups, and at each age training was completed

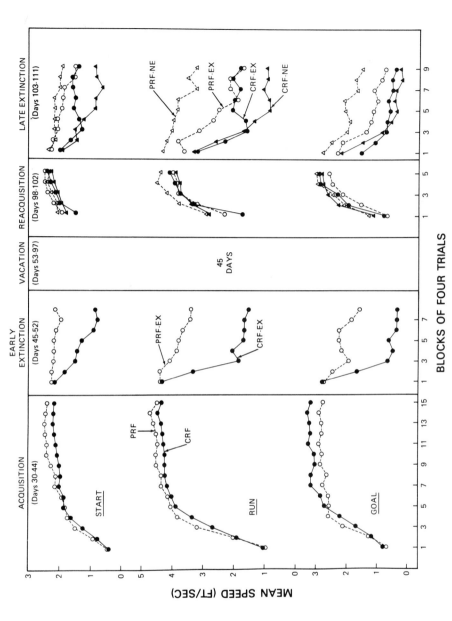

Figure 7.2. Retention and durability of persistence in 30-day-olds: mean speed data across all phases of the experiment. PRF, Partial reinforcement; CRF, continuous reinforcement; EX, extinction; NE, no extinction after original acquisition phase, only in late extinction. (From Chen & Amsel, 1975. Copyright 1975 by the American Psychological Association; reprinted by permission.)

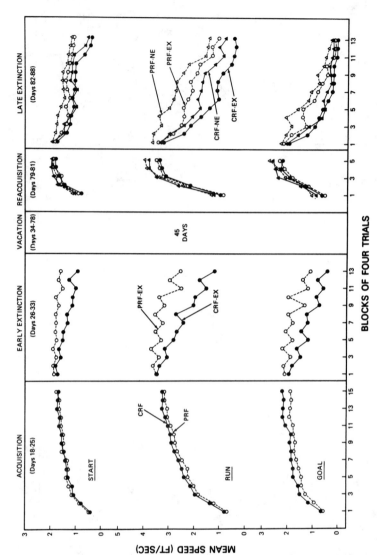

Figure 7.3. Retention and durability of persistence in 18-day-olds: mean speed data across all phases of the experiment. (From Chen & Amsel, 1975. Copyright 1975 by the American Psychological Association; reprinted by permission.)

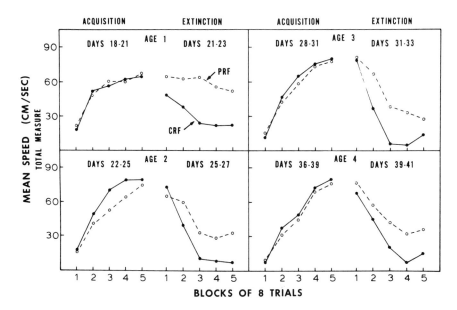

Figure 7.4. Acquisition under CRF and PRF conditions and extinction showing the PREE at four ages from preweanling to juvenile. (From Burdette, Brake, Chen, & Amsel, 1976. Reprinted by permission of the Psychonomic Society.)

in a 3.5-day period. Half the animals at each age were trained under PRF and half under CRF conditions, followed by a 2.5-day extinction period. These experiments revealed remarkable immediate persistence when the PRF treatment was given as early as the 18- to 21-day range, followed by immediate extinction (Figure 7.4). The oldest group, trained from Days 36 to 39, showed relatively less immediate persistence. In a second experiment, another group was added at each age to equate the PRF and CRF groups for rewards rather than trials. (This reward-equated group is designated PRF-R.) Using only the two extreme age groups, trained at 18 to 21 or 36 to 39 days, the result was the same: Whereas persistence was not different between the PRF and PRF-R conditions at either age, the relative persistence due to PRF training was greater in the younger age group of rats when persistence was tested immediately after acquisition, and the overall level of persistence in this younger group was very high (Figure 7.5).

In the experimental study of the ontogeny of learning, the factors of incentive (reward magnitude), level of motivation (hunger, thirst, pain), and their interactions are particularly troublesome because there is no obvious way to hold them equal across ages. Because age-related differences in learning or persistence may be simply a function of such between-ages variation in nondevelopmental factors, it is important to manipulate

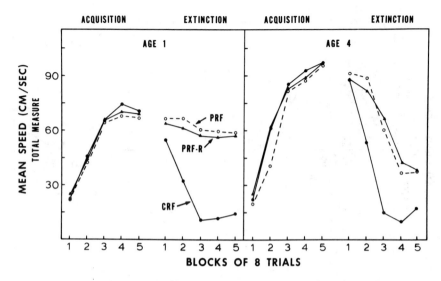

Figure 7.5. Acquisition under three conditions including PRF-R (a PRF condition with rewards equated to the CRF group). Acquisition is at Days 18 to 21 (Age 1) and 36 to 39 (Age 4). Extinction follows immediately in both cases (Days 21 to 23 and 39 to 41, respectively). (From Burdette, Brake, Chen, & Amsel, 1976. Reprinted by permission of the Psychonomic Society.)

the level of such factors and determine if the age effect is eliminated. This is an example of what Bitterman (1975), in his cross-species comparisons, has called "systematic variation."

In two experiments, we examined the effect of size of reward on persistence at various ages. In the first, rats were trained at two levels of reward magnitude and at 18 to 21 or 36 to 39 days of age under CRF conditions (Burdette et al., 1976). The well-established paradoxical effect, the MREE, was shown to operate in preweanling and juvenile rats as it does in adults: Larger reward in acquisition led to faster extinction at both ages. It was also the case that preweanling rats were more persistent than juveniles at both reward levels (45 and 300 mg). This increased the likelihood (a) that the relationships between age and persistence shown in the earlier experiments had general ontogenetic significance and (b) that persistence in rats, whether at about 18 to 21 days of age or adulthood, might involve the same mechanisms.

In the second experiment (Chen, 1978; see also Amsel, 1979), PRF or CRF acquisition was combined factorially with 45-, 97-, or 300-mg rewards in 4-day acquisition phases starting at Day 17, or with 45- and 300-mg rewards starting at Day 30 or 55. Earlier work with adult rats had produced a result that strongly supported a frustration theory interpretation (see Chapter 3). We have seen that, in adults, increasing magnitude of reward not only speeds extinction after CRF training (the MREE), but also retards extinction (makes animals more persistent) after PRF training (Hulse, 1958; Wagner, 1961). This means that the size of the PREE is directly related to size of reward. Chen's data showed this same paradoxical effect of reward magnitude at all three ages; the size of the PREE was greater following training with large than with small reward, though this appeared to be due mainly to greater extinction rates with larger rewards after CRF training (Figure 7.6).

The first four postnatal weeks

There is evidence that ontogenetic changes in a variety of behaviors in the rat are correlated with the rapid development of brain function from the first through at least the fourth postnatal week. The burst in the number of differentiated granule cells of the dentate gyrus of the hippocampus in the rat begins to level off at about 30 days of age (Altman & Das, 1965); spectral composition of the EEG does not approximate that of the adult until the rat is between 25 and 30 days of age (Deza & Eidelberg, 1967); and significant development of cholinergic inhibitory systems in the forebrain is delayed until rats are 20 to 25 days old (Campbell, Lytle, & Fibiger, 1969). At the behavioral level, experiments reveal different patterns in young, and especially in infant, as compared with adult rats in open-field activity (Bronstein, Neiman, Wolkoff, & Levine, 1974), in head-poke habituation (Feigley, Parsons, Hamilton, & Spear, 1972); in spontaneous alternation (Bronstein, Dworkin, & Bilder, 1974; Douglas, Peterson, & Douglas, 1973; Kirkby, 1967); in arousal (Campbell & Mabry, 1972); in the suppressive effects of extinction (Ernst, Dericco, Dempster, & Niemann, 1975); and in conditioning of passive avoidance (Riccio & Marrazo, 1972), to give examples from some of the earlier experiments.

These kinds of considerations, along with our own experiments, led us to focus more minutely on the behavior of animals in the age range comprising the first four postnatal weeks (along with somewhat older animals). Consequently, in two further experiments we included six age groups for comparison, the first three spanning only four days of development (Amsel & Chen, 1976). In the first experiment, these six groups were trained to run for reward in an alley under either PRF or CRF conditions, and there was a clear PREE at all ages. Moreover, particularly after PRF training, an inverse relationship existed between immediate resistance to extinction and age (Figure 7.7): the younger the animal, the greater the persistence. In the second experiment, short- and long-term retention of persistence and durability of persistence were tested following acquisition at three different ages (Figure 7.8). The short-term extinction tests confirmed our earlier results. In the long-term retention-of-persistence extinction tests, which occurred after 53 days and a 24-trial CRF reacquisition phase, the

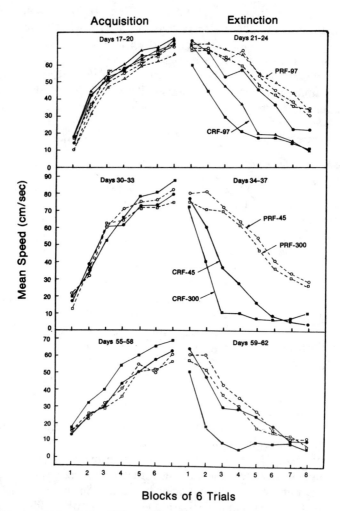

Figure 7.6. Acquisition under CRF and PRF and various reward-magnitude conditions followed by extinction. These data demonstrate not only the PREE but also the paradoxical MREE at all three ages. (From Amsel, 1979. Reprinted by permission of Lawrence Erlbaum Associates, Inc.)

PREE was found in all three age groups. This was also the case in the durability test, which added to the long-term persistence manipulations an early-extinction phase, a 50-day "vacation," and a 24-trial CRF reacquisition before the late-extinction test occurred. In neither test was there an effect of age at original acquisition on the magnitude of adult persistence. These experiments confirmed the earlier findings of remarkable persistence

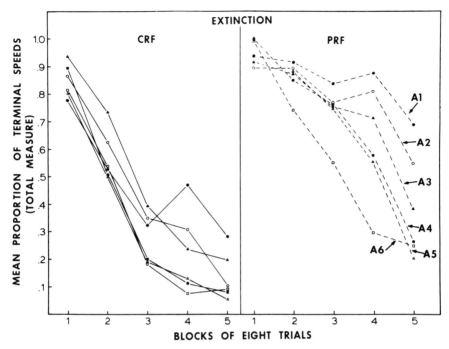

Figure 7.7. Extinction curves for the CRF and PRF conditions shown separately with age in acquisition (A) as the parameter in each case: A_1, 17 to 18 days; A_2, 18 to 19; A_3, 21 to 22; A_4, 28 to 29; A_5, 35 to 36; A_6, 65 to 66. (From Amsel & Chen, 1976. Copyright 1976 by the American Psychological Association; reprinted by permission.)

in rats trained and tested at weanling age. They showed that learned persistence was even greater in preweanlings than in weanlings or young adolescents, and that it was retained into young adulthood, even when a vacation and a CRF reacquisition were interpolated between the original PRF acquisition and the final extinction test.

It might bear reemphasizing that, in these experiments, we found good retention of persistence learned at preweanling age, and not the kind of "infantile amnesia" that was seen by other investigators when retention was of fear conditioning based on a particular CS and on electric shock as the UCS (see, e.g., Campbell, 1967; Campbell & Coulter, 1976). To repeat, this difference may be due to one or more of the following: (a) appetitive versus aversive motivation, (b) many learning trials in our appetitive case versus relatively few in the shock-conditioning case, and/or (c) general temperamental or dispositional learning (learning to persist) versus learning of a specific fear in relation to a specific environmental (episodic) stimulus. There appears to be *no infantile amnesia for learned persistence.*

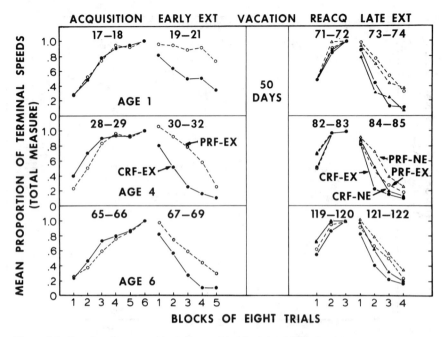

Figure 7.8. Results of an experiment on retention and durability of persistence as a function of age. The prevacation panel represents mean proportion of terminal speeds in the initial acquisition. The postvacation panel represents mean proportion of terminal reacquisition speeds. PRF, Partial reinforcement; EX, extinguished; CRF, continuous reinforcement; NE, not extinguished after original acquisition phase, only in late extinction. (From Amsel & Chen, 1976. Copyright 1976 by the American Psychological Association; reprinted by permission.)

The ten- to fifteen-day age range and later

As we have seen, there is some neuroanatomical and behavioral support for the hypothesis advanced by Jackson (1931) that, during ontogenesis, neural centers responsible for inhibition mature later than centers responsible for excitation, and that the later-developing, phylogenetically "higher" neural centers regulate (inhibit?) the excitation of the "lower" centers. In altricial mammalian species like the rat, functional maturity of the hippocampus and neocortex comes relatively late in development, and the hippocampus, in particular, has been implicated in the expression of behavioral inhibition or (as I prefer) suppression (Altman, Brunner, & Bayer, 1973; R. J. Douglas, 1967; D. P. Kimble, 1968). Preweanling rats, in the third and fourth weeks of life, appear to lack maturity for tasks requiring suppression of responding; for example, they seem particularly deficient in passive avoidance learning and in habituating to novel stimuli (Bronstein et al., 1974; Feigley et al., 1972). These task deficits resemble those shown in hippocampally lesioned adult animals (McCleary, 1966). If appetitive extinction is analogous to passive avoidance of shock and if extinction and habituation involve common processes (Thompson & Spencer, 1966), we may well wonder at what age rats (younger than 17 to 18 days of age) show differential persistence in extinction after CRF and PRF training, and whether, and for how long, this differential persistence (reflected in the PREE) is retained.

A number of observations suggest that the 10- to 15-day age range may be an important transitional period for persistence. Infant rats spend at least 12 hours of each day attached to a nipple, but receive milk in discrete episodes following the milk ejection reflex triggered by the release of the hormone oxytocin (Wakerley & Lincoln, 1971). Hall, Cramer, and Blass (1975) have reported two changes in the suckling behavior of rat pups 12 to 14 days of age: (a) There is a sharp increase in the number of pups detaching from a nipple immediately after milk ejection and scrambling for another nipple (cf. Drewett, Statham, & Wakerley, 1974). (b) Beginning at this age, nipple-attachment latency is inversely related to duration of food deprivation. At around 14 days of age, pups first open their eyes, gain the ability to thermoregulate, begin to leave the nest, and begin to meet their nutritional needs in ways other than suckling (Bolles & Woods, 1964). In addition, this age range marks the appearance of a presumptive maternal pheromone that attracts infants back to the nest (Holinka & Carlson, 1976; Leon, 1974; Leon & Moltz, 1972).

It seems to be the case, then, that in the rat the mechanisms responsible for maintaining the mother–infant bond undergo a change at about 2 weeks of age. Up to that age, suckling, even during the long no-milk intervals, and contact with the mother may be viewed as involving a kind of built-in persistence, essential for survival at this pretransitional stage. This suggested the possibility that as feeding comes more and more under direct instrumental control of the pup, externally imposed differential-reward schedules may become more and more effective determinants of persistence.

Adopting such an ontogenetic perspective gave rise to several questions: (a) Are there transitional ages for learned persistence (the PREE), learned suppression (the MREE, SNC), and other paradoxical effects? (b) If there are, what is the order of their first appearance? (c) Do the approximate ages of first appearance of these effects correspond to periods during which other significant behavioral and neurobiological changes are occurring? (The answers to the first two questions are summarized in Table 7.1.)

The procedures we use with younger infants involve the apparatus pictured in Figure 7.9. Pups are culled to litters of eight at 3 days of age; and at times appropriate to the deprivation conditions called for in the experiments, they are placed in a plastic chamber maintained for the youngest pups at 32°C, the average temperature in an undisturbed nest. Experi-

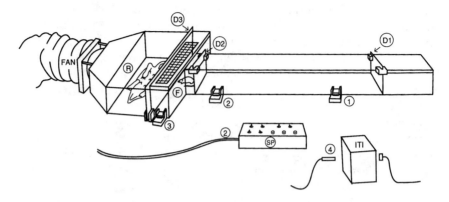

Figure 7.9. Clear Plexiglas pup runway. Photocells are placed at Positions 1, 2, 3, and 4 to time start, run, and goal speeds and ITIs. D1, Door from start box to alley; D2, door from alley to goal box; D3, automated door controlled by Photocell 3 dividing goal box into two compartments. (For older animals, the start box and photocells can be moved to extend the runway, and Photocell 3 controls the delivery of a pellet into a cup.) An anesthetized dam in the rear compartment (R) is accessible on rewarded trials when D3 is raised. On nonrewarded trials, pups remain in the front compartment (F) with D3 in lowered position. On reward-milk trials, milk is delivered via an infusion pump (not shown) into the oral cannula of the pup while it is attached to a nipple and suckling. On reward-dry trials, the procedure is the same, including the activation of the pump, but milk is not delivered. A thermometer records the temperature of the heated runway, which is varied according to age of pups (32°C for youngest). Heating is provided by a thermostatically controlled, circulating hot-water system connected to special plastic heating pads under the floor of the runway apparatus. Soundless digital timers (more recently, computers) record times in three runway segments. A switch panel (SP) with indicator lights (more recently, a computer keyboard) makes possible remote control of reward–nonreward, operation of D3, milk infusion parameters, and timing of the ITI with Photocell 4 mounted on the ITI box. An exhaust system (FAN) removes odors from the goal box through a duct and out a window so that the pup is never downwind of the dam's odor.

mental training takes place in a runway, usually 30 cm long for the youngest pups and longer for the oldest, and 8 cm wide, also maintained at 32°C or less. Approximately 20 minutes before the first trial, a lactating dam receives an injection of a general anesthetic, producing a surgical level of anesthesia and blocking milk release (Lincoln, Hill, & Wakerley, 1973). Preliminary training consists of one or more "priming" trials in which the pup is placed directly against the dam and allowed to attach to a nipple and suckle for a short period of time. At the end of this time, the pup is gently detached from the nipple, moved to the opposite end of the alley, and allowed to approach the dam. On reward trials, the pup is permitted to attach to a nipple, usually for 30 seconds. If no milk is delivered ("dry suckling"), the reward is small. In the case of a larger reward, the pup gets a milk delivery, either from the dam who has had an injection of oxytocin to induce a milk letdown, or through an oral cannula attached to an infusion pump to control the amount and rate of delivery of milk. The

Ontogeny of dispositional learning 155

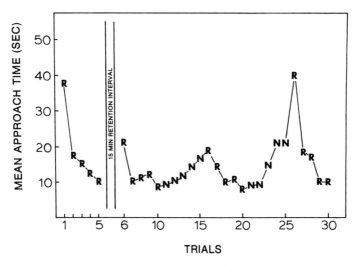

Figure 7.10. Acquisition, reacquisition after a 15-minute interval, and five-trial alternations of reward (R) and nonreward (N) in 10-day-old rat pups run to an anesthetized dam with nonlactating suckling as reward. (From Amsel, Burdette, and Letz, 1976. Copyright 1976 by Macmillan Magazines Ltd.; reprinted by permission of *Nature*.)

"milk" reward is either light cream (half-and-half), if this is a 1-day experiment and nutritional status does not matter, or a diet prepared in the laboratory that is nutritionally similar to mother's milk, if the experiment is over several days or a retention interval is involved. On nonreward trials, the pup is prevented by a barrier from reaching the dam. When postweanlings, adolescents, or adults are run, the apparatus can be lengthened and the anesthetized dam is replaced by milk in a little cup or an automated feeder that delivers a food pellet after the last photobeam is broken.

Here is a sample of results from our work on infant rats with these methods. We began by looking at five-trial alternations of reward and nonreward in 10-day-olds, the reward being dry suckling on an anesthetized dam (Amsel, Burdette, & Letz, 1976). We were able to demonstrate appetitive learning with retention after a 5-minute interval. The five-trial alternations resulted in successive acquisitions and extinctions, a nonparadoxical level of simple instrumental performance, akin to Schneirla's (1959) levels of approach and withdrawal, that can be seen in animals at virtually any level (Figure 7.10). Indeed, Johanson and Hall (1979) later demonstrated that 1-day-old rats learn, at a very slow rate, a simple operant (nudging a paddle) for milk reinforcement, and that even a kind of trial-and-error learning may occur at this age.

In another experiment, we measured runway performance and at the same time recorded ultrasounds (Amsel, Radek, Graham, & Letz, 1977). Ultrasounds in several rodent species have been taken to be an indicant

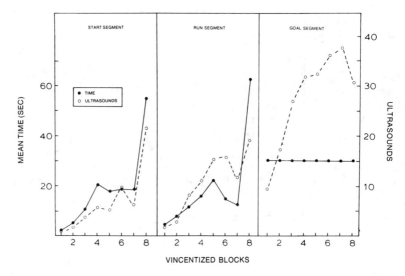

Figure 7.11. Ultrasounds during extinction in the start, run, and goal segments of the runway, with corresponding time spent in each segment. (From Amsel, Radek, Graham, & Letz, 1977.)

of stress and arousal, and they are thought to evoke retrieval behavior in mothers (e.g., Bell, 1974; Noirot, 1972). In this study, 24 rewarded trials were followed by an 8-trial extinction session. Again the reward was dry suckling for 30 seconds. As these 11-day-old pups learned to approach, ultrasounding declined. In extinction, ultrasounding increased with response time, but it also increased fourfold in a constant 30-second period of goal confinement (Figure 7.11). There is at this age, then, evidence of increased arousal (primary frustration). There is also at this age very good evidence for the direct utilization of the frustration-produced feedback cue, as inferred from the presence of a kind of memory-based discrimination learning, patterned alternation.

Patterned alternation occurs reliably and robustly with an 8-second intertrial interval (ITI) at both 11 and 14 days of age (Stanton, Dailey, & Amsel, 1980; see Figure 7.12). The pups are somehow able to respond appropriately on the basis of Trial N to the nature of the goal event on Trial $N + 1$. (Indeed, in our experiments they show this patterning in fewer trials than adults.) And there is evidence at this age not only of patterning on the basis of reward and nonreward, but also of patterning to differential reward magnitudes – suckling with milk versus dry suckling. (Nonnutritive suckling must be regarded as rewarding at this age because, as we have seen, it can serve as the basis for appetitive learning.) We will come back to the patterned alternation phenomenon later in a more theoretically sophisticated connection, and we will have reason to employ

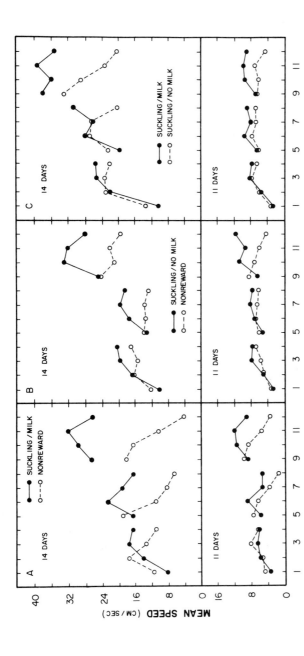

Figure 7.12. Mean alley running speeds on three kinds of single-alternated rewarded (R) and nonrewarded (N) trials. (Adapted from Stanton, Dailey, & Amsel, 1980. Copyright 1980 by the American Psychological Association; reprinted by permission.)

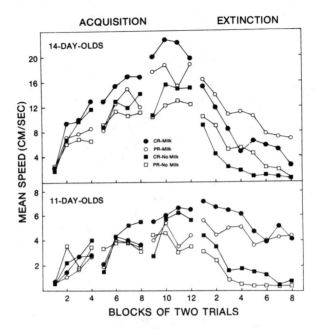

Figure 7.13. Mean approach speeds in acquisition and extinction for continuous- (CR) and partial-reinforcement (PR) groups at 11 and 14 days of age tested with milk or dry suckling as reward. (From Letz, Burdette, Gregg, Kittrell, & Amsel, 1978. Copyright 1978 by the American Psychological Association; reprinted by permission.)

these different reward magnitudes to study other reward-schedule phenomena.

In terms of our earlier definition, none of the above-mentioned effects need be called "paradoxical." The first paradoxical effect to appear is the PREE (Letz, Burdette, Gregg, Kittrell, & Amsel, 1978). The transitional period for this effect was demonstrated, with milk suckling or dry suckling as reward, to lie between 11 and 14 days of age (Figure 7.13). The PREE appeared at 14 days but not at 11 days of age whether the ITI was 10 to 12 minutes or 5 seconds. There was no evidence of the MREE at either 11 or 14 days, which is to say that at neither age did the larger reward result in faster extinction than the smaller reward following CRF training. The PREE was not seen on Day 14 when CRF/PRF training was on Day 11 with CRF reacquisition for both groups on Day 14, again at 10- to 12-minute ITI. In another experiment (Chen & Amsel, 1980b, Exp. 1), the PREE was shown to be present when training was on Days 12 and 13 and extinction was on Day 13, but not when training was on Days 10 and 11 and extinction was on Day 11 (Figure 7.14). The ITI was 8 to 10 minutes. This experiment suggested that the transitional age for the differential-acquisition effects that result in the PREE with our kind of procedure may

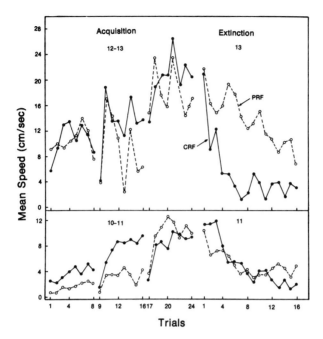

Figure 7.14. Acquisition and extinction speed curves showing the PREE at 12 to 13 days of age but not at 10 to 11 days. (From Chen & Amsel, 1980b. Copyright 1980 by John Wiley and Sons, Inc.)

fall, on the average, between 11 and 12 days. To confirm that this difference was not due to age at extinction, we conducted a second experiment (Chen & Amsel, 1980b, Exp. 2), which showed that the PREE was present on Day 13 after training on Days 11 and 12 and CRF reacquisition on Day 13, but was not present on Day 13 after training on Days 10 and 11 and reacquisition on Day 13.

Parenthetically, the precocial guinea pig, which, unlike the rat, has a well-developed brain and almost mature septohippocampal system at birth (Gillespie, 1974), shows the PREE at 4 to 5 days (Dailey, Lindner, & Amsel, 1983) and would probably show it shortly after birth if it would perform for us at this age. The difficulty is that a certain amount of handling and adjustment to the reward (pureed lettuce, slightly chilled, was our recipe) is necessary. Cross-species comparisons such as this one between the precocial infant guinea pig and the altricial infant rat, and of some of these reward-schedule effects among adult fishes, turtles, pigeons, and rats (see the next section, and the reviews by Bitterman, 1965, 1975), can be thought to lend a degree of phylogenetic validity to our developmental schema of these effects in the infant rat.

In another series of experiments, also conducted at 8 to 10 minutes ITI

(Chen & Amsel, 1980a), we showed that persistence acquired at 14 to 15 days, with suckling from the dam as reward, was retained after 2 weeks and an interpolated period of CRF reacquisition with food pellets as a reward; that is, persistence learned in the suckling system was retained and was expressed in a more adult feeding system. Indeed, persistence learned in a PRF treatment at an age shortly after its first appearance in ontogeny is *remarkably durable*. Recall that we had shown that persistence acquired at 17 to 18 days survived an extinction treatment at 19 to 21 days, a 50-day retention interval, and reacquisition under CRF conditions at 71 to 72 days, to reemerge in a second extinction test at 73 to 74 days (Amsel & Chen, 1976; see Figure 7.8). There is the suggestion here that this learned disposition, this temperamental characteristic of persistence, is retained from infancy to adulthood. (This will be important for our later discussion.) And as I have already emphasized, the retention of such frustration-related dispositional learning from infancy to adulthood seems greater in degree than retention of infantile experiences of episodic fear reported in the literature (e.g., Campbell, 1967; Coulter, 1979).

In contradistinction to work on the PREE, we have found no evidence for SNC (also known as the depression effect) at 11, 14, or 16 days (Stanton & Amsel, 1980). Again the contrasting reward magnitudes were nutritive (milk) and nonnutritive (dry) suckling, and there was at each age an effect of reward magnitude on terminal level of acquisition: The pups ran faster for the milk reward. When they were shifted to a reduced reward, from milk to dry suckling, their response level declined to the level of the unshifted (dry suckling) control, *but not below that level:* There was no SNC effect. In another series of experiments, SNC showed up clearly at Days 25 to 26, marginally at Days 20 to 21, and not at Days 16 to 17 (Figure 7.15). The rewards in these experiments were larger and smaller amounts of milk in a cup (Chen, Gross, & Amsel, 1981).

Not having found paradoxical suppression of performance at 16 to 17 days with two amounts of milk in a cup (a more adult reward system) as the contrasting rewards, we tested for a number of other effects at this age within this reward system (Chen et al., 1981). The PREE was, of course, substantial at Days 16 to 17 under either of two levels of reward magnitude (Figure 7.16), and again there was in acquisition a direct effect of magnitude of reward on runway speed. There were also at this age, as we have seen, clear effects of partial delay and variable magnitude of reward (the PDREE and VMREE). We also found a larger PREE with the larger of the two reward magnitudes. This is "adult functioning" – the Hulse (1958) and Wagner (1961) effect – but there was still no SNC effect at this age. Comparing the extinction curves following the two CRF reward-magnitude treatments, there was no MREE at this age (nor was there one at 11 or 14 days of age) – extinction was not faster in the group that had the larger reward; so the Hulse–Wagner effect in these experiments cannot be at-

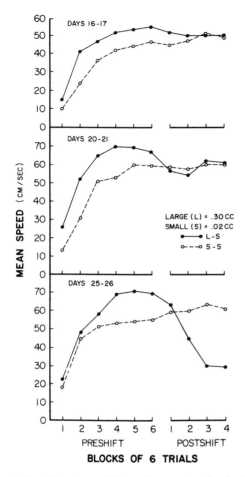

Figure 7.15. Preshift and postshift speed curves showing level of SNC at 16 to 17, 20 to 21, and 25 to 26 days of age. (The milk rewards were large [L] = .30 cc) and small [S] = .02 cc.) (From Chen, Gross, & Amsel, 1981. Copyright 1981 by John Wiley and Sons, Inc.)

tributed simply to faster extinction after large-reward CRF training, that is, to a differential MREE related to different reward magnitudes.

Our ontogenetic data appeared to go against the idea that, in infant rats at least, the PREE is purely a case of SNC in the extinction of the continuously reinforced response (Gonzalez & Bitterman, 1969). The reasoning in this interpretation of the *adult* PREE has been that the switch from CRF acquisition to extinction yields a successive-contrast effect that suppresses performance to a suboperant or subzero level, and therefore to a level below extinction performance after PRF acquisition. However, at an age (16 to 17 days) when there is a clear, adult-like PREE in these milk-in-cup experiments, there is no evidence of the intense frustrative suppres-

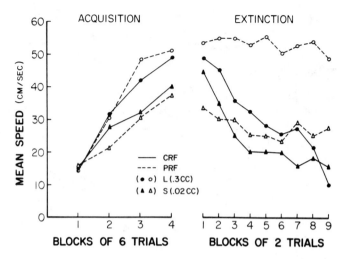

Figure 7.16. PREE as a function of reward magnitude. Mean speeds in acquisition and extinction factorially combining two levels of reward magnitude (large [L] = .30 cc and small [S] = .02 cc) with CRF and PRF at 16 to 17 days of age. (From Chen, Gross, & Amsel, 1981. Copyright 1981 by the American Psychological Association; reprinted by permission.)

sion characteristic of incentive contrast, of either the SNC or MREE variety. In experiments involving both suckling with milk and dry suckling as rewards, the PREE occurs in the 12- to 14-day age range, and the MREE is seen about a week later, but we have never observed frustrative depression (SNC) in approach speeds in the suckling reward system.

Another interesting set of findings in this regard involves the two other persistence effects, variable magnitude of reinforcement (VMREE) and partial delay of reinforcement (PDREE). The VMREE (e.g., Logan, Beier, & Kincaid, 1956) involves the quasi-random intermixing of large and small rewards in acquisition. In the case of our experiments with infants, these large and small reward sizes are defined by milk and dry suckling, respectively. In preweanlings we have used different amounts of milk in a cup. In the PDREE (e.g., Crum, Brown, & Bitterman, 1951), reward is given on every trial, but is delayed on half of the trials, again quasi-randomly. We found a VMREE at 16 to 17 days, but still no SNC at this age with the two magnitudes of reward that, presented intermittently, are sufficient to produce the VMREE. The PDR effect was also present in extinction at 16 to 17 days. As one might expect intuitively, the degree of learned persistence at this age, related to reward schedule, was ordered PRF–VMR–PDR–CRF (Chen et al., 1981; see Figure 7.17). The disposition to persist in infant rats is related, then, to the size of the difference between the reward and the extent of the degraded goal event.

The paradoxical PRAE (e.g., Goodrich, 1959; Haggard, 1959) is defined

Ontogeny of dispositional learning

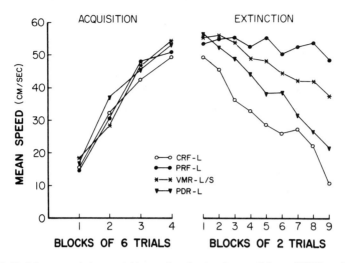

Figure 7.17. Mean speeds in acquisition and extinction for conditions of PDR and VMR in comparison with the appropriate (PRF-L and CRF-L) partial- and continuous-reinforcement groups at 16 to 17 days of age. L, Large; S, small. (Adapted from Chen, Gross, & Amsel, 1981. Copyright 1981 by the American Psychological Association; reprinted by permission.)

by more vigorous performance in PRF than in CRF acquisition in the early segments but not in the goal segment of an instrumental response. (Indeed, it is typically reversed close to the goal.) This remarkable effect, which has been taken to reflect increased activation resulting from anticipatory frustration and uncertainty, followed by suppression as a result of the same factor, all within a second or two, is quite robust at 18 to 20 days (Figure 7.18).

So by the third week postpartum in the rat, all of the intermittent reinforcement effects are present – the PREE, PRAE, VMREE, PDREE, and greater PREE with a large magnitude of reward than with a small magnitude of reward; but contrast-like effects are not observed. Another effect, which, like SNC and the MREE, depends on intense suppression, is slow responding on a schedule of discontinuously negatively correlated reinforcement (DNC). It occurs at 60 to 63 but not at 18 to 21 days (Figure 7.19).[11] Recall that a DNC schedule (Logan, 1960) is one in which the animal must not complete the defined response before a certain amount of time has elapsed, in our experiments 5 seconds. If the animal takes 5 seconds or more to complete the response, it is rewarded. This kind of intense suppression of responding *in the face of reward* is obviously a characteristic of more mature brains (see Chapter 8).

---

11. Some recent unpublished work in our laboratory by Michael Lilliquist has shown that DNC is present in 35-day-old rats.

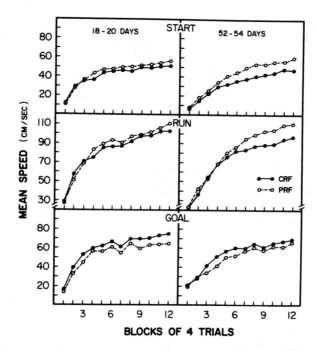

Figure 7.18. Acquisition of a runway response on a CRF or PRF schedule showing the PRAE in preweanling (18- to 20-day-old) and young adult (52- to 54-day-old) rats. At each age, running speeds are shown in the start, run, and goal segments of the runway. Relative positions of CRF and PRF curves in these segments define the PRAE. (From Chen, Gross, Stanton, & Amsel, 1980. Reprinted by permission of the Psychonomic Society, Inc.)

Comparisons with across-species results in vertebrates

Most of the work from experiments similar to ours conducted as comparative investigations in lower vertebrates and in an invertebrate species has been carried out by M. E. Bitterman and his colleagues. An extensive series of experimental findings from the vertebrate work appeared in the literature from about 1959 to 1965. These experiments, summarized in Gonzalez and Bitterman (1965) and in Gonzalez, Behrend, and Bitterman (1965), were conducted under a number of conditions: trials-equated versus reinforcements-equated between CRF and PRF conditions; discrete-trials versus free-operant conditions; spaced trials (24-hour ITI) versus massed trials (2-second ITI); partial or partial delay of reinforcement versus continuous reinforcement. After a series of meticulous investigations on the PREE in African mouthbreeders and goldfish, these investigators concluded that the basis for the PREE in these fish, when it occurs, is the sensory-carryover interpretation of V. F. Sheffield (1949) and the later form of this interpretation, the sequential hypothesis of Capaldi (e.g.,

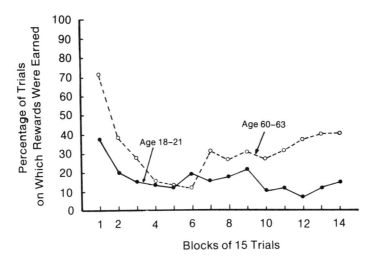

Figure 7.19. Percentage of trials on which rewards were earned under DNC conditions during acquisition training at two ages. The reward criterion was 5.0 seconds or greater for the first 122 cm of the runway. (From Chen, Gross, Stanton, & Amsel, 1981. Copyright 1981 by John Wiley and Sons, Inc.; reprinted by permission.)

1967). The PREE occurs in the fish only under conditions of ITIs in the few-seconds range. As we have seen, when this effect is shown for the first time in the ontogeny of the infant rat, at 12 to 14 days of age, the ITI has been either a few seconds or 8 to 10 minutes. It is therefore possible that, at this longer ITI, what Bitterman et al. call "reinstatement" or "learning about reward" (reward expectancy, r_R) is already operating. A compelling piece of evidence that a mechanism like reinstatement is present in the infant rat at about this age is that two other intermittent-reward extinction effects, the PDREE and the VMREE, are present at 16 to 17 days of age, as is the Hulse–Wagner effect – greater PREE with large than with small magnitude of reward (Chen et al., 1981).

In two experiments with the turtle, there is evidence that could be regarded as conflicting for the presence of the PREE, except that the experiments were run under widely different conditions of trial spacing. Murillo, Diercks, and Capaldi (1961) showed the effect at six trials per day over 14 days at a 30-second ITI, whereas it failed to be shown at an ITI of 24 hours (Eskin & Bitterman, 1961). In this experiment, both groups were given CRF training for 20 days followed by a 56-day period of differential training for the CRF and PRF groups. The conditions of the experiments being so different, it is difficult to pinpoint the reason for the difference in results, but a possibility is that the PREE in the turtle, as in the fish, is a sensory-carryover effect that depends on the kind of intertrial spacing in the experiment of Murillo et al.

The later appearance of SNC than of simultaneous negative contrast (SimNC) in our ontogenetic sequence (Stanton, Lobaugh, & Amsel, 1984) corresponds rather well to experimental results in which the same effects are studied across species, and both sets of results speak to the questions regarding levels of functioning raised earlier. SimNC refers to the fact that performance to the lesser magnitude of reward in a discrimination experiment, in which each magnitude is cued by a different stimulus, is lower (slower, less vigorous) than in a control condition in which separate animals respond to the greater and lesser reward magnitude.

SNC has failed repeatedly to appear in the instrumental responses of goldfish (Gonzalez, Ferry, & Powers, 1974; Gonzalez, Potts, Pitcoff, & Bitterman, 1972; Lowes & Bitterman, 1967; Mackintosh, 1971; Raymond, Aderman, & Wolach, 1972; but see Breuning & Wolach, 1977, with an activity-conditioning procedure). A recent experiment employing the consummatory contrast procedure of Flaherty (1982) as a test of SNC in goldfish also found no evidence for this effect (Couvillon & Bitterman, 1985). Nor is this effect shown in toads (Schmajuk, Segura, & Ruidiaz, 1981) or painted turtles (Pert & Bitterman, 1970; Pert & Gonzalez, 1974). However, SimNC does occur in goldfish (Burns, Woodard, Henderson, & Bitterman, 1974; Gonzalez & Powers, 1973), as does an activating effect of nonreward analogous to the FE (Scobie, Gold, & Fallon, 1974); and SimNC has been shown in painted turtles (Pert & Gonzalez, 1974). Eleven-day-old rat pups show primary frustration in the form of increases in ultrasounding on nonrewarded trials (Amsel, Radek, Graham, & Letz, 1977), but not, as we have seen, the PREE or other paradoxical effects. Couvillon and Bitterman (1984) point to two possibilities to account for the different results for SimNC and SNC across species and in ontogenetic studies: (a) that the mechanisms for simultaneous and successive contrast are different, and (b) that there is a single common mechanism that is better developed in some species than in others and at later ontogenetic stages. In regard to the latter possibility they write as follows:

Just such an interpretation of the ontogenetic data has been offered by Stanton, Lobaugh, and Amsel (1984). According to Amsel's frustration theory, an animal that experiences the preferred reward in a training situation comes in some sense to expect it, and is frustrated upon encountering the less preferred reward instead; the frustration tends to be conditioned to sensory antecedents of the less preferred reward and therefore to interfere with subsequent instrumental performance. If frustration is better conditioned in simultaneous experiments (where presentations of the preferred and less preferred rewards are intermixed) than in successive experiments (where, typically, there is but a single transition from the preferred to the less preferred reward), and if the frustration-conditioning mechanism is not very well developed in young rats, they may be expected to show simultaneous, but not successive, contrast. Especially persuasive is the evidence that certain responses are more sensitive than others to contrast effects: 17-day-old rats trained in a runway with milk versus dry suckling (less preferred) as reward show successive negative contrast in latency of nipple-attachment after reaching the goalbox – where conditioned frustration should be strong – but not in speed of running to the goalbox. Perhaps successive negative contrast has not yet been found in goldfish because conditioned frustration in goldfish is as weak as it has been assumed to be in young rats, and because the responses measured – directed swimming and target-striking – are not very sensitive to frustration. (p. 433)

Reward-schedule effects in the honeybee

In what is clearly the most extensive, if not the only, body of experimentation available on these phenomena in any invertebrate animal, recent work from Bitterman's laboratory has included an experimental analysis of what we have called the paradoxical reward-schedule effects in the honeybee. In these experiments the procedure is to pretrain individual bees to fly from the hive to a shelf in an open window in the laboratory and drink from sucrose solutions placed on colored targets. With this procedure, Bitterman and his colleagues are able to test not only for simple associative learning, but also for the appearance of the aforementioned reward-schedule effects in this animal. These effects include the PREE, SNC, and the overlearning extinction effect (OEE).

One of the very interesting general findings of this work is that there appears to be more expression of these paradoxical effects in the honeybee than in the very young infant rat or in the nonmammalian vertebrates that have been tested. As we have seen, experiments from Bitterman's laboratory have established that SNC is not shown in goldfish, toads, or painted turtles, results that are reminiscent of the relatively reflexive programming ascribed to "fish and reptiles and most birds" in the quoted passage from Hebb that opened this chapter. In more recent collaborative work between laboratories in Argentina and Hawaii, SNC has been shown to occur in two species of marsupials in the consummatory contrast procedure involving a shift from 32% to 4% sucrose solution (Papini, Mustaca, & Bitterman, 1988). The authors suggest that this result "fit[s] the hypothesis, based on comparative work with descendants of older vertebrate lines, that the mechanism of successive negative incentive contrast evolved in a common reptilian ancestor of birds and mammals" (p. 53). In a follow-up experiment, Papini and Ramallo (1990) have provided evidence for primary frustration in one of these species of marsupials, the red oppossum, in a situation analogous to the double runway.

Whereas SNC is not, then, a general phenomenon of vertebrate learning, something like it does occur in honeybees (Couvillon & Bitterman, 1984). And the same can be said in the honeybee for the OEE, which has been shown in a number of experiments (e.g., Couvillon & Bitterman, 1980,

1984; Shinoda & Bitterman, 1987). Employing several ingenious experimental manipulations, Bitterman and his colleagues have concluded that the OEE and SNC in the honeybee are not artifacts of nutritive effects related to prolonged training with high concentrations of sucrose, either preceding extinction or in the shift from higher to lower concentrations. They are now willing to entertain the hypothesis that the rapid extinction in the overlearning condition of the OEE and the "depression" effect in SNC involve "something akin to frustration in vertebrates" (Shinoda & Bitterman, 1987, p. 93), that performance in the honeybee in extinction after prolonged acquisition, and following a shift from high to low levels of reward, may be attributable to emotional responses to the absence of or reduction in the food reward. However, in the case of these effects in the honeybee, they caution against an interpretation in terms of "stronger expectation and greater consequent frustration" (Shinoda & Bitterman, 1987, p. 95; see also Couvillon & Bitterman, 1984), and they propose a "disruptive frustration-like process... generated as a function of the discrepancy between the existing approach tendency and that which the new reinforcer (if any) is capable of supporting" (p. 95). They make no assumption that the honeybee actually remembers and compares the reinforcers, and argue that the OEE and SNC in this animal "can be understood in terms of [the disruptive effects of] unconditioned frustration alone" (p. 95). They also point to their experiments on the PREE in the honeybee, designed specifically to evaluate explanations in terms of associatively reinstated (memorial) effects as opposed to sensory carryover, from which they have concluded that there is no evidence for "remembered nonreward" in the explanation of this effect in this species (Ammon, Abramson, & Bitterman, 1986).

Ammon et al. also note that in our developmental experiments in the infant rat we have seen the PREE at an earlier age than SNC, and that we ourselves point out that frustration theory offers a more straightforward explanation for the reverse (SNC earlier than the PREE): SNC involves only conditioned frustration, whereas according to the theory, the PREE requires the additional mechanism of counterconditioning. Our explanation for this apparent reversal in the expected ontogenetic sequence makes use of the fact that in PRF training there are many transitions between reward and nonreward, whereas in SNC there is just one. Ammon et al. suggest another possibility – that, as in the case of their explanation of the honeybee result, the PREE in the 12- to 14-day-old rat is not based on expectancy or reinstatement but on a sensory-carryover mechanism. What saves the explanation of the PREE as a "reinstatement" or "learning about frustration" effect in these young animals is the kind of evidence offered earlier in this chapter – for example, that persistence learned as a consequence of PRF training, and acquired at 14 to 15 days of age with suckling from the dam as reward, is retained after 2 weeks even if the PRF training is followed by the interpolation of a phase of CRF reacquisition with food-pellet reward (Chen & Amsel, 1980a). This result is not consistent with an explanation of the infant PREE in terms of sensory carryover.

Ontogeny of habituation and general persistence theory

In the remainder of this chapter, we describe a related body of developmental work in which direct habituation to aversive stimulation (electric shock rather than anticipatory frustration) is offered as a mechanism for inducing appetitive persistence. To review briefly the more general theory of persistence (Amsel, 1972a; see Chapter 4, this volume), habituation is an active process in which persistence develops through the counterconditioning of ongoing activity (R_O) to the originally disruptive stimulus (S_X). Repeated presentations of novel and/or noxious stimuli, conditioned or unconditioned, at first evoke responses (R_X) that interfere with ongoing activity. The habituation of these responses, according to this view, involves counterconditioning of R_O to S_X. Other things equal, animals that have undergone prolonged (and varied) habituation treatments should be more resistant to extinction because such animals are then less susceptible to disruption by stimuli of the class S_X, of which feedback cues (S_F) from anticipatory frustration (r_F) can be regarded as members.

In Chapter 4, we reviewed the evidence from a variety of procedures showing that adult rats, subjected to intense aversive stimulation, are more resistant to appetitive extinction than are controls. Most of these procedures are of the "on the baseline" variety; that is to say, the stimulation is administered in the context of the behavior being investigated. To recapitulate, electric shock combined on some acquisition trials with food reward in a runway increases later resistance to extinction (Banks, 1967; Brown & Wagner, 1964; Fallon, 1971; Terris & Wechkin, 1967); loud tones that disrupt bar pressing make fixed-ratio responding more resistant to extinction (Amsel, Glazer, Lakey, McCuller, & Wong, 1973); and coerced approach to shock and even "free" shocks in a runway enhance resistance to subsequent runway extinction (Wong, 1971a, b). Rosellini and Seligman (1975) have demonstrated that animals treated with inescapable shock, a treatment that is said to induce "helplessness," are less capable of escaping from primary frustration. They have also provided some evidence that shock, whether inescapable or not, introduced during the period of appetitive runway training "marginally increased resistance to extinction" (p. 152).

There is not very much developmental work on general persistence or transfer of persistence (and none that I know of on helplessness), but if we are to begin to comprehend the development of this important temperamental characteristic and product of dispositional learning, we must ultimately turn our efforts in this direction. The questions that come to mind are the same as in the case of the ontogeny of the reward-schedule effects, the only difference being that, instead of appetitive acquisition under various schedules of reward and nonreward, we employ treatments in which habituation occurs to a variety of disruptive events. Again, the reasoning, according to a more general theory of persistence, is that behavioral habituation involves the counterconditioning of ongoing responding to originally disruptive stimulation, and that the PREE is a special case of persistence involving habituation (counterconditioning) to anticipatory frustration.

Research on the ontogeny of habituation

In the early to middle 1970s, research on the ontogeny of habituation in the infant-rat animal model provided evidence that there are two kinds of habituation: a reflex kind that can be shown even in 1-day-olds and a kind of exploratory habituation that occurs for the first time at 2 to 3 weeks of age and is related to reaction to novelty. Williams, Hamilton, and Carlton (1975) referred to these as habituation of elicited and emitted responses and questioned whether they could be attributed to a "single hypothetical process called habituation" (p. 733). The evidence was also that these two early- and late-developing kinds of habituation depend on different neurotransmitter systems, serotonin and acetylcholine, respectively; that the second kind – for example, the kind displayed in open-field activity and spontaneous alternation – appears to be related to maturity of hippocampal function; and that up to 2 weeks of age the infant rat, in this respect, is behaviorally similar to the hippocampectomized adult (Altman, Brunner, & Bayer, 1973). From the point of view of the general theory of persistence, the more mature form, but not the reflexive form, of habituation can be thought of as involving persistence, the reinstatement of ongoing behavior in the face of novel or extraneous stimulation, for which the mechanism of counterconditioning has been my explanation. Here is some further evidence.

Habituation is as strong in preweanling rats (15 to 16 days of age) as in young adults (36 to 39 days of age) in an air-puff startle response, but there is more habituation in the older animals than in the preweanlings in a head-poke exploratory response (Williams et al., 1975); and there is no habituation in head-poke responding in 15-day-olds, but clear habituation of this response in 25-day-olds and adults (Feigley, Parsons, Hamilton, & Spear, 1972). The number of trials to habituation of a head-turn response to an air puff in rat pups increased and then declined until about 14 days of age (File & Scott, 1976). Rapid habituation and normal dishabituation of a leg-flexion response to a mild electric shock to the forelimb is displayed in rat pups as young as 3 days of age, and habituation is not affected at 5 days of age as it is at 20 days by decerebration (Campbell & Stehouwer, 1979). In an open-field apparatus, 15-day-old rats show no habituation of activity, whereas 21-day-olds show good habituation (Bronstein, Neiman, Wolkoff, & Levine, 1974).

In the more precocial chick, habituation of a reflexive eye-opening response to a tone was the same on Day 1 as on Day 4 posthatch (Rubel & Rosenthal, 1975). In the open field, however, chicks show habituation of exploratory responding on Day 4 posthatch, but not on Day 1 (Zolman, Sahley, & Mattingly, 1978).

In an across-species comparison, rats display spontaneous alternation, an indication of reaction to novelty, at 27 to 28 days of age, but not earlier; however, guinea pigs display such habituation in the first week of life (Douglas, Peterson, & Douglas, 1973). This corresponds reasonably well to the differences in age at which rats and guinea pigs display the PREE (see earlier references). In an across-age comparison in the rat, spontaneous alternation was seen in 30-day-olds but not in 15-day olds (Bronstein, Dworkin, & Bilder, 1974).

One possible explanation of the differences in age at which reflex and exploratory habituation occur in the rat is that habituation of simple defensive reflexes is mediated by the neurotransmitter serotonin, whereas habituation of the exploratory kind is mediated by acetylcholine. The support for this explanation is that reflex habituation is unaffected by the anticholinergic drug scopolamine (Williams, Hamilton, & Carlton, 1974), whereas PCPA, an antiserotonergic drug, does attenuate reflex habituation (Carlton & Advokat, 1973); sensitivity to the anticholinergic drug scopolamine could not, however, be shown before about 21 days of age, when exploratory habituation is seen (e.g., Campbell, Lytle, & Fibiger, 1969).

Ontogeny of habituation and appetitive persistence

In an early study in adult rats (Chen & Amsel, 1977; see also Chapter 4, this volume), we demonstrated that habituation to inescapable shocks increases appetitive persistence of an instrumental response. Later (Chen & Amsel, 1982), working in a more developmental framework, we found a clear indication that, compared with a no-shock control, a prolonged shock habituation treatment after appetitive acquisition enhances resistance to normal (nonreward) extinction and to combined nonreward and punishment (punished extinction), both at preweanling (17 to 18 days) and juvenile (41 to 42 days) ages, the effect being greater in preweanlings. Another very interesting finding of these experiments was that shock treatment given in preweanling to weanling age (19 to 24 days) increased resistance to nonreward extinction but not to punished extinction when the persistence test was delayed by 42 days. In preweanling to weanling and

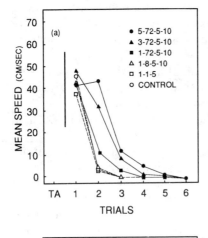
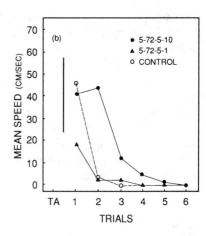
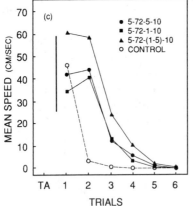

Figure 7.20. (a) Resistance to punished extinction as a function of days over which shock is delivered and total number of shocks. (The first number in the series represents days of shock, the second is number of shocks per day, the third is the duration of shock in seconds, and the fourth is the intershock interval [variable 10 minutes]. Control, no shock; TA, terminal acquisition.) (b) Punished-extinction speed curves as a function of intershock interval. (c) Gradually increasing shock duration (1 to 5 seconds) promotes greater resistance to punished extinction, holding shock duration constant. (From Chen & Amsel, 1982. Copyright 1982 by the American Psychological Association; reprinted by permission.)

juvenile rats, habituation to shock reduced the time taken to recover from suppression of milk intake in the shock situation, and it also reduced latency to eat in a punished-reward test at several retention intervals. Finally, in a series of parametric studies in young adult rats (70 days of age) relating learned persistence to habituation to shock, we found the following (Figure 7.20): (a) The greater the number and the duration of exposures to inescapable shock, the greater was the resistance to punished extinction; (b) the larger the intershock interval, the greater was the resistance to punished extinction; and (c) relative to conditions of constant durations of shock, gradually increasing their duration promoted greater resistance to punished extinction, even when the total amount of shock was held constant. These last experiments, in particular, suggest that the kind of active counterconditioning involved in PRF training and in habituation more generally, and perhaps not some form of passive "helplessness," is an important factor in increasing generalized persistence (Amsel, 1972a; Levis, 1976). In the next chapter, we will review some recent work by Gray and his associates that, in the context of direct septal driving of the hippocampal theta rhythm, points to a nonassociative interpretation of these generalized persistence phenomena.

Summary

In this chapter we have reviewed experiments that point to a basis in early development for dispositional learning and memory in the rat. They suggest a correspondence between the ages of first appearance of a number of reward-schedule effects between infancy and juvenile age and the rapid development of hippocampal and cholinergic function over the same period of development. (Some of these effects appear to have a similar sequence of emergence in experiments performed from a phylogenetic, or at least an across-species, perspective.) In Chapter 8 we will provide a more detailed account of converging evidence that supports this suggestion and the relation of this evidence to a number of theories of hippocampal function.

8 Toward a developmental psychobiology of dispositional learning and memory

For some years, theorists have believed that the explanation of the PREE and some of the other reward-schedule effects involves common underlying mechanisms, though the nature of the specific mechanisms that have been offered has differed from one theory to another. Our ontogenetic investigations begin to suggest that whereas these earlier explanations may or may not be reasonable for adult rats (and other mammalian species), they may not be entirely satisfactory for the infant to weanling rat. This is particularly clear if we compare the ages of appearance of a number of the reward-schedule effects, presented in the last chapter as Table 7.1.

Do theories based on adult behavior explain the order of appearance of the reward-schedule effects in infants?

According to frustration theory, the PREE, first seen in the 12- to 14-day range, involves a more complex chain of associations than the MREE or SNC, which are seen at successively later ages: The latter two effects involve the emergence of primary frustration as a result of extinction or reduced reward, the subsequent conditioning of anticipatory frustration, and the action of this conditioned frustration to strongly (paradoxically) suppress instrumental responding. In contrast, the PREE involves not only the first three stages – the conditioning of reward expectancy, primary frustration, and the conditioning of frustration – but also a fourth, the counterconditioning of feedback stimulation from anticipatory frustration to instrumental responding. In our research, however, the order of appearance of these effects in ontogeny may cause us to doubt that the PREE is a more advanced effect than, say, SNC. In fact, by 16 to 17 days, all of the persistence effects that depend on intermittency are present: not only the PREE, but also the VMREE and the PDREE. Still, SNC is not seen until much later. This pattern of results suggests that the original version of frustration theory that integrates the various reward-schedule effects in the adult animal be revised or at least supplemented to account for the order of emergence of these effects in the infant.

The same can be said for Capaldi's (1967) very influential sequential theory. Effects like the PREE, SNC, and the MREE would seem in this theory to depend on sequential interactions of reward and nonreward, on the kinds of short-term memorial effects ("working" or "carryover" memory) that can be inferred from the occurrence of patterned alternation (PA). Indeed, in Capaldi's more recent theorizing (Capaldi, Verry, Nawrocki, & Miller, 1984), these traces or memories of previous rewarding or nonrewarding events could be long-lasting, up to at least 24 hours. This would suggest that what we call the more advanced effects depend only on the level of mechanism that exists in PA. But sequential theory, like frustration theory, has difficulty with the order of appearance in ontogeny of the reward-schedule effects: If PA, the PREE, the MREE, and SNC do depend on the same sequential, short- and even long-term carryover-memory mechanisms, then how do we explain the following results?

1. As we have seen, at 11 days of age, at 8-second ITI, rat pups can easily discriminate between – and pattern their behavior to – milk suckling versus nonreward, dry suckling versus nonreward, and even milk suckling versus dry suckling (Stanton, Dailey, & Amsel, 1980); they are tremendously good at using the carryover kind of memorial cues at an age at which they manifest no PREE.
2. At 14 days, we again find clear PA in all three conditions and the PREE occurs as well, but not the MREE (Letz et al., 1978) or SNC (Stanton & Amsel, 1980). SNC at this age involves *paradoxical* suppression resulting from detecting a difference between two reward levels, milk suckling and dry suckling, something the pups can do well in PA learning at 11 days; and the MREE involves similar suppression, in this case in relation to the comparison of a large or a small reward in acquisition with nonreward in extinction. Some mechanisms beyond those involved in PA must be required for these effects that depend on intense suppression, effects that are not seen clearly until about Day 20 (the MREE) and Day 25 (SNC).
3. Work by Stanton (1982, 1983) reveals that at 18 days there is no evidence of patterning at the 105-second ITI, though PA does occur at 8-, 30-, and 60-second ITI; that is to say, the duration of this kind of working memory seems to increase with age. However, in the absence of PA at 105-second ITI, a clear PREE does occur when we compare extinction after CRF acquisition and after training on single alternations of reward and nonreward. These results based on infant learning are entirely analogous to some earlier work with adults, in which single alternations of reward and nonreward at 24-hour ITI produced no evidence of patterning but great persistence (virtually no extinction), compared with extremely rapid post-CRF extinction, over 44 extinction trials (Surridge & Amsel, 1966). In these cases, the *expression* of a sequential mechanism appears not to be necessary, let alone sufficient, for the production of the PREE.

Many of these results suggest, at the very least and not surprisingly, that the mechanisms proposed in both the frustration and sequential theories for the control of these reward-schedule effects are different in rats in the infant to postweanling age range than in adults, and that we should look at the developing brain, in relation to developing behavior, for clues to mechanisms for this difference.

Reversion of function: similarity of the infant rat to the hippocampally damaged adult

It is hazardous to think that one part of the developing brain plays a much larger role than any of several others in frustration and arousal, in extinctive suppression, and consequently in the emerging reward-schedule effects, but there is a reasonably good basis for implicating the limbic system in such a hypothesis (early summaries of the relevant work in adult rats can be found in Altman, Brunner, & Bayer, 1973; Douglas, 1967; Gray, 1970; Isaacson & Kimble, 1972; Kimble, 1968). So I proceed in the spirit of Francis Bacon's famous aphorism, paraphrased thus: Truth emerges more readily from error than from confusion.

Angevine and Cotman, in their *Principles of Neuroanatomy* (1981), write that "the hippocampus holds the secret to limbic system functions, or at least a large part of it.... Lesions of the hippocampus in experimental animals make it more difficult for those animals to change an ingrained response to a new one" (p. 261). The hypothesis that has guided our developmental work is that the infant rat is like the hippocampally damaged adult, that the fact of the largely postnatal occurrence in the rat of a significant part of hippocampal cell development and circuitry offers the investigator a natural animal model for gradually increasing levels not only of hippocampal structure, but also of function, as reflected in our case of the developing paradoxical effects (Amsel, 1986; Amsel & Stanton, 1980). It can be said that hippocampal insult in the adult rat, like hippocampal immaturity in the infant, reduces the level of functioning from "paradoxical" to "nonparadoxical" – in Schneirla's (1959) terms, from the level of function he described as "seeking and avoidance" to the level he characterized as "approach and withdrawal." A phrase to describe this kind of effect of hippocampal damage on learning and memory in the adult or older infant rat, analogous but opposite to the familiar "recovery of function," might be "reversion of function."

Here, briefly, are a few interrelated pieces of evidence that relate the integrity of the hippocampal formation, defined inclusively, to the reward-schedule effects.[12] The evidence comes from work with *adult* animals, ours

---

12. For much more extensive reviews of much of this work on septal and hippocampal

being the first to investigate these behavioral effects systematically over the infant to postweanling age range and to suggest hypotheses that relate these phenomena to aspects of the neural substrate at these ages. The kind of questions our particular work is designed to ask are these: Is a level of function in dispositional learning and memory that is thought to depend on hippocampal maturity in intact adult animals degraded (or is it spared) when that area of the brain suffers damage or neuronal agenesis that presumably prevents its normal development? Is there in the developing infant rat a "reversion of function" as a result of these kinds of hippocampal intervention? And if such function is degraded in infant animals, is there recovery of that function in adults? We begin with a very brief summary of the relevant work on hippocampal neurogenesis and neuroanatomy.[13]

Hippocampal cells and their development

The three panels of Figure 8.1 show, respectively (A) a schematic orienting view of the position of the hippocampus in a sagittal section of the limbic system of the mammalian brain; (B) a schematic of the horizontal aspect of cell types and synaptic connections in the hippocampal formation of the rat; and (C) a photomicrograph of a horizontal section of the hippocampal formation, including the dentate gyrus, of the rat, showing again the location of the two main cell types, the granule cells of the dentate gyrus and the pyramidal cells of the hippocampus proper.

In the rat, the pyramidal cells are almost completely laid down by the time of birth. However, much of the neurogenesis, differentiation, and synaptogenesis of the granule cells in the dentate gyrus of the hippocampal formation of the rat occurs between about embryonic Day 14 and postnatal Day 21 (Altman & Das, 1965; Bayer, 1980a,b; Cotman, Taylor, & Lynch, 1973; Schlessinger, Cowan, & Gottlieb, 1975). About 90% of granule cell development takes place postnatally; Bayer (1982) has shown that granule cells continue to be formed throughout life. The greatest developmental spurt is between about 5 and 30 days of age, when close to adult levels are reached (Altman & Bayer, 1975). Altman and Das (1965) have shown that a particularly strong period of cell development is between 15 and 30 days of age, following a peak of "precursor" cell number at 15 days (see Figure 8.2). The 12- to 14-day and 25- to 30-day ranges correspond to the ages of our early-appearing and late-appearing paradoxical effects.

involvement in reward-schedule effects and other phenomena of learning in adult animals, the reader is referred to Gray (1982) and Gray and McNaughton (1983). See also the end of this chapter for comments on the relation of this kind of emphasis to the more recent emphases on interpretations in terms of cognitive maps and working memory.
13. For a more detailed account of septohippocampal anatomy and neurophysiology as they relate to our present interests, see Gray (1982, chaps. 3 and 4).

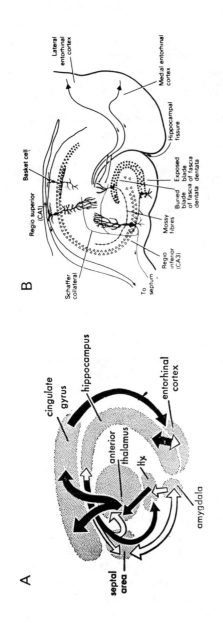
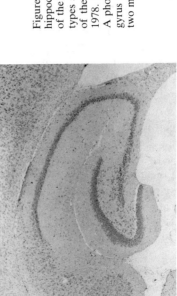

Figure 8.1. (A) A schematic view of the position of the hippocampus in a saggital section of the limbic system of the mammalian brain. (B) A schematic view of cell types and synaptic connections in a horizontal section of the right hippocampus. (From O'Keefe & Nadel, 1978. By permission of Oxford University Press.) (C) A photomicrograph of a coronal section of the dentate gyrus and the hippocampus proper, showing again the two main cell types, granule and pyramidal.

Psychobiology of dispositional learning

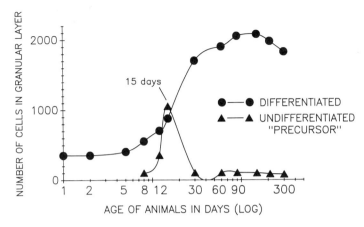

Figure 8.2. Ontogeny of undifferentiated (precursor) and differentiated cells in the granular layer of the dentate gyrus. (From Altman & Das, 1965. Reprinted by permission of Wiley-Ligg, a division of John Wiley and Sons, Inc.; © 1965 by John Wiley and Sons, Inc.)

Hippocampal factors and dispositional learning in adult animals

There is a basis for thinking that there may be a connection between the developing hippocampal formation and its inputs from entorhinal cortex and septum, on the one hand, and our developing reward-schedule effects, on the other. The following selective lines of evidence from work with adult animals point in this direction.

Lesion effects

The results of two studies in adult rats are that hippocampal but not cortical lesions reduce, but do not eliminate, the capacity of adult rats to pattern their responses to single alternations of reward and nonreward (Cogan, Posey, & Reeves, 1976; Franchina & Brown, 1970). Franchina and Brown find this result in simple single-alternation patterning; Cogan et al. find no differences in the simple case, but only in the case of patterning to single alternations of immediate and delayed rewards.

The weight of the evidence in adult rats is that nonspecific lesions in the septum, fornix, and hippocampus reduce or eliminate a number of effects based on reward elimination or reduction in frequency or magnitude. It was shown some time ago that hippocampal lesions in adult rats reduce the capacity to alter responding in line with changed aversive-reinforcement contingencies (Ellen & Wilson, 1963). An example closer to our present interests is that such lesions increase resistance to extinction after appetitive CRF training (Jarrard, Isaacson, & Wickelgren, 1964; Kimble & Kimble, 1965; Winocur & Mills, 1969). It is not surprising, then, that the PREE is reduced or eliminated when hippocampal or septal lesions increase persistence in extinction after CRF training and, depending on the ITI, reduce or have no effect on persistence after PRF training (Feldon & Gray, 1979a,b; Henke, 1974, 1977; Rawlins, Feldon, & Gray, 1980; see Gray & McNaughton, 1983, for an extensive review).

Hippocampal lesions have been shown to eliminate incentive contrast (SNC) in a runway in adult rats (Franchina & Brown, 1971). There is a very extensive body of work on the psychobiology of SNC in which "consummatory contrast" is defined by lick rate, related to shifts in sucrose concentration, usually from 32% to 4%. In this work, a question frequently raised by Flaherty[14] and his co-workers is whether contrast, in this case, is a sensory or an emotional effect, or a combination of both. This question arose in an early examination of the effect on consummatory contrast of septal lesions (Flaherty, Capobianco, & Hamilton, 1973). Flaherty et al. demonstrated that whereas control-operated rats showed retention of contrast 1, 4, 5, and 7 days following the shift from 32% to 4% sucrose, rats with septal lesions showed the effect only at the 1-day interval. Their hypothesis was that, on the first postshift day, contrast is a sensory effect, determined by stimulus change, whereas the emotional effects (and/or conflict related to reward reduction) develop on the second day. In more recent experiments, they have shown that corticosterone level is not elevated on the first day after the incentive shift, but that it is elevated on the second postshift day and that the level of elevation is positively correlated with the degree of response suppression (Flaherty, Becker, & Pohorecky, 1985).

Hippocampally lesioned adult rats cannot maintain low rates of responding on a differential reinforcement for low rates of responding (DRL) schedule (Clark & Isaacson, 1965; Nonneman, Voigt, & Kolb, 1974); nor in a runway analog to DRL (discontinuously negatively correlated reinforcement, DNC), can they adjust to running slow as a condition of receiving reward (Rickert & Bennett, 1972). Adult cats are relatively unimpaired in DRL if neocortical or hippocampal lesions had been made at or before 4 months of age; cats lesioned in the first few days of life are later unimpaired in the extinction of a runway response for food reward, whereas those receiving lesions at 6 weeks of age or later are significantly impaired (Nonneman & Isaacson, 1973). Recall that in our developmental research, DNC (DRL-like) performance in a runway is not observed in 21-day-old rats but is present in young adults (see Table 7.1).

Radiation and alcohol effects

X-irradiation at 2 to 15 days of age in the rat retards the formation of dentate granule cells measured at 30, 60, or 90 days (Altman & Bayer, 1975) or at 21 days (Diaz-Granados, Greene, & Amsel, 1991a). There is

---

14. See Flaherty (1982) for a review of work in this field.

no direct effect on the already developed hippocampal pyramidal cells; however, the granule cells serve an important "gating" function in the trisynaptic pathway from entorhinal cortex to the CA3 and CA1 pyramidal cells. X-irradiation in infancy, with its effect on granule cell development, reduces the size of the adult PREE; however, neonatal X-irradiation does not eliminate or even reduce *adult* single-alternation patterning except under the very special conditions of two trials per day (Brunner, Haggbloom, & Gazzara, 1974). As we shall see, our own work suggests that the effects of infant X-irradiation on PA and the PREE *in infants* are at least as severe. On the other hand, there is some evidence that exposure to ethanol in utero affects the laying down of the prenatally arising pyramidal cells (Barnes & Walker, 1981) and mature, but not immature, granule cells (Wigal & Amsel, 1990). Until recently it was not known how exposure to ethanol in utero might affect these same behavioral reward-schedule effects in adult or infant rats. Work of this kind will be described later in this chapter.

To summarize the last two sections, the lesser effects in these early studies of X-irradiation in infancy and hippocampal lesions in adults on PA than on the PREE, SNC, and DNC (and DRL) *in adults* correspond to the order of appearance of these behavioral effects in ontogeny. Again, it would seem that the reward-schedule effects that are the latest to appear in ontogeny are the most strongly affected in adults by damage to the septohippocampal region of the brain.

As we shall see, however, a caveat from our developmental work is in order: Even PA, the earliest-appearing of our reward-schedule effects ontogenetically, is affected in infants by hippocampal lesions (Lobaugh, Greene, Grant, Nick, & Amsel, 1989) and by postnatal exposure to X-irradiation (Diaz-Granados, Greene, & Amsel, 1991a) and ethanol (see the next section) if the ITI reaches about 1 minute.

Drug effects

There were early demonstrations in adult rats that sodium amobarbital and minor tranquilizers eliminate the PREE but do not affect – or have much effect on – single-alternation patterning (for references see the Appendix, and for a review of some of the more recent work see Gray, 1982). This, too, suggests that PA, particularly at short ITIs, is more "primitive" than the other effects: It is shown in the goldfish under conditions in which the PREE is not shown (Couvillon & Bitterman, 1981) and is attributed to "carryover" effects based on short-term or "working" memory, rather than to "reinstatement" based on longer-term or "reference" memory. The PREE is attenuated following injection of the benzodiazepine chlordiazepoxide (Feldon & Gray, 1981) but also, paradoxically perhaps, by amphetamine (Weiner, Bercovitz, Lubow, & Feldon, 1985). Negative incentive contrast, a reward-schedule effect that is more ontogenetically advanced in our work than the PREE, is virtually eliminated with the administration of chlordiazepoxide and is reliably reduced by ethanol (Becker & Flaherty, 1983). A more recent finding is that in incentive contrast of the kind studied in Flaherty's laboratory (consummatory contrast reflected in lick rate), the sequential processes of "detection, evaluation, and conflict" are involved over the postshift period and that chlordiazepoxide attenuates the contrast effect only when the conflict stage is reached (Flaherty, Grigson, & Rowan, 1986). When SNC experiments are conducted in runways, and the effect is seen on approach-to-the-goal behavior (instead of the consummatory behavior in Flaherty's case), the suppression of behavior characterizing contrast can also be shown to coincide with the appearance of conflict-like behavior (retracing, urination, etc.).

As in the case of lesions, one might be led to conclude that drugs like the benzodiazepines and ethanol eliminate only the "more advanced" (declarative, paradoxical) effects in adults, but again, at least in the case of ethanol, one would have to include PA when the ITI is extended, a kind of learning that requires anticipations of reward and nonreward (reinstatement; see Greene, Diaz-Granados, & Amsel, 1991a). This would exclude learning in which reward does not serve a reinstatement function, but only a direct reinforcing (stamping-in) function, such as PA learning at short ITI in which the sequential, carried-over consequences of stimuli produced by reward and nonreward are sufficient.

Hippocampal EEG

Hippocampal EEG, specifically in the rhythmic slow activity or theta range immediately surrounding 7.7 Hz, has been linked to anticipatory frustration and persistence in Gray's early experiments (Gray, 1972; Gray, Araujo-Silva, & Quintao, 1972) and in my own laboratory in Glazer's (1972, 1974a, b) work. As we shall see, Gray's (1982) more recent theory of the behavioral inhibition system (BIS) that relates "anxiety" to septohippocampal activity, linking 7.7-Hz theta to the inhibitory effects of nonreward, novelty, punishment, and subsequent persistence, is in some respects a neurophysiological analog of the general theory of persistence (Amsel, 1972a). (The specific range of EEG defining theta is apparently important, and other ranges have been linked to other behavioral functions – for example, to voluntary action [e.g., Vanderwolf, 1971] and to spatial learning [O'Keefe & Nadel, 1978].)

There is evidence that the theta rhythm of the hippocampal EEG is first seen in the rat pup at around 11 to 12 days of age (Gillespie, 1974), and that the minimum in the threshold-frequency function at 7.7 Hz, characteristic of adult male rats, appeared at 15 to 16 days (Lanfumey, Adrien, & Gray, 1982), which is about the age at which we have observed the earliest paradoxical effects (e.g., the PREE) in rat pups. As further, albeit indirect, evidence linking theta to the paradoxical effects, Rose (1983) has demonstrated that the granule cells of the dentate gyrus, which as we have seen are in a stage of rapid development at this age, are theta-producing cells, driven by pacemaker cells in the septum.

Another, more recent body of results on hippocampal EEG from the laboratory of Gray and his associates is particularly interesting in connection with general persistence theory and has potentially important tie-ins to the developmental work, reported in Chapter 7, relating "off the baseline" habituation to shock and appetitive persistence (Chen & Amsel, 1982). This research addresses the important subject of tolerance and cross-tolerance to stress, and raises the possibility that the phenomena we have called "general persistence effects" and have attributed to the associative mechanism of counterconditioning are in fact nonassociative in origin. Holt and Gray (1983a) describe this work as "deriving directly from a paper by Glazer (1974a) . . . based on Amsel's theory of the generalized PREE" (p. 99), and they relate it specifically to the experiment by Chen and Amsel (1977) reported in Chapter 4. Glazer had shown that rats can be instrumentally conditioned, with appetitive reward, to produce theta (C theta) in the 7.5- to 8.5-Hz band. As we have seen, this is a frequency range identified with anticipatory frustration in Gray's (1970, 1972) theorizing. Furthermore, Glazer showed that such animals, compared with controls conditioned to produce nontheta in the 4- to 5-Hz band (or to a nonconditioned control group), were more resistant to extinction of a subsequent food-rewarded bar-press response.

Whereas both Gray and Glazer interpreted these early results in terms of the counterconditioning mechanism of frustration theory (they could also be accommodated by general persistence theory), the experiment of Chen and Amsel (1977) suggested to Holt and Gray another possible interpretation: that repeated direct elicitations of 7.7-Hz theta exert a proactive nonassociative effect on resistance to extinction and on toleration of stress in general. In a series of experiments, Holt and Gray (1983a,b, 1985) have shown that septal driving of the hippocampal theta rhythm in this band does produce long-term proactive effects on resistance to extinction and to punishment. They suggest that if theta driving produces its effects nonassociatively, then perhaps anticipatory frustration does the same – "toughening-up," they call it. Whether the ultimate explanation of this kind of result is associative, nonassociative, or both, it is seen by Gray and his colleagues (e.g., Williams, Gray, Snape, & Holt, 1989) as central to our understanding not only of the management of habituation to or tolerance for a particular kind of stress, but of cross-tolerance for stress, a general dispositional characteristic of the individual.

If we now return to a consideration of the evidence in the rat pup that the earliest appearance of hippocampal theta is at 11 to 12 days of age and that this coincides with or precedes the age at which the PREE is first observed, it becomes obvious that an important next step would be to initiate a developmental investigation of the proactive effects shown in adults by Holt and Gray. The two central questions would be: (a) Does the buildup of such proactive effects follow an ontogenetic sequence in infancy (starting, perhaps, in the 11- to 12-day range)? (b) Are these effects long lasting enough to survive and affect appetitive persistence and general tolerance to stress in adulthood? That is to say, does such off-the-baseline "toughening-up" treatment in infancy affect dispositional learning and the adult "personality"?

Sensory evoked potentials

Other compelling evidence, based on evoked potentials, for the involvement of the hippocampal formation in schedules of reward and frustrative nonreward comes from two sets of experiments by Deadwyler, West, and Lynch (1979a,b). They recorded sensory (auditory) evoked potentials (AEPs) from an electrode in the perforant-path synaptic zone of the dentate molecular layer in adult rats during differential operant conditioning (see Figure 8.1B). The AEPs did not occur in response to a single auditory stimulus and were not present before discriminative responding; but the AEPs developed during the acquisition of discriminative behavior and disappeared in extinction (Deadwyler et al., 1979a). During the formation of the discrimination and in subsequent discrimination reversal (Deadwyler et al., 1979b), potentials from the perforant-path zone were the same in response to positive $(S+)$ and negative $(S-)$ tones. When the electrode was subsequently lowered into the granule cell layer, the neurons did not respond to a tone signaling the water reward $(S+)$; but if a second tone signaling nonreward $(S-)$ was added, both tones evoked responses in the entorhinal zone and consistently drove the granule cells. In a subsequent phase, as the rats' differential performance improved, the discharge patterns of the granule cells changed, to quote Deadwyler et al., "dramatically": The positive cue elicited a long burst of firing, whereas the negative cue elicited a short burst. On discrimination reversal, the firing patterns were also reversed. During the transition phase of reversal learning, behavioral responding occurred in the presence of both tones, and as the authors point out, "the dentate unit discharge patterns were not statistically distinct and resembled the negative rather than the positive tone unit discharge pattern" (p. 39). The description in these experiments of the stages of nondifferential followed by differential firing of the dentate granule cells in response to $S+$ and $S-$ and their reversal is in striking parallel to the frustration theory analysis of discrimination learning. In terms of our behavioral constructs, it appears to correspond to the last two stages in discrimination learning (and its reversal): In the third (conflict) stage, there is nondifferentiated anticipatory reward (r_R) and anticipatory frustration (r_F) in response to both $S+$ and $S-$, whereas in the fourth stage, responding is differentiated: Anticipatory reward (r_R is elicited to $S+$, and anticipated frustration (r_F) is elicited to S−. Reversal, of course, sets up a PRF-like transitional stage in which r_R and r_F are again elicited nondifferentially, and this is followed, in turn, by differential evokation of r_R and r_F (Amsel, 1962; Amsel & Ward, 1965; see Chapter 5, this volume), analogous to the renewed differential firing of the granule cells.

The behavioral inhibition system

As we have seen, some of the early hypotheses of hippocampal function (e.g., R. J. Douglas, 1967; Douglas & Pribram, 1966) have centered around the idea that "the hippocampus [could be] the site or organ of [Pavlovian] interval inhibition, ... might function to inhibit stimulus–response bonds" and might serve a specific gating function "which acts to inhibit reception of specific stimuli which have been associated with nonreinforcement" (R. J. Douglas, 1967, pp. 435–6). Such a gating function is also a feature of the working/reference-memory hypothesis (Olton, Becker, & Handlemann, 1979), in that irrelevant stimuli are filtered out whereas pertinent stimuli are not. Damage to the hippocampal formation should then result in a weakening of capacity to withhold a response or, in a working-memory paradigm, should lead to revisiting arms in a radial-arm maze. The essential features of Douglas's model have been elaborated in a neurological model, the *behavioral inhibition system* (BIS), a set of hypotheses relating inhibition and persistence to the hippocampus and related structures (Gray, 1982; Gray, Feldon, Rawlins, Owen, & McNaughton, 1978). A feature of this theory is that it lends itself to much more specific predictions than the previous ones, particularly about the effects on the PREE of damage to the various neural components of the BIS.

The BIS and frustration theory

The BIS is in most ways the neurobiological theory that best encompasses and integrates the reward-schedule effects featured in this book and, in general, the mechanisms of dispositional learning. It is a theory that involves septohippocampal function and, more broadly, other limbic system elements – hippocampus proper, dentate gyrus, thalamus, medial and lateral septum, subiculum – and cortical structures, specifically entorhinal, prefrontal, and cingulate cortex. According to Gray's theory, these structures are interconnected in such a way as to produce behavioral inhibition and facilitate behavioral change in the face of conflicting stimulus inputs. Figure 8.3 is a schematic of the theory.

The end product of the BIS action is the comparison of expected and the actual incoming stimuli by the *subicular comparator*. According to this theory, expected stimuli are relayed to the subiculum by the cingulate cortex and the anteroventral thalamus, the former representing motor

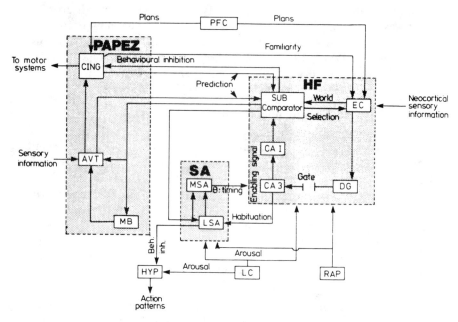

Figure 8.3. A schematic of BIS theory. The three major building blocks are shown in boldface: HF, the hippocampal formation, made up of the entorhinal cortex, EC, the dentate gyrus, DG, CA3, CA1, and the subicular area, SUB; SA, the septal areas, containing the medial and lateral septal areas, MSA and LSA; and the Papez circuit, which receives projections from and returns them to the subicular area via the mammillary bodies, MB, anteroventral thalamus, AVT, and cingulate cortex, CING. Other structures shown are the hypothalamus, HYP, the locus coeruleus, LC, the raphe nuclei, RAP, and the prefrontal cortex, PFC. Arrows show direction of projection; the projection from SUB to LSA lacks anatomical confirmation. Words in lower case show postulated functions; beh. inh., behavioral inhibition. (Figure and caption from Gray, 1982; reprinted by permission of Oxford University Press.)

plans and the latter representing sensory input. The merging of these two inputs accommodates the fact that future movements (motor plans) are affected by incoming stimuli. Processed information about current sensory stimulation comes to the subiculum via the entorhinal cortex, which receives sensory input from all sensory modalities. The BIS, according to Gray, is normally in a "checking" mode. Only when a mismatch occurs does the system take control of behavior, to halt the ongoing response and initiate a "search" for alternative responses. In comparing actual with expected stimuli, the subiculum can be more efficient if unimportant stimuli can be filtered out. This filtering is done in the hippocampus, which serves as a kind of secondary comparator in the system. Through the hippocampus and a septal loop, irrelevant information is filtered out. This septohippocampal loop is central to the BIS and, as we shall see, is important in the expression of the PREE.

The septohippocampal loop is presented by the theory as an "importance detector." Instead of filtering out irrelevant information, it tags information that is important. This "tagging" process begins in the dentate gyrus, which receives the same sensory information as the subiculum from the entorhinal cortex. The granule cells serve as a gate of access for this sensory information to the hippocampus proper. The action of this gate is facilitated by inputs from brain areas associated with arousal, such as the raphe nucleus and the locus coeruleus, which provide a mechanism for entry of motivational inputs to the BIS. On arrival at the CA3 neurons from the granule cell gate in the dentate gyrus, the information passes through the septal loop and back to CA3. In this way, the septum receives input about mismatches from the subicular comparator. According to Gray (1982), the septal loop is responsible for behavioral habituation. By virtue of input from CA3 through CA1 to the subiculum, only sensory stimuli tagged as important are compared with expected stimuli. Habituated sensory input is not tagged by the hippocampus. Though BIS theory is more general, there is obvious similarity between its septohippocampal loop and the action of a specific gate in Douglas's theory that blocks the reception of stimuli associated with nonreinforcement.

In its more general aspects, Gray's neurobiological theory of "anxiety," the BIS theory of habituation, retains features of my own behavioral-construct theory of habituation (counterconditioning) and general persistence (Amsel, 1972a) and its more special form, frustration theory and the counterconditioning of anticipatory frustration-produced suppression. As we have seen, counterconditioning can be thought of as a special case of habituation, and according to Gray, the septal habituation loop operates as follows in PRF training: After a number of reward trials, the conditioned expectancy of reward is fed into the subicular comparator through the anteroventral thalamus and the cingulate cortex. For example, during acquisition in a straight alley, the rat exhibits goal-directed behavior so that the stimuli pertinent to reward are "tagged as important" by the hippocampal formation. If the approach response is subsequently not rewarded, this mismatch engages the BIS and frustration results. On subsequent trials, the BIS initiates a "search" for alternative responses (e.g., freezing, retracing), and there are now conflicting expectancies. Information about the mismatch is then sent from the comparator in the subiculum to the septum. Habituation/counterconditioning occurs in the septal loop, renders the feedback stimuli from anticipatory frustration no longer "important," and these stimuli are "ignored" by the subicular comparator. Habituation to such stimuli (counterconditioning, in frustration theory terms) then accounts for the persistent responding of PRF-trained animals in extinction. CRF-trained animals are relatively desistant because, in this case, the septal loop does not play an important role in acquisition. Following CRF training, the subicular comparator detects a mismatch on the first trial of extinction, and the BIS stops the ongoing response and initiates a search for alternatives. In this case, the alternatives are usually freezing or repetitive retracing (extinction). The PREE is a powerful test of Gray's BIS because it requires consideration of all its components.

A rather important difference between BIS theory and frustration theory that is embedded in the preceding paragraph is one that also exists between frustration theory and other theories of the PREE (see Chapter 6): To paraphrase, it lies in statements such as: In PRF training, habituation in the septal loop renders feedback stimuli from anticipatory frustration no longer important, so that they are ignored by the subicular comparator. This suggests a passive process leading to persistence, a kind of neutralizing of the effects of anticipatory frustration in PRF training. All of the behavioral evidence (see Chapter 4) is that *anticipatory frustration (r_F-s_F) is not neutralized*, and that active counterconditioning is a better description than passive habituation for the action of anticipatory frustration after PRF training in the formation of behavioral rituals, in the transfer of persistence effects from one response system to another, and in other respects.

Behavioral tests of the BIS

A number of techniques have been used to assess the effects on the PREE of damage to neural components of the BIS. Electrolytic and aspiration lesions have been quite common (see Gray & McNaughton, 1983, for a review); and more specific excitotoxins that leave fibers *en passage* intact, such as ibotenic acid and kainic acid, are now frequently used (e.g., Coffey et al. 1989; Jarrard, Feldon, Rawlins, Sinden, & Gray, 1986; Sinden, Jarrard, & Gray, 1988). Excitotoxic damage can be confined to the structure in question because only cell bodies are affected. Knife cuts, particularly of the fimbria-fornix are another method of interfering with normal BIS functioning (e.g., Feldon, Rawlins, & Gray, 1985; Rawlins, Feldon, Tonkiss, & Coffey, 1989). These cuts interrupt the septal loop and so, according to BIS theory, should have an effect on the expression of habituation/counterconditioning.

Subicular lesions

Lesions in different components of the BIS should produce differential effects on the PREE. Lesions of the subiculum should damage the basic comparator, the predicted outcome being, at the very least, the attenuation of mismatches between expected and actual inputs. After CRF training, the result should be increased resistance to extinction, because the mismatch, involving expectancy of reward in acquisition and actual nonreward extinction, will not be detected. The effect of subicular lesions on extinction following PRF training is more difficult to rationalize. Attenuation of the comparator in acquisition would seem to predict decreased resistance to extinction after a large number of acquisition trials because the effect of nonreward trials on subsequent persistence would be attenuated; however, like the lesion-CRF condition, if PRF acquisition is now more CRF-like, there is by the same token no comparator to detect a mismatch from acquisition to extinction, and this factor in the lesion-CRF condition would result in the opposite effect: increased resistance to extinction after PRF training following subicular lesions. This latter half of the effect would be a case of persistence without frustration, a case in which PRF-trained animals with subicular lesions might be no more or less persistent than lesioned CRF-trained animals. PRF-trained animals with subicular lesions would then be more persistent than unlesioned CRF counterparts and less persistent than their unlesioned PRF counterparts, but the persistence observed in both PRF conditions would have different derivations – the former from the absence of counterconditioning but failure of the comparator to detect a mismatch from acquisition to extinction, the latter from normal counterconditioning in acquisition and a weakened comparator function in the transition to extinction.

So far the results have been inconsistent. Jarrard et al. (1986) find that ibotenic acid-induced subicular damage, combined with either hippocampal or entorhinal cortex damage, produces increased resistance to extinction (persistence) in CRF-trained animals and decreased resistance to extinction in PRF-trained animals. In a more definitive study, Sinden et al. (1988) applied ibotenic acid only to the subiculum and found no effect on the PREE, a troublesome result for BIS theory. Subsequently, Rawlins and his colleagues (1989) severed two subicular output pathways that course through the septum. They argued that the subiculoaccumbens pathway is critical for the expression of learned persistence and found decreased resistance to extinction after PRF training (CRF performance was not affected) following knife cuts of this pathway. This result conflicts with that of Sinden et al., because the ibotenic acid lesion made in that study would also have effectively attenuated output through the subiculoaccumbens pathway. An unlikely reconciliation of these results is that outputs from the subiculum have conflicting functions, and so by lesioning the entire subiculum, there is a kind of cancellation of effects. Such conflicting results regarding specific extinction patterns under two different schedules of reinforcement reflect the specificity of prediction that a theory with the complexity of the BIS commands.

Hippocampal lesions

According to BIS theory (Gray, 1982), the hippocampal formation is the "importance detector." Only those stimulus inputs that are tagged as important produce a reaction in the subicular comparator. If the hippocampus were sufficiently damaged, nothing would be tagged (or inputs would be tagged incorrectly), and so there would be no basis for the subicular comparator to react differentially to matches and mismatches. Another possibility is that mismatches without counterconditioning could continue to occur in PRF acquisition. In either case, PRF extinction should be rapid. The basis for counterconditioning might exist at the neural level in PRF training if the habituation loop between the septum and hippocampus remained intact, but this would not necessarily show up as persistence at the behavioral level if hippocampal damage prevented the message from being conveyed to the subiculum; mismatches would continue to occur in the subicular comparator, and normal PRF-induced persistence would not occur. In the transition from CRF acquisition to extinction, persistence would increase to the extent that the inhibitory aspect of the BIS was affected.

In the bulk of studies in adult rats on the effects of hippocampal lesions on the PREE, the PREE is abolished after hippocampal aspirations (e.g., Rawlins, Feldon, & Gray, 1980; Rawlins, Feldon, Ursin, & Gray, 1985); and the absence of this effect in these experiments, as predicted from the preceding analysis, is characterized by increased resistance to extinction after CRF training and decreased resistance to extinction after PRF training. We have found the same pattern of results after electrolytic hippocampal lesions in adults, both in a runway apparatus and procedure (Lobaugh, Bootin, & Amsel, 1985) and in a discrete-trial lever-box apparatus (Amsel et al., 1973). The PRF result agrees with the predictions from BIS theory, as does the CRF result, unless one additional factor is involved: If, in fact, the lesion eliminates not only the mechanism for behavioral inhibition (r_F) resulting from mismatches but also the mechanism for reward expectancy (r_R), the decline in performance following PRF training might be expected to be faster, as animals so treated have received fewer rewards and more nonrewards in acquisition than the corresponding CRF group; the lesion-PRF group might actually appear to be more desistant than the lesion-CRF group in extinction. In fact, some of the data (Rawlins et al., 1989) appear to support this suggestion. This recent result is contrary to both the theory (Gray et al., 1978; Gray, 1982) and the aspiration results of Rawlins et al. (1980). The reason for this discrepancy is not clear. Aspiration lesions damage more of the hippocampus than electrolytic lesions do, and may also damage more of the surrounding structures, like the subiculum. As we have seen, damage to more than one structure can be a complicating factor for interpretations in terms of BIS theory (Jarrard et al., 1986).

Septal lesions

The septum is divided into medial and lateral components. It is the lateral septum and its input from the subicular comparator that, according to Gray's model, are the final mediators of habituation/counterconditioning (see Figure 8.3). Septal function is typically disrupted by either septal lesions or knife cuts of the fimbria–fornix, which is made up of fibers coursing between hippocampus and septum. Disruption of septal function by lesion or fimbria–fornix section should prevent counterconditioning or prevent its expression. According to BIS theory, without input from the septum the hippocampus should continue to tag as important stimuli to which habituation has or should have occurred (e.g., conditioned frustration during acquisition on a PRF schedule). Following septal damage both PRF and CRF training should be followed by mismatches during extinction, and rapid extinction in both would be expected under both conditions.

Electrolytic lesions of the medial (Gray, Quintao, & Araujo-Silva, 1972), lateral (Feldon & Gray, 1979a,b) and total septum (Coffey et al., 1989; Henke, 1974), and fimbria–fornix section (Feldon et al., 1985) have all been shown to abolish the PREE, in the form of increased persistence in CRF-trained and decreased persistence in PRF-trained animals. The CRF result may reflect the fact that all of these lesions also sever axon tracts extending from the subiculum through the septum to the nucleus accumbens. Medial septal lesions that spare these axons also spare the PREE (Feldon & Gray, 1979a,b). When those axons are selectively cut, persistence in extinction is reduced in PRF-trained animals and increased following CRF training, and the PREE is abolished (Rawlins et al., 1989). This result is exactly the prediction for the PREE following septal damage. The problem is that this subiculoaccumbens tract, coursing through the septum, has very little to do with septal function per se and everything to do with subicular function. Consequently, the precise role of the septum in learned persistence remains unclear.

Conclusions concerning the BIS

Even in this simplified version of Gray's BIS theory, there are additional problems that should be noted. No particular function is given to region CA1. Information tagged as "important" by CA3 and the septal loop simply passes through CA1 on its way to the subiculum. This seems unlikely, given that the CA1 pyramidals are the cells of choice in investigations of long-term potentiation, a cellular analog of the study of memory. The BIS theory is also not explicit about the function of the dentate gyrus-to-CA3 gate. As I indicated earlier, this gate could have a motivational function, but lacking that, its raison d'être is unclear. What is also unclear is what anatomical structure(s) are involved in mediating the search for alternative responses or how the BIS facilitates this search. One possibility is that alternative responses might involve cortical structures that have been implicated in the storage of habits (Mishkin & Petri, 1984). Finally, there is no reference in BIS theory to the amygdala. The amygdala seems to be important for the expression of primary frustration (the FE; eg., Henke, 1973; Mabry & Peeler, 1972), and it has been implicated in the conditioning of fear (Davis, Hitchcock, & Rosen, 1987), so perhaps it might also be a factor in conditioned frustration. Any such function, however, is missing from the BIS.

The trend of the work on the BIS appears to have become progressively less theoretical. This is especially clear in the recent work with knife cuts of subicular outputs (Rawlins et al., 1989). The goal of the work seems to have evolved toward finding a persistence *center* rather than a persistence *system*. The PREE can certainly be regarded as a particularly powerful behavioral assay of BIS function, since PRF performance is specifically related to different structures and functions of the system. The substrate of the PREE can also be localized, as in the case of Rawlins et al. (1989), to one structure (in this case an output from the subiculum) that mediates persistence. If such a single "persistence center" can be identified, it perhaps reduces the necessity for a more complex neurobiological theory of inhibition and persistence such as the BIS. However, this seems unlikely given the complexity of the hippocampal formation and its interconnections with other structures, including, but not limited to, the septum and the subiculum.

It is clear that because it has ventured to be highly specific in relating structure to function in the extended hippocampal formation, the BIS has pointed the way to a great deal of research on the dispositional mechanisms of inhibition (or suppression) and persistence.

Prenatal and postnatal damage to hippocampus and dispositional learning and memory in infants and adults

There are two distinct ways to study in development the relationship between brain structure and function and the behavior they are thought to control. The first requires a knowledge of the ontogeny of some specific features of brain development, so that day-to-day development of that feature in early life can be related to the day-to-day emergence of the behaviors that that part of the brain is thought to control. An example of this is the case of the ontogeny of the reward-schedule effects detailed in Chapter 7. The second way is to interfere with – presumably retard – those features of brain development and look for appropriate retardation in the development of those same behaviors. The present section of this chapter begins to address the second method.

The burden of the psychobiological and neurobiological evidence led us to begin to examine some of our developmental patterns of learning and memory in relation to degraded hippocampal function due to permanent damage caused by (a) lesions, (b) prenatal and postnatal exposure to ethanol, and (c) X-irradiation. This section is a partial review of the status of this ongoing work. Whereas most of the tests of learning and memory following these hippocampal interventions have been for reversion of function in the infant to weanling-age rat, another objective is to test adults that have suffered hippocampal damage as infants, that is, to test for the possibility of recovery of function.

Two kinds of reversion and recovery of function

There are two quite different kinds of age-related hypotheses of the effects of lesions (or neurotoxins or X-irradiation) on behavior. The first and more familiar one compares the effects of early and late damage on one specific behavioral function. This still-controversial hypothesis is that the functional deficit is smaller, and the recovery of function greater, when the brain damage occurs at an early age than when it occurs later (see Stein, Finger, & Hart, 1983).

The second kind of age-related hypothesis, largely untested, is one that our developmental work, summarized in Table 7.1, allows us to begin to consider. It has to do with whether behavioral effects that appear early in ontogeny are less severely affected by damage to the brain than later-appearing effects. (A corollary of this hypothesis is that the earlier the appearance of a reward-schedule effect in infancy, the easier it is to demonstrate and the more robust it is in adults.) To test this kind of hypothesis, one must first establish a family of learned behavioral effects (a kind of behavioral assay), preferably with some common underlying conceptualization, in which the learned behaviors have a particular order of first appearance in ontogeny. The following specific question can then be asked: Is a learned reward-schedule effect that makes its first appearance earlier in ontogeny (e.g., simple acquisition and extinction; patterned alternation) more or less resistant to the effects of brain damage (a lesion, a neurotoxin, X-irradiation) than one that appears only later (e.g., the PREE; SNC)? Few, if any, studies address the question of how damage to and recovery of a member of such a *family* of learned effects may be related to the age at which that effect matures. (The work of Teitelbaum, Cheng, & Rozin

[1969] on the parallels between development of feeding and recovery of feeding following hypothalmic damage is an example of recovery related to successive developmental stages that are apparently unrelated to learning.)

What follows are four kinds of predictions about the appearance of reward-schedule effects and their relation to hippocampal development and to damage to the hippocampus in infancy. We have begun to test these predictions by the following means: infant hippocampal lesions, prenatal and early postnatal exposure to ethanol, and X-irradiation of the hippocampus over the first 2 weeks of life.

Prediction 1: Interventions producing hippocampal damage *in early infancy* will have a greater effect *in infants* on paradoxical learning, such as the PREE, than on nonparadoxical learning, such as PA. They will also have a greater effect on the later-appearing (e.g., SNC) than on the earlier-appearing (e.g., PREE) paradoxical effects.

Prediction 2: The degree to which these effects are eliminated *in adults* by hippocampal interventions that occurs *in infancy* will inversely recapitulate the order of their appearance in ontogeny; that is, there will be relative sparing of function in adults from damage that occurs in infancy if the function – the behavioral "effect" in our case – occurs early in infancy.

Prediction 3: Hippocampal damage will cause a greater reduction in the size of a reward-schedule effect in adults if that effect is a later-occurring one in ontogeny. (A juxtaposition of my partial review, in the present chapter, of the adult-lesion literature on hippocampal interventions and the results of our developmental behavioral work in Chapter 7 would support this prediction for at least a few of the effects, specifically PA, the PREE, and SNC.)

Prediction 4: The order of recovery of these reward-schedule effects *in adults*, following hippocampal interventions *in adults* that reduce or eliminate these effects, will recapitulate the order of their appearance in ontogeny; that is, an effect that is reduced or eliminated by hippocampal damage in the adult will recover more quickly if the effect appears earlier in infancy.

Experiments involving hippocampal damage

Infant lesions

In the first study of the effects of lesions on the reward-schedule effects in infant rats, we (Lobaugh, Bootin, & Amsel, 1985) investigated the effects of hippocampal lesions made at Day 10 on (a) PA and the PREE at Day 16 and (b) on PA and the PREE in adults; we also examined (c) the effects on PA and the PREE when rats were lesioned and tested as young adults

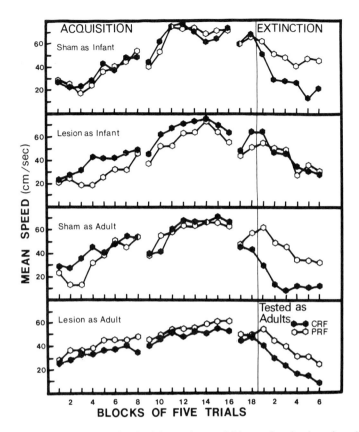

Figure 8.4. Mean running speeds of adult rats in acquisition and extinction after sham or hippocampal lesions made in infancy or in adulthood. (From Lobaugh, Bootin, & Amsel, 1985. Copyright 1985 by the American Psychological Association; reprinted by permission.)

(60 to 65 days). We found, using the runway apparatus already described, with nutritive suckling as the reward and an 8-second ITI, that (a) among subjects lesioned and tested as infants, the lesion had no effect on PA learning, but eliminated the PREE (Prediction 1); (b) among subjects lesioned as infants and tested as adults or (c) lesioned and tested as adults, there was virtually no effect on PA; patterning was just about the same as in the shams. The PREE in adults, however, was eliminated by both the infant and the adult lesions (Figure 8.4; Predictions 2 and 3). The major new finding here is that, at 8-second ITI, two reward-schedule effects that appear a day or two apart in ontogeny are differentially affected by lesions of the hippocampus, the lesion affecting the later, paradoxical effect and not the earlier, nonparadoxical one.

The results of a subsequent experiment with infants (Lobaugh, 1986) suggested, however, that the absence of the PREE following the infant hippocampal lesion may not be an all-or-none effect. This work showed that if the CRF/PRF training was extended to 140 trials (55 trials were given in Lobaugh et al., 1985), a late-developing, weak PREE was seen in infancy following the lesion, because the rate of extinction following CRF training was somewhat greater than it was after the smaller number of acquisition trials in the earlier work; that is to say, a kind of overtraining extinction effect following CRF training appears to occur in these lesioned infant rats tested at about 17 days of age.

There is no other body of evidence from lesions *in infants* with which we can compare these results, though they are not out of line with the results from adults cited earlier. A fair summary statement of the adult-lesion work would be that nonspecific lesions in the septum or hippocampus reduce or eliminate the PREE, usually by increasing persistence in extinction after CRF training and by reducing or having no effect on persistence after PRF training (reviewed by Gray & McNaughton, 1983; Henke, 1974, 1977). These results, and ours with infants, are very much in line with the following long-standing generalization based on adult-lesion work: In one way or another, hippocampal lesions reduce the detection of changed reinforcement contingencies (Ellen & Wilson, 1963; Franchina & Brown, 1971; Jarrard et al., 1964; Kimble & Kimble, 1965; Winocur & Mills, 1969). What our more recent work adds to this picture is that infant lesions affect the PREE in infancy, mainly by increasing persistence after CRF training, reverting the 16-day-old rat pup to the 11-day-old extinction pattern (as we shall see, postnatal X-irradiation appears to have the same effect on the PREE [Diaz-Granados, Greene, & Amsel, 1991b]); that in infants, as in adults (Henke, 1974), this effect of the lesion is attenuated when the acquisition phase is extended; and, perhaps most important, that the infant lesion has a lasting effect on the PREE, showing up, as it does, in adulthood in the adult form – increased persistence after CRF training and reduced persistence after PRF training.

In the earlier-cited experiment on PA by Lobaugh et al. (1985), training was at 8-second ITI. In some recently completed work (Lobaugh, Greene, Grant, Nick, & Amsel, 1989), we showed that, like hippocampal lesions, amygdala lesions on Day 10 have no effect on patterning on Day 16 at 8-second ITI. However, if combined lesions of the hippocampus and amygdala were made, they delayed the appearance of PA, even at an 8-second ITI, but did not eliminate it. Mishkin and his colleagues have shown in monkeys that such combined lesions have a strong effect on what they call the "limbic-dependent memory system," but not (or to a much lesser degree) on the habit system (e.g., Mishkin, Malamut, & Bachevalier, 1984). They have shown that this dissociation also occurs when nonlesioned infant or adult monkeys are compared on the tasks defining these two systems: non-matching-to-sample (memory) and 24-hour concurrent discrimination learning (habit) (Bachevalier & Mishkin, 1984).

We have seen that PA, a kind of memory-based discrimination learning, appears to be a primitive capability that, at an 8-second ITI, does not depend, in infants or adults, on an intact hippocampus or amygdala, but is affected somewhat when both structures are damaged. It had seemed obvious, particularly because of the ease with which it can be demonstrated in infants (who learn it *much faster* than do adults at an 8-second ITI), that PA learning is based on the mechanisms that Schneirla (1959) called "approach and withdrawal" rather than "seeking and avoidance" – the kind of retention system Mishkin and Petri (1984) refer to as involving "knowing how" rather than "knowing that" – the kind of level of functioning that, in our reward-schedule experiments, we call "nonparadoxical" rather than "paradoxical." However, Lobaugh et al. (1989) showed that lesions of the hippocampus alone do retard PA learning when the ITI is extended from 8 to 30 to 60 seconds. (As we shall see, virtually identical results are shown following postnatal exposure to ethanol [Greene, Diaz-Granados, & Amsel, 1992a] and to X-irradiation [Diaz-Granados, Greene, & Amsel, 1992a].) This suggests an involvement of hippocampal function, even in 16-day-olds – that this kind of learning may depend on more than just simple reward and nonreward feedback cues from the previous trial; it may involve a kind of "intermediate-term" memory (Rawlins, 1985), a possible mechanism being the prolongation of the carryover cues by differential conditioned expectancies (r_R and r_F in our notation). Other possible characterizations of such a result in infant rats are that, at very short ITIs, the operating system is "approach and withdrawal," whereas at longer ITIs, it is "seeking and avoidance," and that, in Mishkin's terms, at short ITIs it is "S–R habit formation," whereas at longer ITIs it involves the "memory system."

Prenatal exposure to ethanol

Another means of producing hippocampal (as well as cerebellar and neocortical) damage, in this case to the *prenatally* arising pyramidal cells (see Figure 8.1), is prenatal administration of ethanol (e.g., Barnes & Walker, 1981; Hammer & Scheibel, 1981; Stoltenburg-Didinger & Spohr, 1983; but see Pentney, Cotter, & Abel, 1984). An important consequence of this pyramidal cell agenesis in the hippocampus is that the postnatally developing mossy fibers of the dentate granule cells, most of which terminate in area CA3, are hypertrophied and permanently disarranged in relation to their normal target cells (West, Hodges, & Black, 1981; West & Hodges-Savola, 1983); however, the actual number of granule cells, which are formed primarily postnatally (see Figure 8.2), is not affected (Barnes & Walker, 1981). This finding, that prenatal ethanol administration permanently affects hippocampal structure, raises the possibility that the appearance of the various forms of paradoxical responding may be delayed or reordered in ethanol-treated pups, and that there may be differences in the extent to which there is recovery in these effects in adults.

The literature tells us that there are four main effects of prenatal ethanol on behavior and dispositional learning: (a) hyperactivity, (b) hyperemotionality, (c) abnormalities in perseverance or persistence, and (d) inhibition deficits (the first two of these are often confused). We have been attempting to determine if one or more of these four abnormalities really exist in our infant rats and, if more than one, whether they are "symptoms" reflecting the same mechanism or different ones. In our normal pups, we have seen direct emotional consequences of *primary* frustration as early as 11 days of age, in terms of increased ultrasonic vocalizations to frustrative nonreward (Amsel, Radek, Graham, & Letz, 1977). Loss of persistence (or increased suppression) resulting from exposure to a CRF schedule, and the first appearance of the PREE, is seen in the 12- to 14-day range. (The two other effects of inconsistent reward, the VMREE and the PDREE, show up a few days later.) The more intense suppression ("inhibition"), evidenced by the MREE and SNC, appears, successively, even later in ontogeny. In general terms, then, it appears reasonable to regard delays in, or rearrangement of, the order of the first appearances of a number of reward-schedule effects as indicants of teratogenic effects in the rat pup of exposure to ethanol in utero. Early postnatal exposure in the rat is said to correspond to the third trimester of brain development in the human. Therefore, the information we have acquired on the ontogeny of – and specifically on the transitional periods for – a number of paradoxical effects in rats can, perhaps, serve as a behavioral assay for evaluating the effects of exposure to fetal alcohol in the first two trimesters of pregnancy in humans. The PREE can be taken as an indicant of perseverance or persistence; the MREE and SNC can be taken as indicants of inhibition or suppression. In addition to these paradoxical effects, PA can be taken as an indicant of memory-based learning, and we can employ measures of general activity and ultrasonic cries as indicants of hyperactivity and hyperemotionality, respectively.

In a series of recent experiments, we examined the effects of prenatal exposure to ethanol on PA and on the paradoxical PREE. Testing 15-day-olds exposed to ethanol in utero via a diet fed to the dam throughout gestation, we (Wigal, Lobaugh, Wigal, Greene, & Amsel, 1988) found the following: *At this age,* prenatal ethanol treatment had no effect on the level of general activity in a stabilimeter (no "hyperactivity"). Nor was there at this age any evidence from ultrasounding of differences in distress in this novel environment on first removal from the nest. There was no significant effect of fetal alcohol exposure on primary frustration in infants as measured by ultrasonic cries in the runway on nonrewarded trials, that is, no evidence of "hyperemotionality" *at this age and in this context.*

At 15 days of age, the ethanol treatment had no effect on PA at an 8-

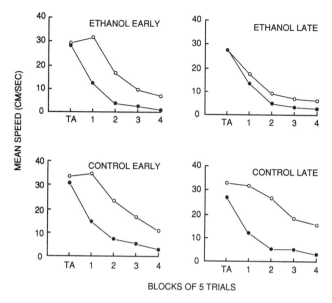

Figure 8.5. Mean running speeds in extinction following conditions of CRF (solid circles) and PRF (open circles) of 15-day-old preweanling rats whose mothers were intubated with ethanol at two different gestational periods: Early, G7 to 9; late, G14 to 16; TA, terminal acquisition. (From Wigal, Greene, & Amsel, 1988. Copyright 1988 by the American Psychological Association; reprinted by permission.)

second ITI, but it had a clear effect on the PREE; in this case, there was an apparent failure of PRF training to induce persistence; that is, the ethanol-exposed PRF infants were less persistent than controls. They reacted to discrepancy, to the onset of extinction, sooner than controls. This result was replicated under conditions of direct intubation of ethanol into the dam's stomach at gestational Days 14 to 16, a period during which there is rapid accretion of hippocampal pyramidal cells. However, when intubation was at gestational Days 7–9, the PREE at 15 days of age was normal (Figure 8.5). The 15-day-old ethanol-exposed infant is in some behavioral respects, then, similar to the normal 11-day-old: At an 8-second ITI, patterning is normal but there is no PREE. However – and this may be important – the ethanol-exposed pup appears to differ from normal 11-day-olds and the hippocampally lesioned 16-day-olds in that, in the latter two preparations, the absence of the PREE reflects persistence after CRF training rather than lack of persistence after PRF training. In the pups that had been exposed to ethanol in utero, the normal increased persistence following PRF training did not occur, and this lack of persistence in extinction appeared to be correlated with an extension during PRF training of the "conflict" stage (Stage 3 in frustration theory). That is to say, there is behavioral variability and urination and defecation in the runway in the

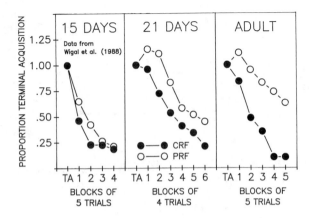

Figure 8.6. Extinction speeds expressed as a proportion of terminal acquisition (TA) for animals prenatally exposed to ethanol and tested at 15 days, 21 days, and as adults. (From Lobaugh, Wigal, Greene, Diaz-Granados, & Amsel, 1991. Reprinted by permission of Elsevier Science Publishing Co., Inc.)

ethanol-exposed pups after the manifestations are no longer seen in controls. Theoretically, there is, concomitant with the extended conflict stage, a failure to reach Stage 4 – a failure of $s_F \rightarrow$ counterconditioning. (Data collected later showed that if rats exposed to ethanol in utero were tested as adults, the PREE was quite normal – indeed, that persistence after PRF training was very strong [Lobaugh, Wigal, Greene, Diaz-Granados, & Amsel, 1991; see also Figure 8.6], suggesting either recovery or maturation of function.)

Postnatal exposure to ethanol and to X-irradiation

One might expect the effects in infancy of postnatal ethanol and X-irradiation to resemble those of hippocampal lesions in infancy (and the effects of exposure to prenatal ethanol in humans, since hippocampal maturity at birth is greater in humans than in rats), and, indeed, this appears to be the case. In this work, we administered ethanol and X-irradiation during the first 2 weeks of postnatal life, the rough equivalent of the third trimester of gestation in humans.

Effects on PA

Postnatal exposure to ethanol from Day 4 to Day 10 has the same effect on PA as infant lesions if the exposure results in a high blood ethanol content, but not if it is low (Greene et al., 1992a); the same is the case for exposure to X-irradiation from Day 2 through Day 15, when a high

Psychobiology of dispositional learning 201

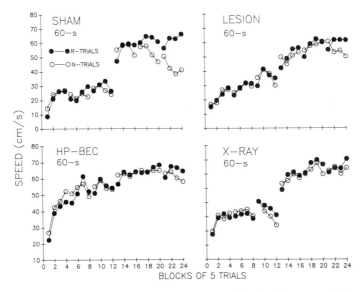

Figure 8.7. Patterned alternation learning at 60-second ITI after hippocampal electrolytic lesions, postnatal exposure to ethanol producing a high-peak blood ethanol content (HP-BEC), and postnatal exposure to X-irradiation. (Data from Diaz-Granados, Greene, & Amsel, 1991a; Greene, Diaz-Granados, & Amsel, 1992; Lobaugh, Greene, Grant, Nick, & Amsel, 1989.)

dose results in an 80% granule cell agenesis (Diaz-Granados et al., 1992a). Plotting the data from these three treatments on the same graph yields Figure 8.7. The effect in all three cases is to greatly retard memory-based (PA) learning in the preweanling rat.

Effects on the PREE

In one experiment involving postnatal exposure to ethanol, pups were exposed either in utero, postnatally, or both. Ethanol was administered postnatally by intubation via intragastric cannulation (see Diaz & Samson, 1980; Hall, 1975); it was administered prenatally via an ethanol-adulterated diet given to the mother. The behavioral results of this experiment from tests for the PREE at weanling age (21 days) are shown in Figure 8.8 (Wigal & Amsel, 1990). First of all, if alcohol is administered only prenatally and the rat is tested at 21 days of age, the effect on persistence following PRF training is somewhat less than that following prenatal administration and testing at 15 days of age (cf. Figure 8.5). If alcohol is administered only postnatally, the PREE is eliminated at 21 days of age; in these weanling rats, extinction is slower after CRF and faster after PRF training, a result much like the effects of lesions in adults and of infant lesions when

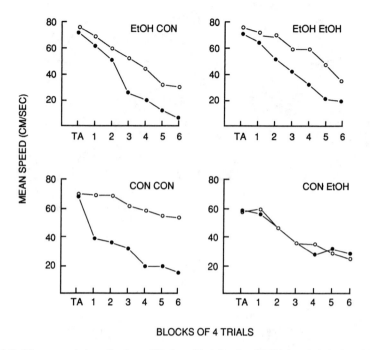

Figure 8.8. Mean speeds in extinction of 21-day-olds following CRF (closed circles) and PRF (open circles) acquisition for the four prenatal/postnatal treatment groups. CON, Control; EtOH, ethanol; TA, terminal acquisition. (From Wigal & Amsel, 1990. Copyright 1990 by the American Psychological Association; reprinted by permission.)

the test for PREE is in adults (cf. Figure 8.4). It does not seem unreasonable that the effect of postnatal exposure to alcohol should produce a behavioral effect closer to a postnatal lesion than to prenatal exposure to alcohol. If the administration of alcohol was both pre- and postnatal, the effect on the PREE at 21 days of age was less than in the postnatal-only condition, as if the prenatal exposure somehow protects or "immunizes" the pup against the postnatal effect. In a recent experiment in which the postnatal ethanol yielded an even higher level of blood alcohol, this combined pre- and postnatal effect was seen again (Greene, Diaz-Granados, & Amsel, 1992b). The latter findings are reminiscent of the many results from studies of one-stage versus two-stage brain damage in which the effect on function of the two-stage or serial lesion is frequently less than the one-stage effect. In one example, entorhinal cortical lesions caused undamaged fibers in the deafferented hippocampus to sprout and form new connections within 4 to 7 days, but when a partial lesion preceded a more complete one by a few days, the comparable axon sprouting occurred in just 2 days (Scheff, Benardo, & Cotman, 1977).

In an experiment on the effects of X-irradiation (Diaz-Granados et al.,

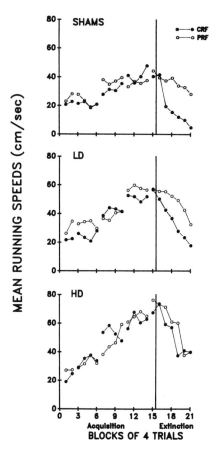

Figure 8.9. Mean speeds in extinction following CRF (closed circles) and PRF (open circles) acquisition at 21 days of age for controls and for pups exposed to high and low levels of X-irradiation from Day 2 to Day 15. (From Diaz-Granados, Greene, & Amsel, 1991b.)

1992b), infant rats were exposed from postnatal Days 2 to 15, a treatment that produces moderate to severe agenesis of the postnatally arising granule cells of the hippocampus depending on dose level. Following a high dose (HD) of irradiation that resulted in about 80% agenesis of cells, rats tested at 21 days of age showed virtual elimination of the PREE compared with shams; following a low dose (LD) the effect was intermediate between HD and the sham condition (Figure 8.9). Again, the absence of the PREE at 21 days of age in the HD condition can be described as being the adult-lesion pattern: reduced persistence in the PRF and increased persistence in the CRF conditions. What is also the case here, as in the PA data, is that absolute running speeds in the irradiated weanlings were considerably higher than in controls, the HD speeds being higher than the LD speeds, suggesting that degree of hyperactivity or suppression deficit is related to the level of cell agenesis from the irradiation treatment.

Summary

In this chapter I began by asking whether either frustration theory or Capaldi's sequential theory, based as they are on the behavior of the adult rat, can provide a consistent account of the order of appearance of the reward-schedule effects in the infant to weanling rat, and concluded that both fell short of doing so in some respects. It was then proposed that clues to this shortcoming might come from attempts to study in parallel the developing brain and the ordering of the behavioral effects. The working hypothesis proposed in this regard was that the infant rat is like the hippocampally damaged adult, and beginning with a brief review of hippocampal morphology, evidence was adduced in support of this hypothesis. This was followed by a brief overview of the work in the adult rat on the effects on dispositional learning of lesions, X-irradiation, alcohol, and drugs; on the relation of hippocampal EEG (theta) to learned persistence; and on hippocampal sensory-evoked potentials during discrimination learning. There followed an extensive discussion of the behavioral inhibition system, a circuit proposed by Gray, following the early accounts of Douglas, Simonov, Vinogradova, and others, to account for the mechanisms of inhibition and persistence, two of the central factors in frustration theory. Behavioral tests of the effects of interventions in the BIS in the form of subicular, hippocampal, and septal lesions were described, and the strengths and weaknesses of this theory were noted.

The last part of the chapter was based on the ontogeny of the reward-schedule effects described in Chapter 7, a body of experimental results that a developmental theory of the BIS (like a developmental frustration theory) must integrate. The point made was that the relationship between the reward-schedule effects and the brain can best be understood in two interlocking ways. One is to study the relationships that exist between the developing behavioral effects and the intact developing brain; the other is to interfere with features of that brain development, particularly those that are implicated in dispositional learning in the adult rat, and to look for predicted retardations in the development of that kind of learning. The chapter closed with a review of the experimental work now being undertaken in this regard – on infant lesions, prenatal and postnatal exposure to ethanol, and postnatal exposure to X-irradiation as they affect PA and the PREE in infant and weanling rats.

9 Summing up: steps in the psychobiological study of related behavioral effects

In Table 1.2 (p. 11), I outlined a six-step strategy that might be followed in an attempt to understand the development of what I have called dispositional learning, as it is studied in the context of the reward-schedule effects. Let us sum up by returning briefly to this strategy and asking to what extent it can be said to have been followed in this book.

Step 1: Observe a number of related effects

The observation and description of a family of related behavioral effects make up a large part of Chapters 3 and 5, are a basis for the developmental work in Chapter 7, and are a feature of the Appendix. The important feature of these reward-schedule effects is that all of them are manipulations of reward and nonreward (or reduced or delayed reward) in some kind of sequence, each of them different from all the others. They are then a family of related effects in this sense, but perhaps more important, also in the sense that each of them is a kind of learning that combines simple classical and instrumental conditioning, that involves many experiences or "trials," and that governs long-term tendencies or dispositions – for example, to approach or avoid, to persist or desist.

Step 2: Develop a conceptualization of the effects

These effects (including discrimination learning) are conceptualized in terms of the body of empirical constructs that comprise frustration theory, and this is the principal subject matter of Chapters 3 and 5. Chapter 4 provides the main body of evidence for the more general theory and for the fact that these effects are very durable, perhaps permanent in the life of the animal. This is the explanatory domain of frustration theory as it applies to dispositional learning and to its long-time retention. The particular long-term dispositions that are involved in discrimination learning – tendencies to form discriminations faster and slower as a function of the nature of earlier exposure to the discriminanda – are shown to be deducible from, and an integral part of, frustration theory. As the review of alternative conceptions in Chapter 6 attempts to make clear, the virtue of our approach relative to others is not simply that it provides a theoretical account of a particular phenomenon (in this case the PREE), but that it integrates a reasonably large number of phenomena; that even in the case of the PREE, it offers mechanisms to account not only for the learning of persistence, but also for its expression in a variety of forms and for its durability in the face of intervening time and events.

Step 3: Study the behavioral effects ontogenetically

The ontogenetic investigation of nine of the reward-schedule effects is, along with other developmental investigations, the subject matter of Chapter 7. The move toward a developmental investigation of the order of appearance of these behavioral effects is now so obvious that, in hindsight, it virtually imposes itself on the theorist. Dispositions are learned (to the extent they are not inherited), and they involve simple classical–instrumental interactions, both of which lead to procedural rather than declarative forms of memory (see Table 1.1). There now appears to be common agreement that this kind of learning begins to occur very early in the life of the infant. Our own work shows that in the rat, between approximately 10 and 25 days of age, there is a fixed sequence of first appearance of the reward-schedule effects ranging from simple acquisition and extinction to discrimination based on feedback stimuli from reward and nonreward (PA), through learned persistence based on a number of intermittent schedules (PREE, VMREE, PDREE), to paradoxical faster extinction following large than small rewards in acquisition (MREE), to strong relative suppression when reward magnitude is shifted from large to small (SNC), and finally to the establishment of patterns of ritualized slow responding (DNC) when the requirement is for the animal to take time for reward to occur.

Step 4: Study the effects in relation to neural substrate

The first part of Chapter 8 is a review of what is known about the neural substrate of the reward-schedule effects and dispositional learning in adult animals. This kind of learning appears to depend on the integrity of the hippocampal formation, taken broadly to include the hippocampus proper and the dentate gyrus, the subicular area, the septum and the septohippocampal connections via fimbria and fornix, and inputs via perforant-path fibers from the entorhinal cortex to the dentate gyrus. These various interconnections define what Gray calls the behavioral inhibition system, and the latter part of this discussion revolves around the experimental work, mainly by Gray and his colleagues, to test the theory of the BIS.

Summing up 207

Step 5: Relate the effects to developing neural substrate

The second part of Chapter 8 moves to material relevant to the development of the neural substrate, and to some recent work on pre- and postnatal interventions in the substrate and their effects on two of the reward-schedule effects in the developing rat pup. These effects, PA and the PREE, define memory-based learning and persistence, respectively, and the work reported suggests that these early hippocampal interventions return the preweanling and weanling infant rat to levels of memory and persistence characteristic of animals at younger infant ages; in some cases the effects of these infant treatments survive in the adult animal, and in others the effects seem not so permanent.

Step 6: Modify the theory

This brings us to Step 6, and to the following question: Does what we have learned about the ontogeny of the reward-schedule effects and their relation to their developing neural substrate offer some basis for a possible modification of frustration theory? Does such a modification take into account developmental changes that occur in the substrate, interventions in its development, and the little that we know about how such changes and interventions affect dispositional learning in the infant and young animal? Because the answer to this question is at best provisional, a necessary caveat is that this theoretical exercise is at best tentative and heuristic. It is derived at present only from the purely developmental (noninvasive) experiments on memory-based learning (PA), on learned persistence in the form of the PREE, and on the few experiments of an invasive kind that we have undertaken. Let us first review the results from the invasive case.

Memory-based learning is clearly affected by a number of hippocampal interventions in the developing infant rat. Infant lesions and postnatal exposure to ethanol and X-irradiation appear to affect the performance of the preweanling rat in similar ways, reducing its capacity in each case to learn a discrimination based on internal, memorial cues. On the basis of the purely behavioral developmental work described in Chapter 7, we know that the hippocampally lesioned, the alcohol-exposed, and the X-irradiated preweanling rat reverts to a stage of development in this respect characteristic of the younger infant animal. In these three cases, the deficit in PA learning in the 17-day-old rat is pronounced at an ITI of 60 seconds. When lesions are made in both hippocampus and amygdala, a deficit is apparent even at an 8-second ITI. These ontogenetic reversions do not appear to include motor deficits in the sense of reduction in speed or vigor of performance. Indeed, most clearly in the case of X-irradiation, there is evidence of increased vigor of performance, of a kind of hyperactivity or distractibility that could perhaps interfere with the formation of the memorial discrimination. What these results also suggest is brought out most clearly in a discussion of our infant-lesion work that relates our results on PA to a distinction in levels of functioning made by several investigators (e.g., Mishkin & Petri, 1984; Wickelgren, 1979.) According to Lobaugh et al. (1989):

> PA discrimination appears to involve processes seen both in simple stimulus-response, habit-system learning and those seen in more complex memory-system learning tasks. PA is similar to habit tasks in that it is a kind of simple, successive discrimination: The pup simply has to learn over several trials that nonreward is a cue for reward on the next trial and vice versa. Olton (1983) has argued that this aspect of PA learning makes it more a reference memory task than a pure working memory task. However, unlike the usual successive (e.g., black–white) discrimination, the cue is not simply "there"; the animal must carry it over to the next trial. Consequently, the outcome of Trial n is necessary for appropriate responding on Trial $n + 1$, and this has the feature of working memory. The relative importance of each process for PA learning in the pup probably depends on ITI and the integrity of limbic structures. In short-ITI PA (8–15 s in infant rats), carryover cues of the emotional or nonemotional sort, acting through the habit system, may be sufficient for response patterning. However, if the carryover cue is weak at the beginning of a trial, because, for example, ITI is too long, the carryover cue may engage activity in the more advanced memory system that involves and depends on the integrity of the hippocampus. A new or modified cue is produced (the constructed cue) that enables the animal to bridge the delay between trials. This system may involve the formation of differential classical conditioned anticipations or expectancies, and particularly, perhaps, anticipatory frustration (Gray, 1982). These constructed cues would provide the basis for the kind of intermediate-term memory store (Rawlins, 1985) that is required for PA learning when the ITI cannot be bridged by carryover cues alone. (p. 1166)

The implication we drew from these results was that PA learning at relatively short ITI and/or at immature or degraded levels of hippocampal function depends on the more primitive, sensory-carryover cue function, whereas at relatively longer ITI, the patterned discrimination is formed only if the carryover cue is supplemented by an expectancy-based memorial cue. We assume that this more advanced memorial cue depends on the integrity of the hippocampal formation, at least as it exists in the preweanling rat. Another way of putting this argument is that the intact rat at an earlier (infant) ontogenetic stage or the hippocampally damaged rat at later (weanling) stage is functioning at a level of "approach and withdrawal," whereas the intact weanling rat is already functioning at a level of "seeking and avoidance" (Schneirla, 1959). The latter level is the one to which the name *memory-based* can properly be applied as a mechanism in PA learning.

We have also seen that the PREE in the infant and adult rat is affected

Summing up 209

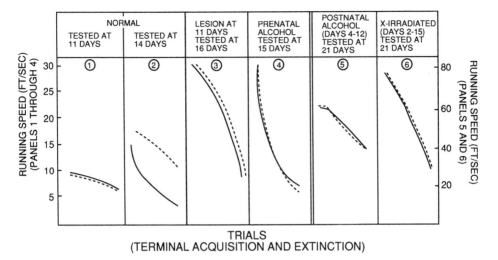

Figure 9.1. Idealized extinction curves showing persistence patterns in normal infant rats (1, 2) at 11 and 14 days of age, (3) at 16 days after lesions at 11 days, (4) at 15 days after exposure to fetal alcohol, (5) at 21 days after postnatal exposure to alcohol, and (6) at 21 days following X-irradiation. Solid lines, CRF; dashed lines, PRF.

by infant hippocampal lesions, prenatal and postnatal exposure to ethanol, and hippocampal X-irradiation in infancy, though in this case not always in the same way. Figure 9.1 presents a partial summary of these differences comparing, in the form of idealized extinction/persistence patterns, the performance of infant rats: normal and tested at 11 and 14 days of age, respectively (Panels 1 and 2), hippocampally lesioned at 11 and tested at 16 days of age (Panel 3), prenatally exposed to ethanol and tested at 15 days of age (Panel 4), postnatally exposed to alcohol and tested at 21 days of age (Panel 5), and postnatally exposed to X-irradiation and tested at 21 days of age (Panel 6). It is easy to jump to the conclusion that the differences in the form of these results after lesions, after fetal and postnatal alcohol, and after postnatal exposure to radiation depend somehow on the fact that the treatments are applied at different stages of development and/or that the different interventions result in different kinds of damage: that lesions produce extensive, generalized postnatal hippocampal damage, that postnatal X-irradiation results in severe and specific granule cell agenesis, and that prenatal and postnatal exposure to alcohol produces more systemic and generalized neurological effects including pyramidal and mature granule cell loss. Why these different kinds of damage should be associated with different forms of the non-PREE is not immediately clear, particularly in the case of the alcohol treatments, in which prenatal exposure is associated with CRF-like extinction at 15 days of age after both CRF and PRF acquisition, whereas after postnatal exposure the somewhat intermediate, adult-like form shows up at 21 days of age.

Nevertheless, a comparison of the data from the noninvasive developmental work in the PREE (Panels 1 and 2 of Figure 9.1), from hippocampal lesions (Panel 3), from the effects of fetal and postnatal alcohol (Panels 4 and 5), and from X-irradiation (Panel 6) suggests the possibility that the current version of the frustration theory of persistence, which was developed out of data from the adult, may not apply without modification to the infant and weanling, at least in the case of altricial mammals like the rat. It now seems possible to advance, quite tentatively, a more inclusive, developmental version of the theory that, in hindsight, is consistent with earlier experimental results (Amsel & Chen, 1976), which showed that there is a PREE at every age in rats from 17 to 65 days but that the *level* of persistence after PRF training, and to a lesser extent after CRF training, gradually declines as the animal gets older (see Figure 7.7, p. 151).

The premise on which the frustration theory of 1958/1962 and most other theories of the PREE were based was that, in adult rats at least (but also in people), nonpersistence (rapid extinction) was the normal or innate state of affairs and that the disposition to persist was learned as a consequence of PRF training – that is, that persistence developed because of a history or schedule of inconsistent reward. Recall that the frustration theory of the PREE postulated the following sequence of stages in the acquisition of persistence: (1) the formation of a conditioned reward expectancy during early PRF training; (2) the emergence during later PRF training of primary (unconditioned) frustration as a result of the occurrence of nonreward in the presence of that reward expectancy; (3) the conditioning of anticipated frustration on the basis of nonreward and primary frustration as the US and UR, respectively, with resultant conflict between $s_R \rightarrow$ approach and $s_F \rightarrow$ avoidance, reflected in increased emotionality and variability in behavior; and (4) the counterconditioning of approach or continued responding (persistence) to the anticipatory-frustration-produced stimuli (s_F).

Based on the rather extensive, purely behavioral developmental work reviewed in Chapter 7 and the more limited developmental psychobiological work reviewed in Chapter 8, both selectively summarized in Figure 9.1, it is now tempting to speculate that there is a kind of inborn persistence, that the mechanisms through which persistence is acquired via PRF training in adults may also serve to *preserve* or *reinstate* the inborn persistence that already exists in infancy and appears to decline in infant development, at least in the altricial rat. (Because the human infant's hippocampal development at birth is said to be comparable to the rat's after the first two postnatal weeks, inborn persistence may be a characteristic more of the infant rat than of the infant human.) Reference to Figure 9.1 will help to make this point. The normal pattern in the 11-day-old rat is equal and strong resistance to extinction following both PRF and CRF training, but by 14 days of age CRF training weakens persistence. If lesions are made at Day 11, the 16-day-old pup appears to revert to something like the normal 11-day-old pattern of persistence. (As we have seen in Chapter 8, in the adult rat the absence of the PREE following infant or adult lesions [not shown in Figure 9.1] is usually an in-between effect.) Fetal alcohol treatment causes 15-day-olds to exhibit *lack of persistence* after both CRF and PRF training; but as we have seen (Lobaugh et al., 1991; see also Figure 8.6), by Day 21 there is some evidence of the beginnings of persistence after PRF training, and by adulthood the PREE is absolutely full blown. In contrast, postnatal exposure to ethanol (Figure 9.1, Panel 5) is associated at Day 21 with an intermediate form of the non-PREE that resembles the adult effect following lesions, as does exposure to X-irradiation from 2 to 15 days of age.

The earlier conjecture was that these developmental data point to the possibility that the rat, perhaps like other altricial mammals, is born persistent: The 11-day-old infant shows strong and equal resistance to extinction following both CRF and PRF training. Another intriguing possibility, however, is that the infant rat acquires persistence as a consequence of the intermittency of milk letdown in nursing, a kind of partial reinforcement of the suckling response. (After all, we do not test for persistence until 11 days of age.) This kind of intermittency is not a characteristic of nursing in all mammals, for example, in the rabbit and guinea pig, and we have seen that in the guinea pig, at least, there is a normal PREE at 4 days of age. As I argued in the case of PA learning, the rat and other altricial mammals may be operating, in Schneirla's (1959) terms, at birth and for some days later at a level of "approach and withdrawal" (Mishkin & Petri's "habit" level) and only some days after that at a level of "seeking and avoidance" (Mishkin & Petri's "memory" level). The habit level is, of course, a level of functioning that continues in rats and humans throughout life, as in Spence's (1966) demonstration that adult humans function "noncognitively" in eye-blink conditioning, in that extinction is the same after CRF and PRF training, slow in both cases.

Arguing in Gray's (1982) terms, the critical factor for the mature PREE is the level of development of the BIS and the subicular comparator. Immaturity in these systems would obviate the suppression that occurs in Stage 3 of PRF training and the learned persistence that follows in Stage 4. In infant rats there is then no PREE, because only innate persistence in extinction follows both PRF and CRF training. An immature BIS would also affect the expectancy-related suppression required for PA at longer ITI and in hippocampally damaged adults (Rawlins, 1985). If the rat is persistent at birth, the intermittency of early feeding experiences may at least serve to preserve this persistence. (If it is not persistent at birth, the intermittency of feeding may be the cause of persistence seen at 11 days of age.) We know that later in life the intermittency of rewarding and frustrating (or rewarding and other punishing) events is an agent in the learning of persistence, and perhaps also in preserving or reinstating persistence, in specific contexts and even more generally. The infant PREE, when it first emerges, may result from the fact that consistent (CRF) training contributes to a *loss of persistence.* (This was a position taken for the PREE in adult rats some years ago [Gonzalez & Bitterman, 1969] – that the PREE is essentially a form of SNC in the transition from CRF acquisition to extinction relative to the transition from PRF acquisition to extinction.) This early PREE in rat pups, seen for the first time at 12 to 14 days of age, may therefore be a case of what happens after CRF, not PRF, training. The hypothesis would be that the basis for the *loss of persistence,* starting at this age, is *a certain degree of hippocampal maturity that is necessary for the detection of discrepancy.* This maturity provides the basis for the shift from Schneirla's passive withdrawal on the basis of gradually weakened associations to the more precipitous, incentive-mediated avoidance in the face of anticipated frustration. (As we know, however, the more intense suppression that defines SNC is not seen until 2 weeks later.) This would explain why extensive lesions of the hippocampus, even at 11 days of age, have the effect of reversion of function, extending the infant persistence pattern and the infant form of PA learning beyond preweanling age in the rat – and even into adulthood.

There is also the puzzling finding from the fetal alcohol work that if hippocampal damage occurs prenatally, the infant persistence pattern fails to appear at 15 days of age and there is, genuinely, no PREE: Both CRF- and PRF-trained pups are nonpersistent, and extinction is rapid in both cases. (Could it be that prenatal exposure to ethanol makes the suckling experience less PRF-like?) Unlike the postnatal treatment, this implies that fetal alcohol-exposed pups are *more* capable of detecting discrepancy than normal pups, despite the fact that exposure to ethanol in utero causes definite, though perhaps subtle, hippocampal damage (pyramidal cell loss and mossy fiber hypertrophy and disorganization) and that hippocampal damage has been said to induce *deficits* in inhibition or suppression. The rapid extinction after PRF training of fetal alcohol-exposed pups surely does not look like such a deficit. If, however, we do set aside the prenatal result as an anomaly and deal only with the results of postnatal treatments as representing a model of third-trimester exposure in the human fetus, we can proceed with the speculative analysis.

In the human, fetal and perinatal exposure to alcohol and to X-irradiation, along with exposure to lead and other neurotoxins, are kinds of treatments that lead to what Altman (1986) calls "microneuronal hypoplasia." One of the possible pathological outcomes of such microneuronal deficits, according to Altman, is increased distractibility and reaction to novelty, which can take the form of the so-called attention deficit–hyperactivity disorders in children. If the deficits in persistence and in intermediate-term memory we have observed in weanling and preweanling rat pups are indeed symptoms of microneuronal hypoplasia, the agenesis, in Altman's terms, of the hippocampal microneurons and the consequent hypertrophy and disorganization of the normal intrahippocampal postnatal connections, these would obviously interfere with the development of R. J. Douglas's "gating function" and with the system through which messages arrive at Gray's subicular comparator.

The set of hypothetical statements that can at least be entertained, then, is as follows:

1. Persistence in infants and adults very likely depends on two quite different mechanisms and neural substrates.
2. In the normal infant rat, there is innate persistence related to immaturity of the BIS, and CRF training at about 2 weeks of age results in a reduction in persistence, whereas at this age PRF training begins to permit a relative conservation of persistence. (The immature substrate may also be in control of the very remarkable early PA at short ITI, another example of the operation of Schneirla's approach–withdrawal function.)
3. In adults, PRF training appears to return the rat to an earlier ontogenic level by somehow overriding mature hippocampal function, the suppressive action of the mature BIS.
4. The mature function, which permits the detection of a discrepancy (Gray, 1982; Simonov, 1974; Vinogradova, 1975), also promotes rapid extinction following CRF training and is involved in the expectancy-based suppressive effects on nonrewarded trials in the more adult ("seeking and avoidance") mode of memory-based (PA) learning.
5. One way to reduce, or even negate, mature hippocampal function is to damage the hippocampal formation, broadly defined to include, in addition to the hippocampus proper, the dentate gyrus, the septum and its cholinergic afferents and efferents, the subiculum, and the entorhinal cortex. Another way is PRF training, either preceding extinction or PA learning.
6. In adults, maximal persistence in a particular context (or even persistence more generally) and optimal memory-based learning at the longer ITIs depend on an intact hippocampal formation and, in Gray's terms, a functioning BIS suppresses behavior but can be overridden by a history of PRF experience in that particular context.

Concluding comments: in the eyes of the beholder

For the past decade, theorizing and experimentation relating hippocampal function to behavior, and particularly to instrumental learning, have been enormously influenced by a book, published more than a decade ago, by O'Keefe and Nadel (1978). This book, as its title plainly announces, presents a theory that likens hippocampal function to a "cognitive map." One result has been a very great amount of experimental research designed to support or disprove this cognitive-map or spatial-memory theory of hippocampal function. Much of the work has been done in the radial-arm maze of Olton (e.g., Olton & Samuelson, 1976) and the Morris water maze (e.g., Morris, 1981). During this period, one of the main opposing views to the spatial-map hypothesis, particularly in its explanation of radial-arm-maze learning, has been that the hippocampus subserves the function of working memory (e.g., Olton, Becker, & Handlemann, 1979; Olton, Wible, & Shapiro, 1986).

In the recent experimental literature that has stemmed from the foregoing analyses, particularly from the O'Keefe and Nadel position, other theories of hippocampal function, particularly the ones featured in this and the preceding chapter as explanations for what I have called dispositional learning, have been more or less set aside. I refer here to conceptualizations of hippocampal function in such terms as novelty, discrepancy, and habituation (e.g., Simonov, 1974; Vinogradova, 1975); suppression and inhibition (e.g., Altman et al., 1973; R. J. Douglas, 1967; Gray, 1982; D. P. Kimble, 1968); neuromodulatory effects (Isaacson, 1980); changing an ingrained response to a new one (Angevine & Cotman, 1981); and in the spirit of most of these analysts, frustration and generalized persistence (Amsel et al., 1973; Glazer, 1974a). Theory, like beauty, is then in the eye of the beholder.

The interesting – and even instructive – thing about this apparent divergence in theorizing about hippocampal function is that it extends beyond the conceptual level and down to the level of the physical arrangement of the apparatus in the experiments involved. If the hippocampus is a "cognitive map," it cannot subserve the non-map-like functions implied in the latter set of theories: It is difficult to adduce the existence of or deficits in cognitive maps or spatial memory from behavior in simple runways and lever boxes (e.g., in extinction and discrimination reversal or in the ability to suppress responding and take time in a DRL or DNC schedule), which almost invariably involve homogeneous, unchanging environments that are impoverished with respect to spatial cues. Gray and McNaughton (1983) have provided a very extensive account of behavioral effects of septal and hippocampal lesions in experiments of this kind, most of which have no explanation in terms of cognitive mapping and/or working memory. Yet it is equally difficult to adduce suppression, inhibition, frustration, and persistence from experiments involving a radial-arm maze or Morris water maze. These are prime examples of how a conceptualization suggests a test situation, to the likely neglect of some other equally plausible conceptualizations. (An extremely even-handed presentation and discussion

Summing up

of results from these alternative perspectives are provided in Gray, 1982, chap. 6.) It is also possible, however, that the different "traditions" of experiments on hippocampal function actually identify different important hippocampal functions; there is no obvious reason that there be just one.

The other major difference between these two traditions of studying hippocampal function is the one to which I alluded in Chapter 1. This is the difference, in the study of animal models of human function, between a cognitive orientation and one cast in terms of associative and nonassociative factors. My own preference for the latter theoretical approach has been obvious; but perhaps this is because in the study of the role of frustration in dispositional learning, and in a theory that attempts to integrate interpretations of arousal, suppression, persistence, and regression, nonassociative as well as associative factors are important and cognitive maps play no necessary role. In O'Keefe and Nadel's (1978) terms, the "taxon" rather than the "locale" system seems to be the one involved in dispositional learning, and within the "taxon" system, perhaps "guidance" rather than "orientation" is the major factor. On the other side, the arguments in favor of a "locale" system can best be made in the context of what is called "spatial learning," and not, in Schneirla's sense and in my own, in terms of simple "approach and withdrawal" or even nonspatial "seeking and avoidance." It should be fairly obvious, then, as an example of this kind of "ships that pass in the night," why a deficit in adult rats with hippocampal lesions in the ability to swim to a safe location in a tank full of opaque water is not at present accounted for by the same theory that accounts for deficits in the PREE in hippocampectomized infant and adult rats, or in memory-based PA in hippocampectomized infant rats when the ITI goes beyond about 1 minute.

10 Applications to humans: a recapitulation and an addendum

Temperamental characteristics of human beings, to the extent that they are not the result of inheritance, depend largely, if not entirely, on dispositional learning, a kind of learning that is, in large part, the sum of many experiences with reward and frustration, and that occurs without specific awareness. Dispositions in humans that depend on a history of inconsistent reinforcement (intermittent frustration) may be subject to the same etiological principles that, as we have seen, operate in animals. I have taken the opportunity in several places in this book to give examples of how such principles derived from animal research may apply to humans.

It now seems safe to assert, on the basis of research not only with animals but also with humans, that a connection exists between negative emotional states and frustrative events; between patterns of inconsistent reward and frustration (or smaller and larger reward or immediate and delayed reward) for what seems to be the same behavior (or effort) and persistence; and between principles of instrumental learning derived from such schedules of reward and frustration and accounts of helplessness and depression. In this book, there has been an attempt to show that there is now some basis in learning theory, and particularly in frustration theory, for coming closer to a comprehension not only of the conceptual mechanisms of these relationships, but also of their sequence in development and, to an early approximation at least, their neural underpinnings.

As should be very apparent by now, the relationship that has been studied more than any other, not only in laboratory animals but also in humans, is the partial-reinforcement extinction effect (PREE). The effect in itself is simply the statement of an experimental outcome; it describes the fact that relatively persistent responding is produced by reinforcing a particular goal-directed behavior randomly and intermittently rather than on every occasion. The treatment of disordered behavior in humans sometimes involves training to persist in the face of outcomes that are at times negative. It is also important that such training result in persistence that is durable. And because the results of psychotherapy transfer so incompletely, as a rule, to other environments, treatment that promotes generalizability and transfer of persistence becomes crucial. Such persistence and generalized persistence training as a part of therapy in humans is an outgrowth in many respects of the animal-learning literature on the PREE, the durable PREE, and the generalized PREE, and such training has been recommended, and in some cases has even been employed, as a strategy in therapy. At the level of dealing with dispositions in humans, Nation and Massad (1978) have described a partial-reinforcement procedure for reversing learned helplessness and depression, and Nation and Woods (1980) have suggested that the stimulus–response analysis of frustration theory, which has helped to uncover and to integrate a number of reward-schedule phenomena, may find some common ground with cognitively oriented behavior therapies, particularly in cases in which these therapies are directed at disorders that can be reduced to descriptions that entail similar phenomena.[15]

Let us now review some of these descriptions that appear here and there in this volume.

Intermittency of reward and persistence

In order for frustration, fear, or, as we have seen, disruptive events in general to lead to persistence, there must be intermittent reward and nonreward (or intermittent reward and punishment) for the same behavior in approximately the same situation. In Chapter 4, an example was given of the application of these simple principles to consistent or inconsistent reinforcement of crying in infants. When an infant cries and is lifted out of the crib, crying is reinforced. If the infant is picked up sometimes, but not always, this inconsistent treatment will produce a persistent crier: When the parents decide in desperation to let the infant "cry it out," crying will continue for a long time. (In contrast, if crying is always rewarded, it will, of course, be established as a very strong response; however, under these circumstances of continuous reward it will also tend to extinguish very quickly.) If such inconsistent treatment is applied to other features of the infant's and then the child's behavior, if he or she is sometimes rewarded and sometimes frustrated (or punished), the child will tend to persist in the face of frustration or punishment. And it is possible, on the basis of a principle of fear–frustration commonality, that such early inconsistent treatment with regard to frustration will result in later persistence in the face of punishment and vice versa. If we can generalize from the animal model, such instances of transfer of persistence would not be unexpected as a result of inconsistency in infancy and childhood. Of course, whether such persistence would be favorable and adaptive later on could be determined only in relation to specific situations and behaviors. One way or another, however, persistence or desistance learned early in life must have important implications for the developmental study of emotion, temperament, and personality.

A complicating factor of some importance for the application of persistence theory to humans is that the strength of counterconditioning of

---

15. See Nation and Woods (1980) for a particularly compelling analysis of the importance of persistence training in psychotherapy.

$s_F \rightarrow$ approach, the mechanism of persistence, must be thought of as increasing gradually during PRF acquisition as r_F-s_F increases in strength. What this means is that, if the intensity of s_F in PRF acquisition becomes very great, and if all intensities of s_F including the greatest are thereby counterconditioned to the approach response, persistence will be maintained in extinction to all of these s_F intensities. However, what if this were not the case? What if, as the experiment of Traupmann et al. (1973) implies, responding can be counterconditioned to specific s_F intensities? In some cases persistence would be evoked only when anticipatory frustration was weak, whereas in other cases persistence would be manifested even when frustration was intense. Is this the difference between good, "wholesome" persistence and maladaptive stubbornness? What implications follow for therapy, education, child rearing? (These questions obviously can be asked for courage in relation to fear. For example, in training recruits to continue moving forward in the face of fear, what is the level of fear to which the behavior should be counterconditioned?)

When we think of learned persistence in a developmental context, it is interesting to speculate that, in the case of the newborn in not very precocial species, including humans, because sensory capacities increase gradually over the first postnatal days and weeks, the strength or intensity of any repetitive, disruptive, external (or internal) event should also increase gradually, providing the optimal conditions for habituation, counterconditioning in our terms (see Figure 4.7). This being the case, one could argue that consistent and inconsistent treatment in humans during the earliest days and weeks may be of critical importance in acquiring lifelong dispositions of persistence and desistance.

Transfer and generality of persistence

If a disposition to persist can be acquired early in life in relation to frustration or other disruptive stimulation, how general can this persistence become? Is there transfer from one persistence system to another? In other words, if a disposition to persist is learned in one context, can it transfer to another? At the end of Chapter 4, I identified eight experimental categories in which persistence and transfer of persistence have been studied, each category representing some degree of difference between the experimental condition under which persistence was acquired and the conditions to which it transferred. Each of these categories, derived from experiments with laboratory rats, has a counterpart in humans. For present purposes I have rearranged the presentation of these categories under three general headings:

1. *Appetitive persistence and transfer of persistence acquired and tested under conditions of PRF, VMR, or PDR.* These appetitive motivational conditions of inconsistency with respect to reward versus nonreward, large

reward versus reduced reward, or immediate reward versus delay of reward are analogous, respectively, to breast feeding in which there is no milk, a reduced flow of mil, or a delay in its availability. If dispositions to persist or desist are learned in these cases of feeding frustration, will they transfer to frustration tolerance or intolerance in eating situations later in life? For example, would the child whose mother had a reliable, abundant milk supply be the least tolerant of delay in adulthood? The experiment by Ross (1964) suggests transfer of persistence across appetitive conditions, for example, for hunger to thirst. In rats, such transfer occurs even when highly distinctive external cues accompany each of the two appetitive systems.

2. *Transfer of persistence from appetitive to aversive motivation.* In one case the transfer of persistence is from conditions based on appetitive frustration to conditions based on pain and fear – to what we call "courage" in humans (Brown & Wagner, 1964). The question is, will the child who has learned to tolerate and persist in the face of anticipated frustration also be disposed to tolerate and persist in the face of anticipated pain, and vice versa? Another case of transfer of persistence is from approach to an inconsistently given reward to approaching and consuming an aversive taste. This case is represented by the experiment of Chen and Amsel (1980). It provides the mechanisms whereby a human patient can be immunized against the aversiveness of the nausea brought on by the consumption of food shortly before a session of radiation or chemotherapy.

3. *Transfer of persistence across very different situations and responses.* Will persistence acquired on a variable ratio schedule of reinforcement, the kind of schedule that is involved in playing a slot machine, transfer to the rather different schedules of reinforcement that are involved at the craps or black-jack tables of a casino? Will persistence acquired in one or several of these gambling situations transfer to the kind of persistence represented by physical courage on the battlefield? If we can generalize from an animal model of such effects, it is not unreasonable to expect such transfer in humans. Chapter 4 provides outlines of many experiments with animals to this effect, some involving interpolated periods of CRF and/or "vacation" between the two phases (Nation et al., 1980; Wong 1971a,b). Still other animal studies of transsituational and transmotivational persistence described in Chapter 4 are perhaps even more analogous to human persistence: They involve the idea that when effort becomes a generalized component of instrumental behavior, the greater the effort the greater the persistence in the face of anticipated frustration.

We have seen that the studies of suppression, persistence, and regression are the end products of dispositional learning. Arousal, whether or not it takes the form of aggression, is more a consequence of the direct activating effects of primary frustration. We have examined these phenomena in transfer experiments in adults (Chapters 4 and 5), and in more developmental and psychobiological ways (Chapters 7 and 8). An exciting possibility is that, when we study persistence in a developmental manner in animals, we may be uncovering the mechanisms controlling this disposition or characteristic of temperament in humans – mechanisms that govern tendencies to persist or to desist. We may also find that learning such dispositions may be related to occasions, perhaps at sensitive early stages of development, on which goal-directed behaviors occur in the face of frustration, physical punishment, or other kinds of disruptive events.

Regression as a transfer-of-persistence phenomenon

A portion of our work can be interpreted as showing that frustration and persistence are implicated in regression, the return to earlier "successful" modes of behavior – that in a sense, persistence can take the form either of increased resistance to extinction or of transfer of persistence and regression.

How do we get from transfer of persistence to regression? As we have seen, persistence acquired in one situation can have effects in a different situation, involving different responses and different motivational-reward conditions. Another way of putting it is that there can be regression to a mode of persistent behavior learned in the context of earlier PRF acquisition, and this regression can be said to be mediated by the internal feedback cues from anticipated frustration. In this case, *to persist is to regress*. A particularly good example of a disposition to regress is the acquisition of idiosyncratic response rituals, superstitions in the human case. In our experiments these rituals were learned under conditions of discontinuously negatively correlated reinforcement (DNC; Logan, 1960), a condition in which animals must learn to run slowly (actually to take time) to be rewarded. Recall that when these same animals then learn normal, fast responding in a new situation on a CRF schedule and are then exposed to extinction, persisting in extinction in the face of anticipated frustration takes the form of returning to the response ritual that was learned in the first situation – the animal "regresses" to a frustration-based ritual or superstition learned much earlier in a situation different from the current one. In the comparable human case, idiosyncratic rituals are observed in both normal and psychopathological conditions, while superstitions are a part of daily life; some people are more prone to them than others. What this means in the present context is that the feedback cues (s_F) from *anticipatory* (conditioned) frustration (r_F) call forth responses that produced reward intermittently in some earlier, perhaps very different situation. This frustration–regression hypothesis should be differentiated from a frustration–aggression hypothesis, in which frustration usually means *primary* (unconditioned) frustration (R_F) and the controlling cue is its feedback (S_F). The difference is in the nature of the response-evoking stimulation that emerges in the former case out of the anticipation of frustration and in the latter case out of its direct experience.

Frustrative and nonfrustrative persistence

I have claimed that there are at least two kinds of persistence, "frustrative" and "nonfrustrative," and that only the first involves intermittent reinforcement. Frustrative (or "emotional") persistence requires that frustration build up in acquisition, that its anticipatory (conditioned) form be acquired, that it evoke avoidance, and that, finally, there be counterconditioning of this conditioned form to continued responding. (Persistence is relatively weaker in the consistently rewarded condition because acquisition involves little or no frustration; the exposure to frustration is only in extinction.) Relative nonpersistence can be achieved even under PRF conditions by holding frustration down in acquisition with a sedative or tranquilizer (see several references in the Appendix). The effect of the drug in acquisition is to suppress the development of frustration during PRF training, thereby preventing the formation of a connection between anticipatory frustration-produced cues and responding.

Persistence of the nonfrustrative or "phlegmatic" kind can be demonstrated by preventing frustrative excitement in either acquisition or extinction. One example of such nonfrustrative persistence would involve acquisition of a continuously rewarded response followed by extinction under the influence of a sedative or tranquilizer. Persistence in this case is due to failure of frustration to operate as an aversive factor in extinction. Another example of phlegmatic persistence occurs in Terrace's work (e.g., 1963a,b,c; see Chapter 5, this volume), in which discrimination training is carried out under conditions of "fading in" the discriminanda so gradually that no "errors" (actually, frustrations) occur, or where frustration can occur in a discrimination but its effects are reduced by drugs. All of these results have been shown in animals, and there is no reason to suppose they do not have their counterparts in humans, since it is clear there are drugs that reduce frustration, or more generally affect, in people, as they appear to do in animals.

Learned persistence and the family

In Chapter 4 and elsewhere in this book we examined results from between-subjects experiments, a model of how separate but similar organisms are affected by different patterns of reinforcement for the same response, and within-subjects experiments, a model for the development within organisms of different dispositional systems relative to different environmental events with their associated reinforcement contingencies. I offered, as a less abstract example of the between- and within-subjects cases the relationships that exist in an ideal family – for experimental purposes, a mother, a father, and identical-twin children. Each child is then a "subject" in both (a) within- and (b) between-subjects experiments to the extent that (a) each of the twins, separately, is on a different reward–frustration schedule in relation to each of the two parents and (b) the two parents exhibit different patterns of inconsistency in their treatment of the behavior of the two children. The questions this situation raises are obvious: In the *within-subjects* case, can a child learn different patterns of vigor–persistence relationships in response to each of the two parents as differential stimuli and reinforcing agents? And assuming that this can occur, to what extent will persistence learned in relation to the inconsistent-reinforcing personality of (say) the father transfer to the mother, who is more consistent? In contrast, the basic *between-subjects* question is: Can the temporally interacting patterns of consistent and inconsistent reward and frustration provided by mother and father together produce, even in twins, differences in disposition, one reflecting more or less persistence than the other?

No infantile amnesia for dispositional learning

As we saw in Chapter 7, there is a remarkable degree of persistence in PRF- relative to CRF-trained rats tested at preweanling age. In fact, learned persistence in preweanlings was greater than in weanlings or young adolescents. What is more striking, however, is that this persistence was retained into young adulthood, even when a "vacation" and a block of CRF reacquisition trials were interpolated between the original PRF acquisition and the final extinction test.

It is important to compare these results with those from experiments on "infantile amnesia." In almost all of these cases, the amnesia is inferred from lack of retention of conditioned fear in which a particular CS was followed by an electric shock UCS in infancy (see Chapter 7). If we can generalize the difference between these two sets of findings to humans, the message is that, whereas there may be infantile amnesia for memories of specific fears in relation to specific environmental (episodic) stimuli, there appears to be much less amnesia for memories based on dispositional learning; for example, learning to persist is not forgotten. (This difference can also be attributed to one or both of the following: appetitive versus aversive motivation and many learning trials in the appetitive case versus relatively few in the shock conditioning case.)

A corollary of the point just made is that reactions to discrepancy, and particularly anticipated frustration, have been underplayed as major factors in emotional development compared with fear based on electric shock, not only in experimental psychology but also in psychopathology. We have seen that the persistent (perseverative, idiosyncratic, fixated) behavior that is a feature of so many emotional disorders is frustration-related; it is a small step to conclude that the debilitating conflicts that are so frequent in psychopathology arise at least as much out of the counteracting expectations of hope and anticipated disappointment as they do out of fear and anticipated relief, to use Mowrer's (1960) terminology.

Toughening up

In Chapters 7 and 8 we considered a case of dispositional learning that involves the effects of simple habituation to electric shocks on appetitive resistance to extinction, and vice versa. This case is different from the others in that the disruptive stimulation of shock is not introduced in the context of appetitive learning, or even in relation to a specific instrumental response; it is introduced as a simple habituation procedure. The main finding in this case was that, compared with unshocked controls, habituation to shock in the first phase appeared to induce resistance to appetitive extinction, and even to punishment, in the second phase.

An explanation by Gray and his colleagues, based on experiments on septal driving of the hippocampal theta rhythm, is that persistence in this case is a matter of toleration of stress, the operation of simple proactive effects of habituation to punishment. They suggest that if theta driving produces its effects nonassociatively, then perhaps anticipatory frustration does the same – "toughening up," they call it, after a term used by N. E. Miller.

An important next step would be to initiate a developmental investigation of these proactive effects. Whether these effects are determined by the mechanisms of general persistence or by "toughening up," there are three central questions for human development: (a) Does the buildup of such proactive effects follow an ontogenetic sequence in infancy, starting in the rat (say) in the 12- to 14-day range? (b) Are these effects long lasting enough to survive into adulthood and affect persistence and general tolerance to stress? (c) Does such toughening-up treatment at a comparable age in human infancy affect the disposition of adults to persist?

Alcohol, X-irradiation, and reversion of function

Memory-based learning and learned persistence are clearly affected by a number of hippocampal interventions in the developing infant rat. Infant lesions and postnatal exposure to ethanol and to X-irradiation appear to affect the performance of the weanling rat in similar ways. They reduce its capacity in each case to learn discriminations on the basis of internal, memorial cues, and they reduce the effect on persistence of a PRF schedule in appetitive learning. On the basis of the purely behavioral developmental work summarized in Chapter 7 (see Table 7.1), it is known that the alcohol-exposed and X-irradiated preweanling rat reverts to a stage of development characteristic of the 10- to 11-day-old infant animal. These ontogenetic reversions do not appear to include motor deficits in the sense of reduction in speed or vigor of performance. Indeed, most clearly in the case of X-irradiation, there is some evidence of increased vigor of performance, of a kind of hyperactivity or distractability that could perhaps interfere with the formation of the memorial discrimination. In our experiments, we expose the pups to alcohol and X-irradiation during the first 10 to 12 postnatal days. Because the rat's gestational period is 22 days and its brain development at birth is said to be equivalent to the human's at the end of the second trimester, these postnatal days of exposure to alcohol and X-irradiation correspond roughly to equivalent exposure in the third human trimester, the period in pregnancy of most intense brain development. The obvious effects of such treatments on memory and persistence functions in the rat, and the corresponding effects on hippocampal neuroanatomy, therefore suggest corresponding deficits from third-trimester exposure to alcohol or X-irradiation in humans. I expand on this possibility in the section that follows.

Addendum: attention deficit–hyperactivity disorder

A major disorder in childhood and adolescence, attention deficit–hyperactivity disorder (ADHD) has been thought to be related to or to involve the kind of minimal brain damage or dysfunction in infant rats that results from perinatal exposure to ethanol and postnatal exposure to X-irradiation. Results from experiments reported at the end of Chapter 8 have suggested an application of frustration theory and the more general theory of persistence to ADHD, a particular problem in human learning, personality, and psychopathology (Amsel, 1990).[16]

Fetal and postnatal alcohol exposure, X-irradiation, and microneuronal hypoplasia

This work on the effects of postnatal exposure to alcohol and X-irradiation was greatly influenced by work from Altman's laboratory and the concept of *microneuronal hypoplasia* in his animal model of minimal brain dys-

---

16. In the 1990 article, however, I did not have the benefit of the results from experiments on early postnatal alcohol exposure and X-irradiation. Postnatal exposure in the first 2 weeks of life in the rat is, in terms of brain development, the approximate equivalent to exposure in the third trimester of pregnancy in the human fetus.

function (Altman, 1986). Microneuronal hypoplasia, in Altman's terms, represents the agenesis of the smallest neurons (they are distinguished from mesoneurons and macroneurons), which are also the last-developing cells, the "local elements" (interneurons) that connect subcomponents of a single structure. Some of these microneurons are scattered throughout the brain, and some are in discrete layers, the layered kind being the granule cells, which, as we have seen, can be found in the olfactory bulb, cerebellar cortex, and the dentate gyrus of the hippocampal formation of the rat. Altman's extensive discussion of theory and experiments, many of them from his own laboratory, has to do with cerebellar and hippocampal hypoplasia. Much of the actual experimental work from Altman's laboratory involves focal X-irradiation over a prolonged period in infancy, a procedure, adopted in our research (see Chapter 8), that prevents the duplication and maturation of cerebellar and hippocampal granule cells and drastically reduces their number.

Altman summarizes the behavioral findings on this subject as follows: Cerebellar granule cell hypoplasia produces minor motor deficits but no apparent learning disabilities in adult rats; however, it does result in some hyperactivity, which Altman links to the concept of decreased response inhibition. In contrast, adult rats with hippocampal granule cell hypoplasia, in addition to being hyperactive, have learning disabilities. These are characterized for the most part by what appears to be a decreased attentional capacity (increased distractibility), exemplified by deficiencies in spontaneous alternation, in discrimination learning in which the cues are very difficult to detect, and in discrimination reversal. These kinds of deficits in Altman's adult rats correspond, as we have seen, to normal behaviors of hippocampally immature infant rats (Chapter 7), of adult rats with hippocampal lesions (Chapter 8; Gray & McNaughton, 1983), and of weanling rats exposed in infancy to alcohol or X-irradiation (Chapter 8). The deficits are in passive avoidance and (much the same thing) in appetitive instrumental extinction, in discrimination reversal, and in the ability to reverse a spatial response (spontaneous alternation) that is, presumably, in the service of novelty seeking. To emphasize the point, these deficits can be said to comprise a kind of behavioral syndrome implicating attentional deficits that can take the form of a less than normal tendency to persist in goal-seeking behavior.

I have pointed to the similarity of behavioral effects from early postnatal exposure to alcohol and X-irradiation. One possible basis for this similarity is suggested in Chapter 8: There is considerable evidence that postnatal exposure to ethanol in the rat, particularly when the exposure results in high-peak blood ethanol levels, results, as in the case of exposure to X-irradiation, in significant hippocampal cell hypoplasia, in this case involving both pyramidal and granule cells. Although the hypoplasia-producing manipulation in the experimental work reported by Altman (1986) was primarily X-irradiation, he emphasized that minimal brain dysfunction and the accompanying hyperactivity and learning deficits in the human child are more often affected by exposure to alcohol during pregnancy than by any other neurotoxin. Supporting Altman's position that exposure to alcohol is a (perhaps the) major cause of attentional disorders and learning disabilities is a recent experimental study in which 20 children with fetal alcohol syndrome (FAS) or fetal alcohol effect (FAE) were compared with 20 children diagnosed as having ADHD and 20 normal controls. The results were summarized as follows: "Although the children with FAS/FAE are significantly more impaired intellectually, their attentional deficits and behavioral problems are similar to those of children with AD[H]D" (Nanson & Hiscock, 1990, p. 656). Our own work with fetal and postnatal exposure to alcohol is more a model of FAE than of FAS, and perhaps, therefore, closer to being an animal model of ADHD.

ADHD and frustration theory

Virginia Douglas has for some years taken the position that hyperactivity and attentional deficits in children are due not to any deficit in information or cognition, but rather to motivational–emotional distress resulting from the unusual sensitivity of such children to rewards and nonrewards. In a series of publications (Douglas, 1980; Douglas & Parry, 1983; Freibergs & Douglas, 1969; Parry & Douglas, 1983), Douglas and her associates have pointed to the importance of "consistent contingent reinforcers" (Douglas & Parry, 1983, p. 324) in guiding and controlling the behavior of hyperactive children, attributing their impulsivity and attention deficits to the energizing effects of primary frustration (Douglas & Parry, 1983) and, in an earlier paper, to the "activating component of r_F [anticipatory frustration] by raising ... [the child's] level of arousal beyond an optimum limit" (Freibergs & Douglas, 1969, p. 394).

There is nothing in this position, so far as it goes, with which our recent research would cause me to differ. In my view, however, taking another step in the theory provides a clearer differentiation between ADHD children and others: The research on the effects of fetal and postnatal alcohol exposure and postnatal X-irradiation on memory-based learning and persistence in weanling rats, described earlier, emboldened me to speculate that a perhaps more important factor that differentiates hyperactive children from others is that in the presence of intermittent rewards and frustrations (or punishments of other kinds), "frustration tolerance" (i.e., persistence) is not achieved. In my terms, counterconditioning of $s_F \rightarrow$ approach is, at best, slower to occur in the hyperactive child, so that the activating effects of frustration and anticipated frustration, to which Douglas refers, and the subsequent approach–avoidance conflict are not blunted.

This analysis in terms of frustration theory suggests that, as an indicant of ADHD, hyperactivity reflects increased responsiveness to stimulation from anticipatory frustration. It can take the form of increased and prolonged conflict of the kind that exists in Stage 3 of PRF training, which in turn gives rise to a *reduced* tendency to persist in the face of continuous frustrative nonreward (extinction). This amounts to an increase in the tendency to be distracted by inconsistency and a greater difficulty in counterconditioning, that is, a greater difficulty in resolving the resulting conflict in favor of persisting in some ongoing goal-directed behavior.

ADHD and the more general theory of persistence

I have proposed that a factor contributing to hyperactivity is the absence of persistence – that in the case of ADHD, counterconditioning of $s_F \rightarrow$ approach is slow to occur in the context of inconsistency of rewards and frustrations in relation to the same behavior, that the child is left in a state of conflict and high arousal (in Stage 3 rather than Stage 4). However, another, more general basis for hyperactivity and lack of persistence is simply failure of habituation (counterconditioning) to occur to external stimuli that are normally disruptive to ongoing behavior. If the hyperactive child were especially sensitive to disruptive stimuli and if each presentation of this kind of stimulus were perceived as novel, a high level of distractibility would result, and persistence in the face of this distractibility would be compromised. The so-called general theory of persistence (Amsel, 1972a), introduced in Chapter 4 (see Figure 4.5), addresses the simple mechanisms involved in this case.

To summarize the earlier discussion, the more general theory encompasses the frustration theory as a special case, in the sense that persistence may result from the counterconditioning of ongoing behavior to stimuli other than the specific frustration-produced stimuli called for in the theory of the PREE. According to this more general view, persistence develops whenever there is learning to approach or maintain a response in the face of any of a class of stimuli (S_X) that elicit competing-disruptive responses (R_X). When S_X is introduced into a situation (S_O), a response (R_X) is at first evoked and it interferes with the ongoing response (R_O); however, as instrumental counterconditioning of R_O to S_X proceeds, R_O comes to be more dominant and more resistant to disruption by stimuli of the class S_X. Generally speaking, S_Xs can be regarded as a class of stimuli to whose disruptive effects organisms are said to habituate. The general position is that feedback stimuli from anticipated frustration are simply one of these classes of stimuli.

This brings us back to an earlier statement to the effect that a possible mechanism in the symptoms of ADHD is a failure of counterconditioning of ongoing behavior to frustrative or other disruptive stimulation to occur, and that the continued disruption resulting from this failure, with the attendant conflict and heightened arousal, is a basis for hyperactivity and for deficits in attention and persistence in certain children. In the animal model, failures to solve simple discriminations, particularly those based on memory when the ITI is reasonably long (equivalent to PA at 60-second ITI in rats), would also implicate deficits in attention or increased distractibility. Animals exposed to alcohol perinatally or to X-irradiation postnatally appear to manifest just these deficits, along with deficits in persistence after PRF training.

Action of stimulants on hyperactive children

If children with ADHD suffer from deficits in persistence, as these experiments in weanling animals lead one to speculate, why should treatment with stimulants, which have a paradoxically calming effect, have an ameliorating effect on these deficits? If we pursue the hypothesis that under conditions of inconsistent reward and frustration, these children remain in conflict (Stage 3 in frustration theory) longer than normal children do, a reason for the therapeutic effects of stimulant medication suggests itself. As we have seen, in frustration theory terms, Stage 3 in the formation of the PREE is a stage of approach–avoidance conflict (Miller, 1944), which precedes the ultimate formation of persistence if training is continued to Stage 4.

My proposal, then, that a deficiency in persistence is operating in cases of ADHD, is actually influenced by two factors: (a) the animal work reported in Chapter 8 on deficits in memory-based learning and persistence related to fetal and postnatal exposure to alcohol and postnatal exposure to X-irradiation, and (b) an attempt to account in terms of Miller's analysis of approach and avoidance conflict for the paradoxical, calming effects of d-amphetamine or other stimulating drugs, such as methylphenidate (Ritalin), on children with ADHD.

If, as frustration theory holds, the $s_F \rightarrow$ avoidance and $s_R \rightarrow$ approach gradients are both excitatory, it would seem to follow, on first consideration, that a stimulating drug such as amphetamine would be indiscriminate with respect to these gradients, raising the level of both of them. If that were the case, the level of conflict after administration of d-amphetamine (or Ritalin) would not be expected to be reduced; indeed, as Miller's analysis of conflict predicts, increasing the strength of both gradients increases conflict.

But what if the drug's effect is, in fact, differential with respect to the approach and avoidance gradients, raising the level of one more than the other, thereby changing the point of intersection of the gradients, the point of maximum conflict, in relation to the goal? The different possible outcomes of such interactions of approach and avoidance gradients and a stimulant drug or a placebo are depicted in Figure 10.1.

Applications to humans

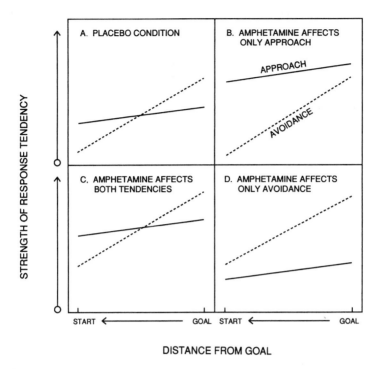

Figure 10.1. Hypothetical gradients of approach and avoidance, based on Miller's (1944) analysis of conflict, as it relates to four alternative hypotheses of the effects of amphetamine-like drugs on conflict and persistence: (A) placebo condition, (B) amphetamine affects the height of only the approach gradient, (C) amphetamine affects the height of both the approach and avoidance gradients, and (D) amphetamine affects the height of only the avoidance gradient.

A. *The placebo condition.* There is no effect on the approach and avoidance gradients. This is a case in which the hyperactive child is left in conflict.

B. *The case of an amphetamine affecting the level of approach but not the avoidance gradient.* In this case, the goal can be reached without strong conflict: There is greater perseveration[17] to the goal. This greater perseveration in the presence of a relatively weaker anticipatory frustration may allow for ultimate counterconditioning – more perseveration leading to more persistence. This outcome would seem to be consistent

---

17. I define *perseveration* as the tendency to complete a task on a given occasion, such as under the special circumstances of the administration of a drug. *Persistence* is defined as a more general disposition to complete tasks, that is, to overcome anticipated frustrations and/or punishments in the completion of a task or approach to a goal.

with the conflict- or hyperactivity-attenuating effect of a stimulant on ADHD.
C. *The case of amphetamine affecting the levels of both the approach and avoidance gradients.* In this case, there would be maximum conflict at the same point in the distance-from-goal space, but the level of conflict would be greater. This outcome would not fit the ameliorating effects of amphetamine on ADHD.
D. *The case of amphetamine affecting the level of only the avoidance gradient.* In this case, the effect of amphetamine is to raise the avoidance gradient, so that conflict between approach and avoidance is absent in the start-to-goal space. Again, this does not fit the effects of amphetamine on ADHD.

The conclusion from this analysis is that Case B would seem to correspond to the effects of d-amphetamine or Ritalin in cases of ADHD if, indeed, the disorder reflects prolonged conflict and a delay in the formation of persistence. The analysis also provides a plausible account for why the effects of stimulating drugs such as amphetamine on ADHD symptoms are only temporary and why, therefore, therapeutic persistence training should be carried out, first, under the drug condition and, gradually, under smaller and smaller doses of the drug.

As I pointed out (Amsel, 1990), some recent evidence of Robbins and his colleagues does not detract from the idea that d-amphetamine, an indirect dopamine agonist, selectively increases positive or approach tendencies (in their terms, "responses providing reinforcers") but has no effect on avoidance tendencies ("responses providing no reinforcers"). The work, which mainly involves operant conditioning procedures, has as one of its major thrusts the investigation of the selective effects of stimulant and nonstimulant drugs on reinforced and nonreinforced responses. In a number of experiments in the past decade, these investigators have shown that amphetamine and pipradrol (but not cocaine or chlordiazepoxide) increase responding that provides conditioned (secondary) positive reinforcers but have no effect or decrease responding for nonreinforcement (Robbins, Watson, Gaskin, & Ennis, 1983); that intracerebral injections of amphetamine, particularly in the nucleus accumbens, produces selective increases in responding in the case of positive, but not negative, reinforcers (Taylor & Robbins, 1984, 1986); that the latter effect is dose dependent (Cador, Robbins, & Everitt, 1989); that food-reinforced tracking of a visual stimulus is facilitated by low and intermediate doses, but not by high doses, of amphetamine; and that the drug produces more switching to the reinforced lever from the nonreinforced lever than vice versa (Evenden & Robbins, 1985).

This and other evidence from Robbins's laboratory is compatible with a generalized raising of the approach, but not the avoidance, gradient by d-amphetamine in Miller's analysis of conflict, and consequently with the idea that amphetamine can reduce or eliminate conflict, making possible counterconditioning of $s_F \to$ approach and thereby promoting increased persistence in children with ADHD.

I have presented a possible application of frustration theory that I think is not intuitively obvious. However, in any attempt to comprehend, on the basis of an animal model, as subtle and as complicated a brain–behavior relationship as is involved in hyperactivity and ADHD in children, we are obviously involved in a case of multiple causation, involving multiple approaches to explanation.

Reprise: the explanatory domain of frustration theory

To return to the theme of the introductory portions of Chapter 3, the experimental study of frustration, not only in animals but also in people, is far from being a new undertaking; it has a very long history, but also a rather unfocused one. This book presents a treatment of frustration in the context of the dynamics of dispositional learning; in such a treatment, *frustration is a factor within the learning itself*. It differs and perhaps goes beyond most previous treatments of frustration, which have tended to be articulated in terms of the consequence of an interaction between something already learned and an external thwarting event. This kind of consequence is at the base of frustration theory and is called *primary frustration*. There is general agreement that this primary kind of frustration has an affective or emotional component and leads, normally, to directed or (more commonly) undirected invigorated behavior. As developed in this book, frustration theory also has a strong Pavlovian flavor; it deals with the consequences of *secondary* and *tertiary* frustration – with the consequences of being exposed to environmental cues that were present during earlier frustration or of anticipating frustration because an ongoing response has led to frustration, consistently or inconsistently, on other occasions. I have summarized the many consequences of these three forms of frustration with the terms *arousal, suppression, persistence,* and *regression* (see Chapter 3); these four terms simply summarize a number of specific behavioral phenomena that can be divided into two kinds of "effects," which I have called nonparadoxical and paradoxical.

Among the nonparadoxical effects of frustration are increased general activity, aggression, escape, invigoration of goal-directed behavior, increased emotionality (in rats, urination, defecation, emission of frustration odor; in infant rats, increased ultrasounding), behavioral suppression and escape, increased response variability, and increased generalized drive level. Among the paradoxical behavioral effects are learned persistence resulting from intermittent reward and nonreward (or reduced or delayed reward), durable and generalized persistence, successive and simultaneous contrast effects, faster extinction as a consequence of larger rewards or a greater number of rewards, and idiosyncratic rituals induced by rewarding responses only if they take time to complete. These are a sample of a larger number of experimental particulars that have been brought more or less under explanatory control or predicted by the small number of first principles that make up frustration theory. A much more inclusive list of such particulars can be found in the Appendix. In this sense, at least, frustration theory fits Hawking's definition of a good theory (p. 122, this volume).

Appendix: some phenomena predicted or explained by frustration theory

This is a partial list of citations in a partial list of categories. In most cases, but not all, the article cites frustration theory as the explanation or a possible explanation of the result. Most of the citations listed are the earliest ones. However, in a few cases later experiments are listed, in addition to or instead of earlier ones, because they represent predictions from frustration theory under special experimental conditions. Recent work discussed in the text is not included.

Frustrative nonreward (FNR)

FNR and general activity	Gallup & Altomari (1969); Dunlap et al. (1971)
FNR and aggression	Gallup (1965); Davis & Donenfeld (1967)
Escape from FNR as reinforcer	Adelman & Maatsch (1955); Daly (1969a, 1974a); Daly & McCroskery (1973)
Invigoration following FNR	Finch (1942); Marzocco (1950); Staddon & Innis, 1969
The double-runway frustration effect (FE)	Amsel & Roussel (1952); Wagner (1959)
FE as a function of time	MacKinnon & Amsel (1964); Davenport et al. (1966); Dunlap et al. (1974)
FE as a function of factors affecting reward size and expectancy	Amsel & Hancock (1957); Amsel et al. (1961); Peckham & Amsel (1967); Daly (1968); Tominga & Imada (1968); Hug (1970b) (but see Jensen & Fallon, 1973)
FE as a function of drive level	McHose & Ludvigson (1964)
FE as a function of number and percentage of rewarded trials	Stimmel & Adams (1969); Yellen (1969); Hug (1970a)
FE in discrimination learning	Amsel & Ward (1965); Hug (1970c); Patten & Latta (1974)
Odors of reward and FNR as cues in learning	McHose & Ludvigson (1966); Ludvigson & Sytsma (1967); Amsel, Hug, & Surridge (1969); Ludvigson, Mathis, & Choquette (1985)

Frustration drive stimulus in selective learning	Amsel & Ward (1954); Amsel & Prouty (1959) (but see Tyler, Marx, & Collier, 1959)

Conditioned (anticipatory) frustration (CAF)

CAF as a suppressor of behavior	Amsel & Surridge (1964)
CAF as a nonspecific energizer	Wagner (1963)
Escape from CAF as reinforcer	Wagner (1963); Daly (1969a); Rilling, Askew, Ahlskog, & Kramer (1969); Terrace (1972)
CAF and response variability in acquisition	Amsel (1958a)
CAF and response variability in extinction	Miller & Stevenson (1936)
CAF and drive level	Cohen (1973)
CAF and magnitude of reward	Amsel & Surridge (1964); Ison & Cook (1964)
CAF in partial or varied reinforcement	Daly (1969b); Ison, Glass, & Daly (1969)

Intermittent reinforcement effects

Partial-reinforcement acquisition effect (PRAE)	Haggard (1959); Goodrich (1959); Wagner (1961); Amsel et al. (1964)
Variability and emotionality in PRF acquisition	Amsel (1958a)
Partial-reinforcement extinction effect (PREE)	Humphreys (1939a); V. F. Sheffield (1949); Rashotte & Surridge (1969)
Partial delay of reinforcement extinction effect (PDREE)	Crum et al. (1951); Donin et al. (1967); Rashotte & Surridge (1969)
Variable magnitude of reinforcement extinction effect (VMREE)	Logan et al. (1956); Chen, Gross, & Amsel (1981)
Factors affecting (generalized or reversed) PREE within subjects	Brown & Logan (1965); Pavlik & Carlton (1965); Amsel et al. (1966); Spear & Pavlik (1966); Boehnert (1973)
PREE as a function of number of nonrewards	Jernstedt (1971)
PREE as a function of reward size	Hulse (1958); Wagner (1961); Jernstedt (1971); Chen, Gross, & Amsel (1981)
Persistence surviving interpolated CRF	Jenkins (1962); Theios (1962); Linden & Hallgren (1973)

Appendix

Durable persistence	Amsel, Wong, & Traupmann (1971); Wong (1977)

Other paradoxical effects of reinforcement

Magnitude of reinforcement extinction effect (MREE) in a runway	Hulse (1958); Armus (1959); Wagner (1961)
Overtraining extinction effect (OEE) in a runway	North & Stimmel (1960); Ison (1962);[18] Theios & Brelsford (1964); Daly (1972)
Successive negative (incentive) contrast (SNC)	Elliott (1928); Black (1968); Crespi (1942); see extensive reviews by Rashotte (1979) and Flaherty (1982)

Discrimination learning (within-subjects) phenomena

Transient positive contrast (induction) in classical conditioning	Pavlov (1927); Senf & Miller (1967); Frey & Ross (1967); Amsel (1971b)
Emotionality in discrimination learning	Amsel & Ward (1965); Terrace (1972)
FE in discrimination formation	Amsel (1962); Amsel & Ward (1965); Nevin & Shettleworth (1966)
Disappearance of FE and emotionality after discrimination is formed	Amsel (1962); Amsel & Ward (1965)
CAF in discrimination learning and reversal	Daly (1971, 1972)
Simultaneous negative contrast (SimNC)	Bower (1961); Gonzalez & Champlin (1974)
No emotionality under "errorless" discrimination	Terrace (1966a, 1972)
Effects of prediscrimination exposure to discriminanda	Amsel (1962); Amsel & Ward (1965); Mandler (1966)
Transient positive (operant) behavioral contrast	Reynolds (1961); Scull, Davies, & Amsel (1970); Gonzalez & Champlin (1974)
Transient and permanent contrast at short intertrial interval (ITI) but not at long ITI	Bloomfield (1967a); Wilton & Clements (1972); Mackintosh, Little, & Lord (1972)
Operant peak-shift effect (OPS)	Hanson (1959); Honig, Thomas, & Guttman (1959); Terrace (1966a)
Magnitude of reinforcement acquisition effect	MacKinnon (1967); McKinnon & Russotti (1967)

18. There are many other studies to this effect, but there are also failures to find it, particularly in operant chambers, but see Senkowski (1978) for a reconciliation of these differences.

PRAE (within-subjects)	Henderson (1966); Amsel et al. (1964, 1966)
Overlearning reversal effect (ORE), with large but not with small reward	Hooper (1967) (see also earlier studies by Reid, 1953; Pubols, 1956)
ORE as a function of amount of discrimination training	Daly (1972)
Differential reward percentage and reward magnitude effects	Henderson (1966); MacKinnon (1967); Galbraith (1971)

Transfer and generality of persistence

From punishment to appetitive extinction and from nonreward to punishment	Miller (1960); Brown & Wagner (1964); Fallon (1968, 1971); Wong (1971a,b); Banks (1973); Linden & Hallgren (1973)
From one motivation/reward system (and one response) to another	Ross (1964); Banks & Torney (1969)
From operant training to runway extinction (and vice versa)	McCuller, Wong, & Amsel (1976) (see also footnote to Lewis, 1964, in Spear & Pavlik, 1966)
From habituation to disruptive stimuli (in context of appetitive acquisition) to extinction	Brown & Bass (1958); Amsel et al. (1973)
From inescapable shock to appetitive extinction	Rosellini & Seligman (1975); Chen & Amsel (1977, 1982)
Idiosyncratic rituals from acquisition to extinction	Rashotte & Amsel (1968); Amsel & Rashotte (1969)
Generalized (within-subjects) PREE	Brown & Logan (1965); Amsel et al. (1966)

Effects of sedative and tranquilizing drugs[19]

Decreased effects of reward reduction	Rosen, Glass, & Ison (1967)
Increased resistance to simple (CRF) extinction	Barry, Wagner, & Miller (1962); Gray (1969); Ison & Pennes (1969)
Reversed PRAE	Wagner (1963)
(a) PREE attenuated	(a) Gray & Dudderidge (1971); Capaldi & Sparling (1971) (see Gray, 1977, for a review)

---

19. For more recent and complete bibliographies, discussions, and summaries of the results in this section and the next, see Gray (1982) and Gray and McNaughton (1983), respectively.

(b) but not FE	(b) Ludvigson (1967); Ison, Daly, & Glass (1967); Gray & Dudderidge (1971)
Errorless and emotionless discrimination under sedatives and tranquilizers	Terrace (1966a)
Reduction or elimination of peak shift under chlorpromazine	Lyons, Klipec, & Eirick (1973); Lyons, Klipec, & Steinsultz (1973)
Disruption of discrimination through increased responding to S−	Wagner (1966); Ison & Rosen (1967)

Relation to brain function

Production of FE with omission of intracranial self-stimulation	Panksepp & Trowill (1968); Johnson et al. (1969)
Enhancement of incentive motivation and FE following septal lesions	Henke (1976)
Mitral cell responses to odors of reward and nonreward (possibly FNR)	Voorhees & Remley (1981)
Reduction of FE and behavioral contrast with lesions of amygdala	Henke (1972); Henke, Allen, & Davison (1972); Mabry & Peeler (1972); Henke (1973a,b, 1977)
Reduction of PREE with lesions of septum or hippocampus	Henke (1974, 1977)
Conditioned frustration, the PREE, and hippocampal theta	Gray (1970, 1972); Gray et al. (1972); Glazer (1972, 1974a, 1974b)
Hippocampal lesion-induced enhancement of operant responding following removal of reinforcement	Swanson & Isaacson (1967)

Mathematical treatments

Arousal-decision model of partial reinforcement	Gray & Smith (1969)
Paradoxical reward effects in adults	Hug (1969); Daly & Daly (1982)
Paradoxical reward effects in developing infants	Daly (1991)
Kinematics of locomotion and the PRAE	Killeen & Amsel (1987)

References

Adelman, H. M., & Maatsch, J. L. (1955). Resistance to extinction as a function of the type of response elicited by frustration. *Journal of Experimental Psychology, 50,* 61–5.
Adelman, H. M., & Rosenbaum, G. (1954). Extinction of instrumental behavior as a function of frustration at various distances from the goal. *Journal of Experimental Psychology, 47,* 429–32.
Altman, J. (1986). An animal model of minimal brain dysfunction. In M. Lewis (Ed.), *Learning disabilities and prenatal risk* (pp. 241–303). Urbana: University of Illinois Press.
Altman, J., & Bayer, S. (1975). Postnatal development of the hippocampal dentate gyrus under normal and experimental conditions. In R. L. Isaacson & K. H. Pribram (Eds.), *The hippocampus: Vol. 1. Structure and development* (pp. 95–119). New York: Plenum.
Altman, J., Brunner, R. L., & Bayer, S. A. (1973). The hippocampus and behavioral maturation. *Behavioral Biology, 8,* 557–96.
Altman, J., & Das, G. D. (1965). Autoradiographic and histological evidence of postnatal hippocampal neurogenesis in rats. *Journal of Comparative Neurology, 124,* 319–36.
Ammon, D., Abramson, C. I., & Bitterman, M. E. (1986). Partial reinforcement and resistance to extinction in honeybees. *Animal Learning & Behavior, 14,* 232–40.
Amsel, A. (1949). Selective association and the anticipatory goal response mechanism as explanatory concepts in learning theory. *Journal of Experimental Psychology, 39,* 785–99.
　(1950a). The combination of a primary appetitional need with primary and secondary emotionally derived needs. *Journal of Experimental Psychology, 40,* 1–14.
　(1950b). The effect upon level of consummatory response of the addition of anxiety to a motivational complex. *Journal of Experimental Psychology, 40,* 709–15.
　(1951, April). *A three-factor theory of inhibition: An addition to Hull's two-factor theory.* Paper delivered at Southern Society for Philosophy and Psychology Meetings, Roanoke.
　(1958a). The role of frustrative nonreward in noncontinuous reward situations. *Psychological Bulletin, 55,* 102–19.
　(1958b). Comment on "Role of prefeeding in an apparent frustration effect." *Journal of Experimental Psychology, 56,* 180–1.
　(1960). Error responses and reinforcement schedules in self-instructional devices. In A. A. Lumsdaine & R. Glaser (Eds.), *Teaching machines and programmed learning* (pp. 505–16). Washington, DC: National Education Association.
　(1961). Hope comes to learning theory [review of O. H. Mowrer's *Learning theory and behavior*]. *Contemporary Psychology, 6,* 33–6.

References

(1962). Frustrative nonreward in partial reinforcement and discrimination learning: Some recent history and a theoretical extension. *Psychological Review, 69,* 306–28.

(1967). Partial reinforcement effects on vigor and persistence: Advances in frustration theory derived from a variety of within-subjects experiments. In K. W. Spence & J. T. Spence (Eds.), *The psychology of learning and motivation: Advances in research and theory* (Vol. 1, pp. 1–65). New York: Academic Press.

(1968). Secondary reinforcement and frustration. *Psychological Bulletin, 10,* 107–8.

(1971a). Frustration, persistence, and regression. In H. D. Kimmel (Ed.), *Experimental psychopathology: Recent research and theory* (pp. 51–69). New York: Academic Press.

(1971b). Positive induction, behavioral contrast, and generalization of inhibition in discrimination learning. In H. H. Kendler & J. T. Spence (Eds.), *Essays in neobehaviorism* (pp. 217–36). New York: Appleton-Century-Crofts.

(1972a). Behavioral habituation, counterconditioning, and a general theory of persistence. In A. H. Black & W. F. Prokasy (Eds.), *Classical conditioning II: Current research and theory* (pp. 409–26). New York: Appleton-Century-Crofts.

(1972b). Inhibition and mediation in classical, Pavlovian and instrumental conditioning. In R. Boakes and S. Halliday (Eds.), *Inhibition and learning* (pp. 275–99). New York: Academic Press.

(1979). The ontogeny of appetitive learning and persistence in the rat. In N. E. Spear & B. A. Campbell (Eds.), *Ontogeny of learning and memory* (pp. 189–224). Hillsdale: Erlbaum.

(1986). Developmental psychobiology and behaviour theory: Reciprocating influences (Daniel E. Berlyne memorial lecture). *Canadian Journal of Psychology, 40,* 311–42.

(1989). *Behaviorism, neobehaviorism, and cognitivism in learning theory: Historical and contemporary perspectives.* Hillsdale, NJ: Erlbaum.

(1990). Arousal, suppression, and persistence: Frustration theory, attention, and its disorders. *Cognition and Emotion, 4,* 239–68.

Amsel, A., Burdette, D. R., & Letz, R. (1976). Appetitive learning, patterned alternation, and extinction in 10-day-old rats with non-lactating suckling as reward. *Nature, 262,* 816–18.

Amsel, A., & Chen, J.-S. (1976). Ontogeny of persistence: Immediate and long-term persistence in rats varying in training age between 17 and 65 days. *Journal of Comparative and Physiological Psychology, 90,* 808–20.

Amsel, A., & Cole, K. F. (1953). Generalization of fear-motivated interference with water intake. *Journal of Experimental Psychology, 46,* 243–7.

Amsel, A., Ernhart, C. B., & Galbrecht, C. R. (1961). Magnitude of frustration effect and strength of antedating goal factors. *Psychological Reports, 8,* 183–6.

Amsel, A., Glazer, H., Lakey, J. R., McCuller, T., & Wong, P. T. P. (1973). Introduction of acoustic stimulation during acquisition and resistance to extinction in the normal and hippocampally-damaged rat. *Journal of Comparative and Physiological Psychology, 84,* 176–86.

Amsel, A., & Hancock, W. (1957). Motivational properties of frustration: III. Relation of frustration effect to antedating goal factors. *Journal of Experimental Psychology, 53,* 126–31.

Amsel, A., Hug, J. J., & Surridge, C. T. (1969). Subject-to-subject trial sequence, odor trails, and patterning at 24-hour ITI. *Psychonomic Science, 15,* 119–20.

Amsel, A., MacKinnon, J. R., Rashotte, M. E., & Surridge, C. T. (1964). Partial reinforcement (acquisition) effects within subjects. *Journal of the Experimental Analysis of Behavior, 7,* 135–8.

Amsel, A., & Maltzman, I. (1950). The effect upon generalized drive strength of emotion-

ality as inferred from the level of consummatory response. *Journal of Experimental Psychology, 40,* 563–9.
Amsel, A., & Prouty, D. (1959). Frustrative factors in selective learning with reward and nonreward as discriminanda. *Journal of Experimental Psychology, 57,* 224–30.
Amsel, A., Radek, C., Graham, M., & Letz, R. (1977). Ultrasound emission in infant rats as an indicant of arousal during appetitive learning and extinction. *Science, 197,* 786–8.
Amsel, A., & Rashotte, M. E. (1969). Transfer of experimenter-imposed slow-response patterns to the extinction of a continuously rewarded response. *Journal of Comparative and Physiological Psychology, 69,* 185–9.
 (1984). *Mechanisms of adaptive behavior: Clark L. Hull's theoretical papers, with commentary.* New York: Columbia University Press.
Amsel, A., Rashotte, M. E., & MacKinnon, J. R. (1966). Partial reinforcement effects within subjects and between subjects. *Psychological Monographs, 80*(20, Whole No. 628).
Amsel, A., & Roussel, J. (1952). Motivational properties of frustration: I. Effect on a running response of the addition of frustration to the motivational complex. *Journal of Experimental Psychology, 43,* 363–8.
Amsel, A., & Stanton, M. (1980). Ontogeny and phylogeny of paradoxical reward effects. In J. S. Rosenblatt, R. A. Hinde, C. Beer, & M. Busnel (Eds.), *Advances in the study of behavior* (pp. 227–74). New York: Academic Press.
Amsel, A., & Surridge, C. T. (1964). The influence of magnitude of reward on the aversive properties of anticipatory frustration. *Canadian Journal of Psychology, 18,* 321–7.
Amsel, A., & Ward, J. S. (1954). Motivational properties of frustration: II. Frustration drive stimulus and frustration reduction in selective learning. *Journal of Experimental Psychology, 48,* 37–47.
 (1965). Frustration and persistence: Resistance to discrimination following prior experience with the discriminanda. *Psychological Monographs, 79*(4, Whole No. 597).
Amsel, A., Wong, P. T. P., & Scull, J. (1971). Transfer of persistence in the domestic chick: Imprinting, punishment and resistance to extinction of a food–reward running response. *Psychonomic Science, 25,* 174–6.
Amsel, A., Wong, P. T. P., & Traupmann, K. L. (1971). Short-term and long-term factors in extinction and durable persistence. *Journal of Experimental Psychology, 90,* 90–5.
Angevine, J. B., Jr., & Cotman, C. W. (1981). *Principles of neuroanatomy.* Oxford: Oxford University Press.
Anisman, H., deCatanzaro, F., & Remington, G. (1978). Escape performance following exposure to inescapable shock: Deficits in motor response maintenance. *Journal of Experimental Psychology: Animal Behavior Processes, 4,* 197–218.
Anokhin, P. K. (1974). *Biology and neurophysiology of the conditioned reflex and its role in adaptative behaviour.* New York: Pergamon.
Armus, H. L. (1959). Effect of magnitude of reinforcement on acquisition and extinction of a running response. *Journal of Experimental Psychology, 58,* 61–3.
Aston-Jones, G., & Bloom, F. E. (1981). Activity of the NE-containing locus coeruleus neurons in behaving rats anticipates fluctuations in the sleep–waking cycle. *Journal of Neuroscience, 1,* 876–86.
Azrin, N. H., Hutchinson, R. R., & Hake, D. F. (1966). Extinction induced aggression. *Journal of the Experimental Analysis of Behavior, 9,* 191–204.
Bachevalier, J., & Mishkin, M. (1984). An early and late developing system for learning and retention in infant monkeys. *Behavioral Neuroscience, 98,* 770–8.

References

Banks, R. K. (1966). Persistence to continuous punishment following intermittent punishment training. *Journal of Experimental Psychology, 71*, 373–7.

(1967). Intermittent punishment effect (IPE) sustained through changed stimulus conditions and through blocks of nonpunished trials. *Journal of Experimental Psychology, 73*, 456–60.

(1973). Generality of persistence: The role of stimulus and response factors in persistence to punishment. *Learning and Motivation, 4*, 218–28.

Banks, R. K., & Torney, D. (1969). Generalization of persistence: The transfer of approach behavior to differing aversive stimuli. *Canadian Journal of Psychology, 23*, 268–73.

Barker, R. G., Dembo, T., & Lewin, K. (1941). Frustration and regression: An experiment in young children. *University of Iowa Studies in Child Welfare, 18*(Whole No. 386).

Barnes, D. E., & Walker, D. W. (1981). Prenatal ethanol exposure permanently reduces the number of pyramidal neurons in rat hippocampus. *Developmental Brain Research, 1*, 333–40.

Barry, H., III, Wagner, A. R., & Miller, N. E. (1962). Effects of alcohol and amobarbital on performance inhibited by experimental extinction. *Journal of Comparative and Physiological Psychology, 55*, 464–8.

Bayer, S. A. (1980a). Development of the hippocampal region in the rat: I. Neurogenesis examined with ^{3}H-thymidine autoradiography. *Journal of Comparative Neurology, 190*, 87–111.

(1980b). Development of the hippocampal region in the rat: II. Morphogenesis during embryonic and early postnatal life. *Journal of Comparative Neurology, 190*, 115–34.

(1982). Changes in the total number of dentate granule cells in juvenile and adult rats: A correlated volumetric and ^{3}H-thymidine autoradiographic study. *Experimental Brain Research, 46*, 315–23.

Becker, H., & Flaherty, C. F. (1983). Chlordiazepoxide and ethanol additivity reduce gustatory negative contrast. *Psychopharmacology, 80*, 35–7.

Bell, R. W. (1974). Ultrasounds in small rodents: Arousal-produced and arousal-producing. *Developmental Psychobiology, 7*, 39–42.

Berger, D. F. (1969). Alternative interpretations of the frustration effect. *Journal of Experimental Psychology, 81*, 475–83.

Berlyne, D. E. (1960). *Conflict, arousal, and curiosity*. New York: McGraw-Hill.

(1964). *Structure and direction in thinking*. New York: Wiley.

(1967). Arousal and reinforcement. In D. Levine (Ed.), *Nebraska symposium on motivation, 1967* (pp. 1–110). Lincoln: University of Nebraska Press.

Bertsch, G. J., & Leitenberg, H. (1970). A "frustration effect" following electric shock. *Learning and Motivation, 1*, 150–6.

Bindra, D. (1969). A unified interpretation of emotion and motivation. *Annals of the New York Academy of Science, 159*, 1071–83.

Birch, D., Ison, J. R., & Sperling, S. E. (1960). Reversal learning under single stimulus presentation. *Journal of Experimental Psychology, 60*, 36–40.

Bitterman, M. E. (1960). Towards a comparative psychology of learning. *American Psychologist, 15*, 704–12.

(1965). Phyletic differences in learning. *American Psychologist, 20*, 396–410.

(1975). The comparative analysis of learning. *Science, 188*, 699–709.

Bitterman, M. E., Fedderson, W. E., & Tyler, D. W. (1953). Secondary reinforcement and the discrimination hypothesis. *American Journal of Psychology, 66*, 456–64.

Black, R. W. (1968). Shifts in magnitude of reward and contrast effects in instrumental and selective learning. *Psychological Review, 75*, 114–26.

Bloomfield, T. M. (1967a). Some temporal properties of behavioral contrast. *Journal of the Experimental Analysis of Behavior, 10,* 159–64.
   (1967b). Frustration, preference, and behavioral contrast. *Quarterly Journal of Experimental Psychology, 19,* 166–9.
Boehnert, J. B. (1973). Extinction of persistence in a within-subject partial reinforcement experiment. *Canadian Journal of Psychology, 27,* 233–46.
Bolles, R. C. (1958). The usefulness of the drive concept. In M. R. Jones (Ed.), *Nebraska symposium on motivation* (Vol. 6, pp. 1–33). Lincoln: University of Nebraska Press.
   (1972). Reinforcement, expectancy, and learning. *Psychological Review, 79,* 394–409.
Bolles, R. C., & Woods, P. J. (1964). The ontogeny of behavior in the albino rat. *Animal Behavior, 12,* 427–41.
Bower, G. H. (1961). A contrast effect in differential conditioning. *Journal of Experimental Psychology, 62,* 196–9.
   (1962). The influence of graded reductions in reward and prior frustrating events upon the magnitude of the frustration effect. *Journal of Comparative and Physiological Psychology, 55,* 582–7.
Boyd, T. L., & Levis, D. J. (1980). Functional depression. In R. Daitzman (Ed.), *Clinical behavior therapy and behavior modification* (pp. 301–50). New York: Garland.
Breuning, S. E., & Wolach, A. H. (1977). Successive negative contrast effects with goldfish (*Carassius auratus*). *Psychological Record, 3,* 565–75.
Brodal, A. (1981). *Neurological anatomy: In relation to clinical medicine* (3rd ed.). Oxford: Oxford University Press.
Bronstein, P. M., Dworkin, T., & Bilder, B. H. (1974). Age-related differences in rats' spontaneous alternation. *Animal Learning & Behavior, 2,* 285–8.
Bronstein, P. M., Neiman, H., Wolkoff, F. D., & Levine, M. J. (1974). The development of habituation in the rat. *Animal Learning & Behavior, 2,* 92–6.
Brown, J. S. (1961). *The motivation of behavior.* New York: McGraw-Hill.
Brown, J. S., & Bass, B. (1958). The acquisition and extinction of an instrumental response under constant and variable stimulus conditions. *Journal of Comparative and Physiological Psychology, 51,* 499–504.
Brown, J. S., & Farber, I. E. (1951). Emotions conceptualized as intervening variables — with suggestions toward a theory of frustration. *Psychological Bulletin, 48,* 465–95.
Brown, R. T., & Logan, F. A. (1965). Generalized partial reinforcement effect. *Journal of Comparative and Physiological Psychology, 60,* 64–9.
Brown, R. T., & Wagner, A. R. (1964). Resistance to punishment and extinction following training with shock or nonreinforcement. *Journal of Experimental Psychology, 68,* 503–7.
Bruce, R. H. (1937). An experimental investigation of the thirst drive in rats with especial reference to the goal-gradient hypothesis. *Journal of General Psychology, 17,* 49–60.
Brunner, R. L., Haggbloom, S. J., & Gazzara, R. A. (1974). Effects of hippocampal X-irradiation-produced granule-cell agenesis on instrumental runway performance in rats. *Physiology and Behavior, 13,* 485–94.
Burdette, D. R., Brake, S., Chen, J.-S., & Amsel, A. (1976). Ontogeny of persistence: Immediate extinction effects in preweanling and weanling rats. *Animal Learning & Behavior, 4,* 131–8.
Burns, R., Woodard, W. T., Henderson, T. B., & Bitterman, M. E. (1974). Simultaneous contrast in the goldfish. *Animal Learning & Behavior, 2,* 97–100.
Bush, R. R., & Mosteller, F. (1951). A mathematical model for simple learning. *Psychological Review, 58,* 313–23.
Cador, M., Robbins, T. W., & Everitt, B. J. (1989). Involvement of the amygdala in stimulus–reward associations: Interaction with the ventral striatum. *Neuroscience, 30,* 77–86.

References

Campbell, B. A. (1967). Developmental studies of learning and motivation in infraprimate mammals. In H. W. Stevenson, E. H. Hess, & H. L. Rheingold (Eds.), *Early behavior: Comparative and developmental approaches* (pp. 43–71). New York: Wiley.
Campbell, B. A., & Coulter, X. (1976). The ontogenesis of learning and memory. In M. R. Rosenzweig & E. L. Bennett (Eds.), *Neural mechanisms of learning and memory* (pp. 209–35). Cambridge, MA: MIT Press.
Campbell, B. A., Lytle, L. D., & Fibiger, H. C. (1969). Ontogeny of adrenergic arousal and cholinergic inhibitory mechanisms in the rat. *Science, 166*, 635–6.
Campbell, B. A., & Mabry, P. D. (1972). Ontogeny of behavioral arousal: A comparative study. *Journal of Comparative and Physiological Psychology, 81*, 371–9.
Campbell, B. A., & Stehouwer, D. J. (1979). Ontogeny of habituation and sensitization in the rat. In N. E. Spear & B. A. Campbell (Eds.), *Ontogeny of learning and memory* (pp. 67–100). Hillsdale, NJ: Erlbaum.
Cannon, W. B. (1932). *The wisdom of the body*. New York: Norton.
Capaldi, E. J. (1966). Partial reinforcement: A hypothesis of sequential effects. *Psychological Review, 73*, 459–77.
  (1967). A sequential hypothesis of instrumental learning. In K. W. Spence & J. T. Spence (Eds.), *The psychology of learning and motivation* (Vol. 1, pp. 67–156). New York: Academic Press.
  (1971). Memory and learning: A sequential viewpoint. In W. K. Honig & H. James (Eds.), *Animal memory* (pp. 111–54). New York: Academic Press.
Capaldi, E. J., & Miller, D. J. (1988a). Counting in rats: Its functional significance and the independent cognitive processes which comprise it. *Journal of Experimental Psychology: Animal Behavior Processes, 14*, 3–17.
  (1988b). The rat's simultaneous anticipation of remote events and current events can be sustained by event memories alone. *Animal Learning & Behavior, 16*, 1–7.
Capaldi, E. J., & Sparling, D. L. (1971). Amobarbital and the partial reinforcement effect in rats: Isolating frustrative control over instrumental responding. *Journal of Comparative and Physiological Psychology, 74*, 467–77.
Capaldi, E. J., & Stevenson, H. W. (1957). Response reversal following different amounts of training. *Journal of Comparative and Physiological Psychology, 50*, 195–8.
Capaldi, E. J., Verry, D. R., Nawrocki, T. M., & Miller, D. J. (1984). Serial learning, interitem associations, phrasing cues, interference, overshadowing, chunking, memory, and extinction. *Animal Learning & Behavior, 12*, 7–20.
Carlton, P. L., & Advokat, C. (1973). Attenuated habituation due to parachlorophenylalanine. *Pharmacology, Biochemistry and Behavior, 1*, 657–63.
Chen, J.-S. (1978). *Mechanisms and limits of durable persistence*. Unpublished doctoral dissertation, University of Texas.
Chen, J.-S., & Amsel, A. (1975). Retention and durability of persistence acquired by young and infant rats. *Journal of Comparative and Physiological Psychology, 89*, 238–45.
  (1977). Prolonged, unsignaled, inescapable shocks increase persistence in subsequent appetitive instrumental learning: General persistence or helplessness? *Animal Learning & Behavior, 5*, 377–85.
  (1980a). Retention under changed-reward conditions of persistence learned by infant rats. *Developmental Psychology, 13*, 469–80.
  (1980b). Learned persistence at 11–12 but not at 10–11 days in infant rats. *Developmental Psychobiology, 13*, 481–91.
  (1980c). Recall versus recognition of taste and immunization against aversive taste based on illness. *Science, 209*, 831–3.
  (1982). Habituation to shock and learned persistence in preweanling, juvenile and adult rats. *Journal of Experimental Psychology: Animal Behavior Processes, 8*, 113–30.

Chen, J.-S., Gross, K., & Amsel, A. (1981). Ontogeny of successive negative contrast and its dissociation from other paradoxical reward effects in preweanling rats. *Journal of Comparative and Physiological Psychology, 95,* 146–59.

Chen, J.-S., Gross, K., Stanton, M., & Amsel, A. (1980). The partial reinforcement acquisition effect in preweanling and juvenile rats. *Bulletin of the Psychonomic Society, 16,* 239–42.

   (1981). Adjustment of weanling and adolescent rats to a reward condition requiring slow responding. *Developmental Psychobiology, 14,* 139–45.

Child, I. L., & Waterhouse, I. K. (1952). Frustration and the quality of performance: I. A critique of the Barker, Dembo, and Lewin experiment. *Psychological Review, 59,* 351–62.

Chomsky, N. (1959). A review of B. F. Skinner's *Verbal Behavior. Language, 35,* 26–58.

Clark, C. V. H., & Isaacson, R. L. (1965). Effect of hippocampal ablation on DRL performance. *Journal of Comparative and Physiological Psychology, 59,* 137–40.

Coffey, P. J., Feldon, J., Mitchell, S., Sinden, J., Gray, J. A., & Rawlins, J. N. P. (1989). Ibotenate-induced total septal lesions reduce resistance to extinction but spare the partial reinforcement extinction effect in the rat. *Experimental Brain Research, 77,* 140–52.

Cogan, D. C., Posey, T. B., & Reeves, J. L. (1976). Response patterning in hippocampectomized rats. *Physiology & Behavior, 16,* 569–76.

Cohen, J. M. (1973). Drive level effects on the conditioning of frustration. *Journal of Experimental Psychology, 98,* 297–301.

Corkin, S. (1968). Acquisition of motor skill after bilateral temporal lobe excisions. *Neuropsychologia, 6,* 225–65.

Cotman, C. W., Taylor, D., & Lynch, G. (1973). Ultrastructural changes in synapses in the dentate gyrus of the rat during development. *Brain Research, 63,* 205–13.

Coulter, X. (1979). The determinants of infantile amnesia. In N. E. Spear & B. A. Campbell (Eds.), *Ontogeny of learning and memory* (pp. 245–70). Hillsdale, NJ: Erlbaum.

Couvillon, P. A., & Bitterman, M. E. (1980). Some phenomena of associative learning in honeybees. *Journal of Comparative and Physiological Psychology, 94,* 878–85.

   (1981). Analysis of alternation patterning in goldfish. *Animal Learning & Behavior, 9,* 169–72.

   (1984). The overlearning-extinction effect and successive negative contrast in honeybees (*Apis mellifera*). *Journal of Comparative Psychology, 98,* 100–109.

   (1985). Effect of experience with a preferred food on consummatory responding for a less preferred food in goldfish. *Animal Learning & Behavior, 13,* 433–8.

Crespi, L. P. (1942). Quantitative variation of incentive and performance in the white rat. *American Journal of Psychology, 55,* 467–517.

Crum, J., Brown, W. L., & Bitterman, M. E. (1951). The effect of partial and delayed reinforcement on resistance to extinction. *American Journal of Psychology, 64,* 228–37.

Dailey, W., Lindner, M., & Amsel, A. (1983). The partial reinforcement extinction effect in the 4–5-day-old guinea pig. *Animal Learning & Behavior, 11,* 337–40.

Daly, H. B. (1968). Excitatory and inhibitory effects of complete and incomplete reward reduction in the double runway. *Journal of Experimental Psychology, 76,* 430–8.

   (1969a). Learning of a hurdle-jump response to escape cues paired with reduced reward or frustrative nonreward. *Journal of Experimental Psychology, 79,* 146–57.

   (1969b). Aversive properties of partial and varied reinforcement during runway acquisition. *Journal of Experimental Psychology, 81,* 54–60.

(1971). Evidence for frustration during discrimination learning. *Journal of Experimental Psychology, 88,* 205–15.

(1972). Hurdle jumping from S+ following discrimination and reversal training: A frustration analysis of the ORE. *Journal of Experimental Psychology, 92,* 332–8.

(1974a). Reinforcing properties of escape from frustration aroused in various learning situations. In G. H. Bower (Ed.), *The psychology of learning and motivation* (Vol. 8, pp. 187–231). New York: Academic Press.

(1974b). Arousal of frustration following gradual reductions in reward magnitude in rats. *Journal of comparative and Physiological Psychology, 86,* 1149–55.

(1991). Changes in learning about aversive nonreward accounts for ontogeny of paradoxical appetitive reward effects in the rat pup: A mathematical model (DMOD) integrates results. *Psychological Bulletin, 109,* 325–39.

Daly, H. B., & Daly, J. T. (1982). A mathematical model of reward and aversive nonreward: Its application in over 30 appetitive learning situations. *Journal of Experimental Psychology: General, 111,* 441–80.

Daly, H. B., & McCroskery, J. H. (1973). Acquisition of a bar-press response to escape frustrative nonreward and reduced reward. *Journal of Experimental Psychology, 98,* 109–12.

Darwin, C. (1859). *The origin of species.* London: John Murray.

Davenport, J. W. (1963). Spatial discrimination and reversal learning based upon differential percentage of reinforcement. *Journal of Comparative and Physiological Psychology, 56,* 1038–43.

Davenport, J. W., Flaherty, C. F., & Dyrud, J. P. (1966). Temporal persistence of frustration effects in monkeys and rats. *Psychonomic Science, 6,* 411–12.

Davenport, J. W., & Thompson, C. I. (1965). The Amsel frustration effect in monkeys. *Psychonomic Science, 3,* 481–2.

Davis, H., & Donenfeld, I. (1967). Extinction induced social interaction in rats. *Psychonomic Science, 7,* 85–6.

Davis, M., Hitchcock, J. M., & Rosen, J. B. (1987). Anxiety and the amygdala: Pharmacological and anatomical analysis of the fear-potentiated startle paradigm. In G. H. Bower (Ed.), *The Psychology of learning and motivation* (Vol. 21, pp. 263–305).

Davis, M., & Wagner, A. R. (1969). Habituation of the startle response under an incremental sequence of stimulus intensities. *Journal of Comparative and Physiological Psychology, 67,* 486–92.

Deadwyler, S. A., West, M., & Lynch, G. (1979a). Synaptically identified hippocampal slow potentials during behavior. *Brain Research, 161,* 211–25.

(1979b). Activity of dentate granule cells during learning: Differentiation of perforant path input. *Brain Research, 169,* 29–43.

Deza, L., & Eidelberg, E. (1967). Development of cortical electrical activity in the rat. *Experimental Neurology, 17,* 425–38.

Diaz, J., & Samson, H. H. (1980). Impaired brain growth in neonatal rats exposed to ethanol. *Science, 208,* 751–3.

Diaz-Granados, J. L., Greene, P. L., & Amsel, A. (1992a). Memory-based learning in the infant rat is affected by x-irradiation-induced hippocampal granule-cell hypoplasia. Unpublished manuscript.

(1992b). Learned persistence in the infant rat is affected by x-irradiation-induced hippocampal granule-cell hypoplasia. Unpublished manuscript.

Dollard, J., Doob, L. W., Miller, N. E., Mowrer, O. H., & Sears, R. R. (1939). *Frustration and aggression.* New Haven, CN: Yale University Press.

Dollard, J., & Miller, N. E. (1950). *Personality and psychotherapy.* New York: McGraw-Hill.

Donin, J. A., Surridge, C. T., & Amsel, A. (1967). Extinction following partial delay of reward with immediate continuous reward interpolated, at 24-hour intertrial intervals. *Journal of Experimental Psychology, 74*, 50–3.

Douglas, R. J. (1967). The hippocampus and behavior. *Psychological Bulletin, 67*, 416–42.

Douglas, R. J., Peterson, J. J., & Douglas, D. P. (1973). The ontogeny of a hippocampus-dependent response in two rodent species. *Behavioral Biology, 8*, 827–37.

Douglas, R. J., & Pribram, K. H. (1966). Learning and limbic lesions. *Neuropsychologia, 4*, 197–220.

Douglas, V. I. (1980). Treatment approaches: Establishing inner or outer control? In C. K. Whalen & B. Henker (Eds), *Hyperactive children: The social ecology of identification and treatment* (pp. 283–317). New York: Academic Press.

Douglas, V. I., & Parry, P. A. (1983). Effects of reward on delayed reaction time task performance of hyperactive children. *Journal of Abnormal Child Psychology, 11*, 313–26.

Drewett, R. F., Statham, C., & Wakerley, J. B. (1974). A quantitative analysis of the feeding behavior of suckling rats. *Animal Behavior, 22*, 907–13.

Dunlap, W. P., Hughes, L. F., Dachowski, L., & O'Brien, T. J. (1974). The temporal course of the frustration effect. *Learning and Motivation, 5*, 484–97.

Dunlap, W. P., Hughes, L. F., O'Brien, T. J., Lewis, J. H., & Dachowski, L. (1971). Goalbox activity as a measure of frustration in a single runway. *Psychonomic Science, 23*, 327–8.

Eisenberger, R., Carlson, J., & Frank, M. (1979). Transfer of persistence to the acquisition of a new behaviour. *Quarterly Journal of Experimental Psychology, 31*, 691–700.

Eisenberger, R., Carlson, J., Guile, M., & Shapiro, N. (1979). Transfer of effort across behaviors. *Learning and Motivation, 10*, 178–97.

Eisenberger, R., Terborg, R., & Carlson, J. (1979). Transfer of persistence across reinforced behaviors. *Animal Learning & Behavior, 7*, 493–8.

Ellen, P., & Wilson, A. S. (1963). Perseveration in the rat following hippocampal lesions. *Experimental Neurology, 8*, 310–17.

Elliot, M. H. (1928). The effect of change of reward on the maze performance of rats. *University of California Publications in Psychology, 4*, 19–30.

Ernst, A. J., Dericco, D., Dempster, J. P., & Neimann, J. (1975). Developmental differences in rats of suppressive effects of extinction as a function of extinction sessions. *Journal of Comparative and Physiological Psychology, 88*, 633–9.

Eskin, R. M., & Bitterman, M. E. (1961). Partial reinforcement in the turtle. *Quarterly Journal of Experimental Psychology, 13*, 112–16.

Estes, W. K. (1950). Toward a statistical theory of learning. *Psychological Review, 57*, 94–107.

(1958). Stimulus–response theory of drive. In M. R. Jones (Eds.), *Nebraska symposium on motivation* (Vol. 6, pp. 35–69). Lincoln: University of Nebraska Press.

(1959). The statistical approach to learning theory. In S. Koch (Ed.), *Psychology: A study of a science* (Vol. 2, pp. 380–491). New York: McGraw-Hill.

Estes, W. K., & Skinner, B. F. (1941). Some quantitative properties of anxiety. *Journal of Experimental Psychology, 29*, 390–400.

Evenden, J. L., & Robbins, T. W. (1985). The effects of d-amphetamine, chlordiazepoxide and alpha-plupenthixol on food-reinforced tracking of a visual stimulus by rats. *Psychopharmacology (Berlin), 85*, 361–6.

Falk, J. L. (1971). The nature and determinants of adjunctive behavior. *Physiology and Behavior, 6*, 577–88.

Fallon, D. (1968). Resistance to extinction following learning with punishment of reinforced and nonreinforced licking. *Journal of Experimental Psychology, 76*, 550–7.

(1971). Increased resistance to extinction following punishment and reward: High frustration tolerance or low frustration magnitude? *Journal of Comparative and Physiological Psychology, 77,* 245–55.
Feigley, D. A., Parsons, P. A., Hamilton, L. W., & Spear, N. E. (1972). Development of habituation to novel environments in the rat. *Journal of Comparative and Physiological Psychology, 79,* 443–52.
Feldon, J., & Gray, J. A. (1979a). Effects of medial and lateral septal lesions on the partial reinforcement extinction effect at one trial a day. *Quarterly Journal of Experimental Psychology, 31,* 653–74.
  (1979b). Effects of medial and lateral septal lesions on the partial reinforcement extinction effect at short inter-trial intervals. *Quarterly Journal of Experimental Psychology, 31,* 675–90.
Feldon, J., Rawlins, J. N. P., & Gray, J. A. (1985). Fimbria-fornix section and the partial reinforcement extinction effect. *Experimental Brain Research, 58,* 435–9.
Festinger, L. (1961). The psychological effects of insufficient rewards. *American Psychologist, 16,* 1–11.
File, S. E., & Scott, E. M. (1976). Acquisition and retention of habituation in the pre-weanling rat. *Developmental Psychobiology, 9,* 97–107.
Finch, G. (1942). Chimpanzee frustration responses. *Psychosomatic Medicine, 4,* 233–51.
Fitzwater, M. E. (1952). The relative effect of reinforcement and nonreinforcement in establishing a form discrimination. *Journal of Comparative and Physiological Psychology, 45,* 476–81.
Flaherty, C. F. (1982). Incentive contrast: A review of behavioral changes following shifts in reward. *Animal Learning & Behavior, 10,* 409–40.
  (In press). *Psychological Reliability.* Cambridge University Press.
Flaherty, C. F., Becker, H. C., & Pohorecky, L. (1985). Correlation of corticosterone elevation and negative contrast varies as a function of postshift day. *Animal Learning & Behavior, 13,* 309–14.
Flaherty, C. F., Capobianco, S., & Hamilton, L. W. (1973). Effects of septal lesions on retention of negative contrast. *Physiology & Behavior, 11,* 625–31.
Flaherty, C. F., Grigson, P. S., & Rowan, G. A. (1986). Chlordiazepoxide and the determinants of negative contrast. *Animal Learning & Behavior, 14,* 315–21.
Franchina, J. J., & Brown, T. S. (1970). Response patterning and extinction in rats with hippocampal lesions. *Journal of Comparative and Physiological Psychology, 70,* 66–72.
  (1971). Reward magnitude shift effects in rats with hippocampal lesions. *Journal of Comparative and Physiological Psychology, 76,* 365–70.
Freibergs, V., & Douglas, V. I. (1969). Concept learning in hyperactive and normal children. *Journal of Abnormal Psychology, 74,* 388–95.
Frey, P. W., & Ross, L. E. (1967). Differential conditioning of the rabbit's eyelid response with an examination of Pavlov's induction hypothesis. *Journal of Comparative and Physiological Psychology, 64,* 277–83.
Galbraith, K. J. (1969). *Mediation in animal discrimination learning.* Unpublished doctoral dissertation, University of Toronto.
  (1971). Differential extinction performance to two stimuli following within-subject acquisition. *Journal of Experimental Psychology, 89,* 343–50.
Gallup, G. G. (1965). Aggression in rats as a function of frustrative nonreward in a straight alley. *Psychonomic Science, 3,* 99–100.
Gallup, G. G., & Altomari, T. S. (1969). Activity as a postsituation measure of frustrative nonreward. *Journal of Comparative and Physiological Psychology, 68,* 382–4.
Garcia, J., & Koelling, R. A. (1966). Relation of cue to consequence in avoidance learning. *Psychonomic Science, 4,* 123–4.

Gillespie, L. A. (1974). *Ontogeny of hippocampal electrical activity and behavior in rat, rabbit and guinea pig.* Unpublished doctoral dissertation, University of Western Ontario.

Glazer, H. I. (1972). Physostigmine and resistance to extinction. *Psychopharmacologia, 26,* 387–94.

(1974a). Instrumental conditioning of hippocampal theta and subsequent response persistence. *Journal of Comparative and Physiological Psychology, 86,* 267–73.

(1974b). Instrumental response persistence following induction of hippocampal theta frequency during fixed-ratio responding in rats. *Journal of Comparative and Physiological Psychology, 86,* 1156–62.

Gonzalez, R. C., Behrend, E. R., & Bitterman, M. E. (1965). Partial reinforcement in the fish: Experiments with spaced trials and partial delay. *American Journal of Psychology, 78,* 198–207.

Gonzalez, R. C., & Bitterman, M. E. (1965). Partial reinforcement and the fish. *American Journal of Psychology, 78,* 136–9.

(1969). Spaced-trials partial reinforcement effect as a function of contrast. *Journal of Comparative and Physiological Psychology, 67,* 94–103.

Gonzalez, R. C., & Champlin, G. (1974). Positive behavioral contrast, negative simultaneous contrast and their relation to frustration in pigeons. *Journal of Comparative and Physiological Psychology, 87,* 173–87.

Gonzalez, R. C., Ferry, M., & Powers, A. S. (1974). The adjustment of goldfish to reduction of magnitude of reward in massed trials. *Animal Learning & Behavior, 2,* 23–6.

Gonzalez, R. C., Potts, A., Pitcoff, K., & Bitterman, M. E. (1972). Runway performance of goldfish as a function of complete and incomplete reduction in amount of reward. *Psychonomic Science, 27,* 305–7.

Gonzalez, R. C., & Powers, A. S. (1973). Simultaneous contrast in goldfish. *Animal Learning & Behavior, 1,* 96–8.

Goodrich, K. P. (1959). Performance in different segments of an instrumental response chain as a function of reinforcement schedule. *Journal of Experimental Psychology, 57,* 57–63.

Gray, J. A. (1967). Disappointment and drugs in the rat. *Advancement of Science, 23,* 595–605.

(1969). Sodium amobarbital and effects of frustrative nonreward. *Journal of Comparative and Physiological Psychology, 69,* 55–64.

(1970). Sodium amobarbital, the hippocampal theta rhythm and the partial reinforcement extinction effect. *Psychological Review, 77,* 465–80.

(1972). Effects of septal driving of the hippocampal theta rhythm of resistance to extinction. *Physiology & Behavior, 8,* 481–90.

(1975). *Elements of two-process theory of learning.* New York: Academic Press.

(1977). Drug effects on fear and frustration: Possible limbic site of action of minor tranquilizers. In L. L. Iversen, S. D. Iversen, & S. H. Snyder (Eds.), *Handbook of pharmacology: Vol. 8. Drugs, neurotransmitters, and behavior* (pp. 433–529). New York: Plenum.

(1982). *The neuropsychology of anxiety: An enquiry into the functions of the septohippocampal system.* New York: Oxford University Press.

Gray, J. A., Araujo-Silva, M. T., & Quintao, L. (1972). Resistance to extinction after partial reinforcement training with blocking of the hippocampal theta rhythm by septal stimulation. *Physiology & Behavior, 8,* 497–502.

Gray, J. A., & Dudderidge, H. (1971). Sodium amylobarbitone, the partial reinforcement extinction effect and the frustration effect in the double runway. *Neuropharmacology, 10,* 217–22.

Gray, J. A., Feldon, J., Rawlins, J. N. P., Owen, S., & McNaughton, N. (1978). The role of the septo-hippocampal system and its noradrenergic afferents in behavioural responses to nonreward. In K. Elliott & J. Whelan (Eds.), *Functions of the septo-hippocampal system*, Ciba Foundation Symposium No. 58 (new series, pp. 275–300). Amsterdam: Elsevier.

Gray, J. A., & McNaughton, N. (1983). Comparison between the behavioral effects of septal and hippocampal lesions: A review. *Neuroscience & Biobehavioral Reviews, 7*, 119–88.

Gray, J. A., Quintao, L., & Araujo-Silva, M. T. (1972). The partial reinforcement extinction effect in rats with medial septal lesions. *Physiology & Behavior, 8*, 491–6.

Gray, J. A., & Smith, P. T. (1969). An arousal–decision model of partial reinforcement and discrimination learning. In R. Gilbert & N. S. Sutherland (Eds.), *Animal discrimination learning* (pp. 243–72). New York: Academic Press.

Greene, P. L., Diaz-Granados, J. L., & Amsel, A. (1992a). Blood ethanol concentration from early postnatal exposure: Effects on memory-based learning and hippocampal neuroanatomy in infant and adult rats. *Behavioral Neuroscience, 106*, 1–10.

(1992b). Effects on learned persistence and hippocampal neuroanatomy of blood ethanol concentration from postnatal exposure in the infant rat. Unpublished manuscript.

Grice, G. R., & Goldman, H. M. (1955). Generalized extinction and secondary reinforcement in visual discrimination learning with delayed reward. *Journal of Experimental Psychology, 50*, 197–200.

Grosslight, J. H., & Child, I. L. (1947). Persistence as a function of previous experience of failure followed by success. *American Journal of Psychology, 60*, 378–87.

Grove, G. R., & Eninger, M. U. (1952, May). The relative importance of approach and avoidance tendencies in brightness discrimination learning. Paper read at the Midwestern Psychological Association, Chicago.

Grusec, T., & Bower, G. (1965). Response effort and the frustration hypothesis. *Journal of Comparative and Physiological Psychology, 60*, 128–30.

Haggard, D. F. (1959). Acquisition of a simple running response as a function of partial and continuous schedules of reinforcement. *Psychological Record, 9*, 11–18.

Hall, W. G. (1975). Weaning and growth of artificially reared rats. *Science, 190*, 1313–15.

Hall, W. G., Cramer, C. P., & Blass, E. (1975). Developmental changes in suckling of rat pups. *Nature, 258*, 318–20.

Hammer, R. P., & Scheibel, A. B. (1981). Morphologic evidence for a delay of neuronal maturation in fetal alcohol exposure. *Experimental Neurology, 74*, 587–96.

Hanson, H. M. (1959). Effects of discrimination training on stimulus generalization. *Journal of Experimental Psychology, 58*, 321–34.

Harlow, H. F. (1949). The formation of learning sets. *Psychological Review, 56*, 51–65.

Hartshorne, M., May, M. A., & Maller, J. B. (1929). *Studies in the nature of character: II. Studies in service and selfcontrol*. New York: Macmillan.

Hawking, S. W. (1988). *A brief history of time*. New York: Bantam.

Hebb, D. O. (1949). *The organization of behavior*. New York: Wiley.

(1955). Drives and the CNS (conceptual nervous system). *Psychological Review, 62*, 243–54.

(1980). *Essay on mind*. Hillsdale, NJ: Erlbaum.

Henderson, K. (1966). Within-subjects partial-reinforcement effects in acquisition and in later discrimination learning. *Journal of Experimental Psychology, 72*, 704–13.

Henke, P. G. (1972). Amygdalectomy and mixed reinforcement schedule contrast effects. *Psychonomic Science, 28*, 301–2.

(1973a). Effects of reinforcement omission on rats with lesions in the amygdala. *Journal of Comparative and Physiological Psychology, 84*, 187–93.

(1973b). Lesions of the amygdala and the frustration effect. *Physiology & Behavior, 10,* 647–50.

(1974). Persistence of runway performance after septal lesion in rats. *Journal of Comparative and Physiological Psychology, 86,* 760–7.

(1976). Septal lesions and aversive nonreward. *Physiology & Behavior, 17,* 483–8.

(1977). Dissociation of the frustration effect and the partial reinforcement extinction effect after limbic lesions in rats. *Journal of Comparative and Physiological Psychology, 91,* 1032–8.

Henke, P. G., Allen, J. D., & Davison, C. (1972). Effect of lesions in the amygdala on behavioral contrast. *Physiology & Behavior, 8,* 173–6.

Henke, P. G., & Maxwell, D. (1973). Lesions in the amygdala and the frustration effect. *Physiology & Behavior, 10,* 647–50.

Hill, W. F. (1968). An attempted clarification of frustration theory. *Psychological Review, 75,* 173–6.

Holder, W. B., Marx, M. H., Holder, E. E., & Collier, G. (1957). Response strength as a function of delay of reward in a runway. *Journal of Experimental Psychology, 53,* 316–23.

Holinka, C. F., & Carlson, A. D. (1976). Pup attraction to lactating Sprague–Dawley rats. *Behavioral Biology, 16,* 489–505.

Holland, P. S., & Rescorla, R. A. (1975). The effect of two ways of devaluing the unconditioned stimulus after first- and second-order appetitive conditioning. *Journal of Experimental Psychology: Animal Behavior Processes, 1,* 355–63.

Holt, L., & Gray, J. A. (1983a). Septal driving of the hippocampal theta rhythm produces a long-term, proactive and non-associative increase in resistance to extinction. *Quarterly Journal of Experimental Psychology, 35B,* 97–118.

(1983b). Proactive behavioral effects of theta-blocking septal stimulation in the rat. *Behavioral Neural Biology, 39,* 7–21.

(1985). Proactive behavioral effects of theta-driving septal stimulation on conditioned suppression and punishment in the rat. *Behavioral Neuroscience, 99,* 60–74.

Honig, W. K., Thomas, D. R., & Guttman, N. (1959). Differential effects of continuous extinction and discrimination training on the generalization gradient. *Journal of Experimental Psychology, 58,* 145–52.

Hooper, R. (1967). Variables controlling the overlearning reversal effect (ORE). *Journal of Experimental Psychology, 73,* 612–19.

Horie, K. (1971). Effects of the length of the detention time in goal box to measure the strength of frustration. *Japanese Journal of Psychology, 42,* 14–23.

Hug, J. J. (1969). *The overlearning extinction effect after partially reinforced acquisition: A quantitative analysis of frustration theory.* Unpublished doctoral dissertation, University of Toronto.

(1970a). Frustration effects after varied numbers of partial and continuous reinforcements: Incentive differences as a function of reinforcement percentage. *Psychonomic Science, 21,* 57–9.

(1970b). Number of food pellets and the development of the frustration effect. *Psychonomic Science, 21,* 59–60.

(1970c). Frustration effect after development of patterned responding to single alternation reinforcement. *Psychonomic Science, 21,* 61–2.

Hug, J. J., & Amsel, A. (1969). Frustration theory and partial reinforcement effects: The acquisition–extinction paradox. *Psychological Review, 76,* 419–21.

Hughes, L. F., & Dachowski, L. (1973). The role of reinforcement and nonreinforcement in an operant frustration effect. *Animal Learning and Behavior, 1,* 68–72.

Hull, C. L. (1930). Knowledge and purpose as habit mechanisms. *Psychological Review, 37,* 511–25.

(1931). Goal attraction and directing ideas conceived as habit phenomena. *Psychological Review, 38,* 487–506.
(1934). The concept of the habit-family hierarchy and maze learning: Part 1. *Psychological Review, 41,* 33–54. (See also Part 2, *Psychological Review,* 1934, *41,* 134–52.
(1938). The goal gradient hypothesis applied to some "field force" problems in young children. *Psychological Review, 45,* 271–99.
(1943). *Principles of behavior.* New York: Appleton-Century-Crofts.
(1950). Simple qualitative discrimination learning. *Psychological Review, 57,* 303–13.
(1952). *A behavior system.* New Haven, CN: Yale University Press.
Hulse, S. H., Jr. (1958). Amount and percentage of reinforcement and duration of goal confinement in conditioning and extinction. *Journal of Experimental Psychology, 56,* 48–57.
Hulse, S. H., Jr., & Stanley, W. C. (1956). Extinction by omission of food as related to partial and secondary reinforcement. *Journal of Experimental Psychology, 52,* 221–7.
Humphreys, L. G. (1939a). The effect of random alternation of reinforcement on the acquisition and extinction of conditioned eyelid reactions. *Journal of Experimental Psychology, 25,* 141–58.
(1939b). Acquisition and extinction of verbal expectations in a situation analogous to conditioning. *Journal of Experimental Psychology, 25,* 294–301.
Isaacson, R. L. (1980). A perspective for the interpretation of limbic system function. *Physiological Psychology, 8,* 183–8.
Isaacson, R. L., & Kimble, D. P. (1972). Lesions of the limbic system: Their effects upon hypotheses and frustration. *Behavioral Biology, 7,* 767–93.
Ison, J. R. (1962). Experimental extinction as a function of number of reinforcements. *Journal of Experimental Psychology, 64,* 314–17.
Ison, J. R., & Cook, P. E. (1964). Extinction performance as a function of incentive magnitude and number of acquisition trials. *Psychonomic Science, 1,* 245–6.
Ison, J. R., Daly, H. B., & Glass, D. H. (1967). Amobarbital sodium and the effects of reward and nonreward in the Amsel double runway. *Psychological Reports, 20,* 491–6.
Ison, J. R., Glass, D. H., & Daly, H. B. (1969). Reward magnitude changes following differential conditioning and partial reinforcement. *Journal of Experimental Psychology, 81,* 81–8.
Ison, J. R., & Krane, R. V. (1969). Induction in differential instrumental conditioning. *Journal of Experimental Psychology, 80,* 183–5.
Ison, J. R., & Pennes, E. S. (1969). Interaction of amobarbital sodium and reinforcement schedule in determining resistance to extinction of an instrumental running response. *Journal of Comparative and Physiological Psychology, 68,* 215–19.
Ison, J. R., & Rosen, A. J. (1967). The effects of amobarbital sodium on differential instrumental conditioning and subsequent extinction. *Psychopharmacologia, 10,* 417–25.
Jackson, J. H. (1931). *Selected writings of John Hughlings Jackson.* London: Hodder & Stoughton.
James, H. (1959). Flicker: An unconditioned stimulus for imprinting. *Canadian Journal of Psychology, 13,* 59–67.
Jarrard, L. E., Isaacson, R. L., & Wickelgren, W. O. (1964). Effects of hippocampal ablation and intertrial interval on runway acquisition and extinction. *Journal of Comparative and Physiological Psychology, 57,* 442–4.
Jarrard, L. E., Feldon, J., Rawlins, J. N. P., Sinden, J. D., & Gray, J. A. (1986). The effects of intrahippocampal ibotenate on resistance to extinction after continuous or partial reinforcement. *Experimental Brain Research, 61,* 519–30.

Jenkins, H. M. (1962). Resistance to extinction when partial reinforcement is followed by regular reinforcement. *Journal of Experimental Psychology, 64,* 441–50.

Jensen, C., & Fallon, D. (1973). Behavioral aftereffects of reinforcement and its omission as a function of reinforcement magnitude. *Journal of the Experimental Analysis of Behavior, 19,* 459–68.

Jernstedt, G. C. (1971). Joint effects of pattern of reinforcement, intertrial interval, and amount of reinforcement in the rat. *Journal of Comparative and Physiological Psychology, 75,* 421–9.

Johanson, I. B., & Hall, W. G. (1979). Appetitive learning in 1-day-old rat pups. *Science, 205,* 419–21.

Johnson, R. N., Lobdell, P., & Levy, R. S. (1969). Intracranial self-stimulation and the rapid decline in frustrative nonreward. *Science, 164,* 971–2.

Keller, F. S., & Schoenfeld, W. N. (1950). *Principles of psychology.* New York: Appleton-Century-Crofts.

Kello, J. E. (1972). The reinforcement-omission effect on fixed-interval schedules: Frustration or inhibition? *Learning and Motivation, 3,* 138–47.

Kendler, H. H. (1946). The influence of simultaneous hunger and thirst drives upon the learning of two opposed spatial responses in the white rat. *Journal of Experimental Psychology, 36,* 212–20.

Kendler, H. H., Pliskoff, S. S., D'Amato, M. R., & Katz, S. (1957). Nonreinforcements versus reinforcements as variables in the partial reinforcement effect. *Journal of Experimental Psychology, 53,* 269–76.

Killeen, P. R., & Amsel, A. (1987). The kinematics of locomotion toward a goal. *Journal of Experimental Psychology, 13,* 92–101.

Kimble, D. P. (1968). Hippocampus and internal inhibition. *Psychological Bulletin, 70,* 285–95.

Kimble, D. P., & Kimble, R. J. (1965). Hippocampectomy and response perseveration in the rat. *Journal of Comparative and Physiological Psychology, 60,* 474–6.

Kimble, G. A. (1961). *Hilgard and Marquis' conditioning and learning.* New York: Appleton-Century-Crofts.

Kirkby, R. J. (1967). A maturation factor in spontaneous alternation. *Nature, 215,* 784.

Klee, J. B. (1944). The relation of frustration and motivation to the production of abnormal fixations in the rat. *Psychological Monographs, 56*(4).

Krippner, R. A., Endsely, R. C., & Tucker, R. S. (1967). Magnitude of G_1 reward and the frustration effect in a between-subjects design. *Psychonomic Science, 9,* 385–6.

Lachman, R. (1960). The model in theory construction. *Psychological Review, 67,* 113–29.

Lambert, W. W., & Solomon, R. L. (1952). Extinction of a running response as a function of distance of block point from the goal. *Journal of Comparative and Physiological Psychology, 45,* 269–79.

Lanfumey, L., Adrien, J., & Gray, J. A. (1982). Septal driving of hippocampal theta rhythm as a function of frequency in the infant male rat. *Experimental Brain Research, 45,* 230–2.

Lawrence, D. H. (1949). Acquired distinctiveness of cues: I. Transfer between discriminations on the basis of familiarity with the stimulus. *Journal of Experimental Psychology, 39,* 770–84.

(1952). The transfer of a discrimination along a continuum. *Journal of Comparative and Physiological Psychology, 45,* 511–16.

Lawrence, D. H., & Festinger, L. (1962). *Deterrents and reinforcement.* Stanford, CA: Stanford University Press.

Lawson, R. (1965). *Frustration: The development of a scientific concept.* New York: Macmillan.

Lawson, R., & Marx, M. H. (1958). Frustration: Theory and experiment. *Genetic Psychology Monographs, 57,* 393–464.
Leon, M. (1974). Maternal pheromone. *Physiology & Behavior, 13,* 441–53.
Leon, M., & Moltz, H. (1972). The development of the pheromonal bond in the albino rat. *Physiology & Behavior, 8,* 683–6.
Leonard, D. W., Weimer, J., & Albin, R. (1968). An examination of Pavlovian induction phenomena in differential instrumental conditioning. *Psychonomic Science, 12,* 89–90.
Letz, R., Burdette, D. R., Gregg, B., Kittrell, M. E., & Amsel, A. (1978). Evidence for a transitional period for the development of persistence in infant rats. *Journal of Comparative and Physiological Psychology, 92,* 856–66.
Levis, D. J. (1976). Learned helplessness: A reply and an alternative S–R interpretation. *Journal of Experimental Psychology: General, 105,* 47–65.
  (1980). The learned helplessness effect: An expectancy, discrimination deficit, or motivation-induced persistence? *Journal of Research in Personality, 14.*
Lewin, K. (1943). Defining the "field at a given time." *Psychological Review, 50,* 292–310.
Lincoln, D. W., Hill, A., & Wakerley, J. B. (1973). The milk-ejection reflex of the rat: An intermittent function not abolished by surgical levels of anesthesia. *Journal of Endocrinology, 57,* 459–76.
Linden, D. R., & Hallgren, S. O. (1973). Transfer of approach responding between punishment and frustrative nonreward sustained through continuous reinforcement. *Learning and Motivation, 4,* 207–17.
Livesey, P. J. (1986). *Learning and emotion: A biological synthesis: Vol. 1. Evolutionary processes.* Hillsdale, NJ: Erlbaum.
Lloyd, D. (1986). The limits of cognitive liberalism. *Behaviorism, 14,* 1–14.
Lobaugh, N. J. (1986). *Expectancies and response inhibition in infant rats following hippocampal lesions.* Unpublished doctoral dissertation, University of Texas.
Lobaugh, N. J., Bootin, M., & Amsel, A. (1985). Sparing of patterned alternation but not partial reinforcement effect after infant and adult hippocampal lesions in the rat. *Behavioral Neuroscience, 99,* 46–59.
Lobaugh, N. J., Greene, P. L., Grant, M., Nick, T., & Amsel, A. (1989). Patterned (single) alternation in infant rats following combined or separate lesions of hippocampus and amygdala. *Behavioral Neuroscience, 103,* 1159–67.
Lobaugh, N. J., Wigal, T., Greene, P. L., Diaz-Granados, J., & Amsel, A. (1991). Effects of prenatal ethanol exposure on learned persistence and hippocampal neuroanatomy in infant, weanling and adult rats. *Behavioral Brain Research. 44,* 81–6.
Logan, F. A. (1960). *Incentive.* New Haven, CN: Yale University Press.
  (1966). Transfer of persistence. *Journal of Experimental Psychology, 71,* 616–18.
  (1968). Frustration effect following correlated nonreinforcement. *Journal of Experimental Psychology, 78,* 396–400.
Logan, F. A., Beier, E. M., & Kincaid, W. D. (1956). Extinction following partial and varied reinforcement. *Journal of Experimental Psychology, 52,* 65–70.
Lowes, G., & Bitterman, M. E. (1967). Reward and learning in the goldfish. *Science, 157,* 455–7.
Ludvigson, H. W. (1967). A preliminary investigation of the effects of sodium amytal, prior reward in G_1, and activity level on the FE. *Psychonomic Science, 8,* 115–16.
Ludvigson, H. W., Mathis, D. A., & Choquette, K. A. (1985). Different odors in rats from large and small rewards. *Animal Learning & Behavior, 13,* 315–20.
Ludvigson, H. W., & Sytsma, D. (1967). The sweet smell of success: Apparent double alternation in the rat. *Psychonomic Science, 9,* 283–4.
Lyons, J., Klipec, W. D., & Eirick, R. (1973). The effect of chlorpromazine on the peak shift in the albino rat. *Physiological Psychology, 1,* 165–8.

Lyons, J., Klipec, W. D., & Steinsultz G. (1973). The effect of chlorpromazine on discrimination performance and the peak shift. *Physiological Psychology, 1,* 121–4.

Mabry, P. D., & Peeler, D. F. (1972). Effects of septal lesions on response to frustrative nonreward. *Physiology & Behavior, 8,* 909–13.

MacKinnon, J. R. (1967). Interactive effects of the two rewards in a differential magnitude of reward discrimination. *Journal of Experimental Psychology, 75,* 329–38.

MacKinnon, J. R., & Amsel, A. (1964). Magnitude of the frustration effect as a function of confinement and detention in the frustrating situation. *Journal of Experimental Psychology, 67,* 468–74.

MacKinnon, J. R., & Russotti, J. S. (1967). An inverse pattern in a discrimination based on differential magnitudes of reward. *Psychonomic Science, 7,* 395–6.

Mackintosh, N. J. (1971). Reward and aftereffects of reward in the learning of goldfish. *Journal of Comparative and Physiological Psychology, 76,* 225–32.

(1974). *The psychology of animal learning.* New York: Academic Press.

Mackintosh, N. J., Little, L., & Lord, J. (1972). Some determinants of behavioral contrast in pigeons and rats. *Learning and Motivation, 3,* 148–61.

Maier, N. R. F. (1949). *Frustration: The study of behavior without a goal.* New York: McGraw-Hill.

(1956). Frustration theory: Restatement and extension. *Psychological Review, 63,* 370–88.

Maier, S. F., & Seligman, M. E. P. (1976). Learned helplessness: Theory and evidence. *Journal of Experimental Psychology: General, 105,* 3–46.

Maltzman, I. (1952). The process need. *Psychological Review, 59,* 40–8.

Mandler, J. M. (1966). Behavior changes during overtraining and their effects on reversal and transfer. *Psychonomic Monograph Supplements, 1,* 187–202.

(1984). Representation and recall in infancy. In M. Moscovitch (Ed.), *Infant memory* (pp. 75–102). New York: Plenum.

Martin, B. (1963). Reward and punishment associated with the same goal response: A factor in the learning of motives. *Psychological Bulletin, 60,* 441–51.

Marx, M. H. (1956). Some relations between frustration and drive. In M. R. Jones (Ed.), *Nebraska symposium on motivation.* Lincoln: University of Nebraska Press.

Marzocco, F. N. (1950). *Frustration effect as a function of drive level, habit strength and distribution of trials during extinction.* Unpublished doctoral dissertation, University of Iowa.

McCain, G., & McVean, G. (1967). Effects of prior reinforcement or nonreinforcement on later performance in a double alley. *Journal of Experimental Psychology, 73,* 620–7.

McCleary, R. A. (1966). Response modulating functions of the limbic system: Initiation and suppression. In E. Stellar & J. M. Sprague (Eds.), *Progress in physiological psychology* (Vol. 1, pp. 209–72). New York: Academic Press.

McCuller, T., Wong, P. T. P., & Amsel, A. (1976). Transfer of persistence from progressive-ratio training to runway extinction. *Animal Learning & Behavior, 4,* 53–7.

McFarland, D. J. (1966). The role of attention in the disinhibition of displacement activities. *Quarterly Journal of Experimental Psychology, 18,* 19–30.

McHose, J. H. (1963). Effect of continued nonreinforcement on the frustration effect. *Journal of Experimental Psychology, 65,* 444–50.

McHose, J. H., & Ludvigson, H. W. (1964). Frustration effect as a function of drive. *Psychological Reports, 14,* 371–4.

(1965). Role of reward magnitude and incomplete reduction of reward magnitude in the frustration effect. *Journal of Experimental Psychology, 70,* 490–5.

(1966). Differential conditioning with nondifferential reinforcement. *Psychonomic Science, 6,* 485–6.

References

Mellgren, R. L., Hoffman, D., Nation, J. R., Williams, J. D., & Wrather, D. M. (1979). Transfer of persistence responding across motivational reward conditions. *American Journal of Psychology, 92,* 95–103.

Miller, N. E. (1944). Experimental studies of conflict. In J. McV. Hunt (Ed.), *Personality and the behavior disorders* (Vol. 1, pp. 431–65). New York: Ronald.

  (1948). Theory and experiment relating psychoanalytic displacement to stimulus–response generalization. *Journal of Abnormal and Social Psychology, 38,* 89–101.

  (1959). Liberalization of basic S–R concepts: Extensions to conflict behavior, motivation, and social learning. In S. Koch (Ed.), *Psychology: A study of a science* (Vol. 2, pp. 196–293). New York: McGraw-Hill.

  (1960). Learning resistance to pain and fear: Effects of overlearning, exposure and rewarded exposure in context. *Journal of Experimental Psychology, 60,* 137–45.

  (1963). Some reflections on the law of effect produce a new alternative to drive reduction. In M. R. Jones (Ed.), *Nebraska symposium on motivation* (Vol. 11, pp. 65–112). Lincoln: University of Nebraska Press.

Miller, N. E., & Dollard, J. C. (1941). *Social learning and imitation.* New Haven, CN: Yale University Press.

Miller, N. E., & Miles, W. R. (1935). Effect of caffeine on the running speed of hungry, satiated, and frustrated rats. *Journal of Comparative Psychology, 20,* 397–412.

  (1936). Alcohol and removal of reward: An analytical study of rodent maze behavior. *Journal of Comparative Psychology, 21,* 179–204.

Miller, N. E., & Stevenson, S. S. (1936). Agitated behavior of rats during experimental extinction and a curve of spontaneous recovery. *Journal of Comparative Psychology, 21,* 205–31.

Mishkin, M., Malamut, B., & Bachevalier, J. (1984). Memories and habits: Two neural systems. In G. Lynch, J. L. McGaugh, & N. M. Weinberger (Eds.), *Neurobiology of learning and memory* (pp. 65–7). New York: Guilford.

Mishkin, M., & Petri, H. L. (1984). Memories and habits: Some implications for the analysis of learning and retention. In L. R. Squire & N. Butters (Eds.), *Neuropsychology of memory* (pp. 287–96). New York: Guilford.

Morgan, C. T., & Fields, P. E. (1938). The effect of variable preliminary feeding upon the rat's speed of locomotion. *Journal of Comparative Psychology, 26,* 331–48.

Morris, R. G. M. (1981). Spatial localization does not require the presence of local cues. *Learning and Motivation, 12,* 239–60.

Morruzi, G., & Magoun, H. W. (1949). Brain stem reticular formation and activation of the EEG. *Electroencephalography and Clinical Neurophysiology, 1,* 455–73.

Mowrer, O. H. (1939). A stimulus–response analysis of anxiety and its role as a reinforcing agent. *Psychological Review, 46,* 553–65.

  (1940). An experimental analogue of "regression" with incidental observations on "reaction-formation." *Journal of Abnormal and Social Psychology, 35,* 56–87.

  (1960). *Learning theory and behavior.* New York: Wiley.

Mowrer, O. H., & Jones, H. M. (1945). Habit strength as a function of the pattern of reinforcement. *Journal of Experimental Psychology, 43,* 293–311.

Murillo, N. R., Diercks, J. K., & Capaldi, E. J. (1961). Performance of the turtle, *Pseudemys scripta troostii,* in a partial-reinforcement experiment. *Journal of Comparative and Physiological Psychology, 54,* 204–6.

Nadel, L., & Zola-Morgan, S. (1984). Toward the understanding of infant memory: Contributions from animal neuropsychology. In M. Moscovitch (Ed.), *Infant memory* (pp. 145–72). New York: Plenum.

Nanson, J. L., & Hiscock, M. (1990). Attention deficits in children exposed to alcohol prenatally. *Alcoholism: Clinical and Experimental Research, 14,* 656–61.

Nation, J. R., & Boyagian, L. G. (1981). Appetitive performance following exposure to inescapable shocks of short or long duration. *American Journal of Psychology, 94,* 605–17.

Nation, J. R., Cooney, J. B., & Gartrell, K. E. (1979). Durability and generalizability of persistence training. *Journal of Abnormal Psychology, 88,* 121–36.

Nation, J. R., & Massad, P. (1978). Persistence training: A partial reinforcement procedure for reversing learned helplessness and depression. *Journal of Experimental Psychology: General, 107,* 436–51.

Nation, J. R., & Woods, D. J. (1980). Persistence: The role of partial reinforcement in psychotherapy. *Journal of Experimental Psychology: General, 109,* 175–207.

Nation, J. R., Wrather, D. M., Mellgren, R. L., & Spivey, M. (1980). Transfer of the partial reinforcement extinction effect between escape (shock) and appetitive (food) conditioning. *Learning and Motivation, 11,* 97–116.

Nevin, J. A., & Shettleworth, S. J. (1966). An analysis of contrast effects in multiple schedules. *Journal of the Experimental Analysis of Behavior, 9,* 305–15.

Noirot, E. (1972). Ultrasounds and maternal behavior in small rodents. *Developmental Psychobiology, 5,* 371–87.

Nonneman, A. J., & Isaacson, R. L. (1973). Task dependent recovery after early brain damage. *Behavioral Biology, 8,* 143–72.

Nonneman, A. J., Voigt, J., & Kolb, B. E. (1974). Comparisons of behavioral effects of hippocampal and prefrontal cortex lesions in the rat. *Journal of Comparative and Physiological Psychology, 87,* 249–60.

North, A. J., & Stimmel, D. T. (1960). Extinction of an instrumental response following a large number of reinforcements. *Psychology Reports, 6,* 227–34.

O'Keefe, J., & Nadel, L. (1978). *The hippocampus as cognitive map.* Oxford: Oxford University Press.

Olds, J. (1953). The influence of practice on the strength of secondary approach drives. *Journal of Experimental Psychology, 46,* 232–6.

  (1956). *The growth and structure of motives.* Glencoe, IL: Free Press.

Olton, D. S. (1983). Memory functions and the hippocampus. In W. Seifert (Ed.), *Neurobiology of the hippocampus* (pp. 335–73). New York: Academic Press.

Olton, D. S., Becker, J. T., & Handlemann, G. E. (1979). Hippocampus, space and memory. *Behavioral & Brain Science, 2,* 313–65.

Olton, D. S., & Samuelson, R. J. (1976). Remembrance of places past: Spatial memory in rats. *Journal of Experimental Psychology: Animal Behavior Processes, 2,* 97–116.

Olton, D. S., Wible, C. G., & Shapiro, M. L. (1986). Mnemonic theories of hippocampal function. *Behavioral Neuroscience, 100,* 852–5.

Panksepp, J., & Trowill, J. (1968). Extinction following intracranial reward: Frustration or drive decay? *Psychonomic Science, 12,* 173–4.

Papini, M. R., Mustaca, A. E., & Bitterman, M. E. (1988). Successive negative contrast in the consummatory responding of didelphid marsupials. *Animal Learning & Behavior, 16,* 53–7.

Papini, M. R., & Ramallo, P. (1990). Primary frustration in the red opossum, *Lutreolina crassicaudata. International Journal of Comparative Psychology, 3,* 235–42.

Parry, P. A., & Douglas, V. I. (1983). Effects of reinforcement on concept identification in hyperactive children. *Journal of Abnormal Child Psychology, 11,* 327–40.

Patten, R. L., & Latta, R. M. (1974). Frustration effect in discrimination: Effect of extended training. *Journal of Experimental Psychology, 103,* 831–6.

Patten, R. L., & Myers, D. B. (1970). Number of training trials and frustration effects of nonzero reward reduction in the double alley. *Psychonomic Science, 18,* 291.

Pavlik, W. B., & Carlton, P. L. (1965). A reversed partial-reinforcement effect. *Journal of Experimental Psychology, 70,* 417–23.

References

Pavlik, W. B., Carlton, P. L., & Hughes, R. A. (1965). Partial reinforcement effects in a runway: Between- and within-Ss. *Psychonomic Science, 3,* 203–4.

Pavlik, W. B., Carlton, P. L., & Manto, P. G. (1965). A further study of the partial reinforcement effect within subjects. *Psychonomic Science, 3,* 533–4.

Pavlov, I. P. (1927). *Conditioned reflexes* (G. V. Anrep, Trans.). Oxford: Oxford University Press.

Peckham, R. H., & Amsel, A. (1964). Magnitude of reward and the frustration effect in a within-subjects design. *Psychonomic Science, 1,* 285–6.

  (1967). Within-subjects demonstration of a relationship between frustration and magnitude of reward in a differential magnitude of reward discrimination. *Journal of Experimental Psychology, 73,* 187–95.

Pentney, R. J., Cotter, J. R., & Abel, E. L. (1984). Quantitative measures of mature neuronal morphology after in utero ethanol exposure. *Neurobehavioral Toxicology and Teratology, 6,* 59–65.

Perry, S. L., & Moore, J. W. (1965). The partial-reinforcement effect sustained through blocks of continuous reinforcement in classical eyelid conditioning. *Journal of Experimental Psychology, 69,* 158–61.

Pert, A., & Bitterman, M. E. (1970). Reward and learning in the turtle. *Learning and Motivation, 1,* 121–8.

Pert, A., & Gonzalez, R. C. (1974). Behavior of the turtle (*Chrysemys picta picta*) in simultaneous, successive, and behavioral contrast situations. *Journal of Comparative and Physiological Psychology, 87,* 526–38.

Premack, D. (1969). On some boundary conditions of contrast. In J. T. Tapp (Ed.), *Reinforcement and behavior* (pp. 120–45). New York: Academic Press.

Pubols, B. H., Jr. (1956). The facilitation of visual and spatial discrimination reversal by overlearning. *Journal of Comparative and Physiological Psychology, 49,* 243–8.

Rashotte, M. E. (1971). Influence of partial reinforcement of running on the extinction of continuously reinforced barpressing in rats. *Psychonomic Science, 25,* 145–7.

  (1979). Reward training: Contrast effects. In M. E. Bitterman, V. M. LoLordo, J. B. Overmier, & M. E. Rashotte (Eds.), *Animal learning: Survey and analysis* (pp. 127–66). New York: Plenum.

Rashotte, M. E., & Amsel, A. (1968). Transfer of slow-response rituals to the extinction of a continuously rewarded response. *Journal of Comparative and Physiological Psychology, 66,* 432–43.

Rashotte, M. E., & Surridge, C. T. (1969). Partial reinforcement and partial delay of reinforcement effects with 72-hour intertrial intervals and interpolated continuous reinforcement. *Quarterly Journal of Experimental Psychology, 21,* 156–61.

Ratliff, R. G., & Clayton, K. N. (1969). Runway extinction as a joint function of acquisition reward precentage and extinction punishment intensity. *Journal of Experimental Psychology, 80,* 574–6.

Rawlins, J. N. P. (1985). Associations across time: The hippocampus as a temporary memory store. *Behavioral & Brain Sciences, 8,* 479–96.

Rawlins, J. N. P., Feldon, J., & Gray, J. A. (1980). The effects of hippocampectomy and of fimbria section upon the partial reinforcement extinction effect in rats. *Experimental Brain Research, 38,* 273–83.

Rawlins, J. N. P., Feldon, J., Tonkiss, J., & Coffey, P. J. (1989). The role of subicular outputs in the development of the partial reinforcement extinction effect. *Experimental Brain Research, 77,* 153–60.

Rawlins, J. N. P., Feldon, J., Ursin, H., & Gray, J. A. (1985). Resistance to extinction after schedules of partial delay or partial reinforcement in rats with hippocampal lesions. *Experimental Brain Research, 59,* 273–81.

Raymond, B., Aderman, M., & Wolach, A. H. (1972). Incentive shifts in the goldfish. *Journal of Comparative and Physiological Psychology, 78,* 10–13.
Razran, G. (1955). Partial reinforcement of salivary CR's in adult human subjects: Preliminary study. *Psychological Reports, 1,* 409–16.
Reid, L. S. (1953). The development of noncontinuity behavior through continuity learning. *Journal of Experimental Psychology, 46,* 107–12.
Rescorla, R. A., & Solomon, R. L. (1967). Two-process learning theory: Relationships between Pavlovian conditioning and instrumental learning. *Psychological Review, 74,* 151–82.
Rescorla, R. A., & Wagner, A. R. (1972). A theory of Pavlovian conditioning: Variations in the effectiveness of reinforcement and nonreinforcement. In A. H. Black & W. F. Prokasy (Eds.), *Classical conditioning: II. Current theory and research* (pp. 64–99). New York: Appleton-Century-Crofts.
Reynolds, G. S. (1961). Behavioral contrast. *Journal of the Experimental Analysis of Behavior, 4,* 57–71.
Riccio, D. C., & Marrazo, M. J. (1972). Effects of punishing active avoidance in young and adult rats. *Journal of Comparative and Psysiological Psychology, 79,* 453–8.
Rickert, E. J., & Bennett, T. L. (1972). Performance of hippocampectomized rats on discontinuous negatively correlated reward. *Behavioral Biology, 7,* 375–82.
Rilling, M. E., Askew, H. R., Ahlskog, J. E., & Kramer, T. J. (1969). Aversive properties of the negative stimulus in a successive discrimination. *Journal of the Experimental Analysis of Behavior, 12,* 917–32.
Robbins, T. W., Watson, B. A., Gaskin, M., & Ennis, C. (1983). Contrasting interactions of pipradrol, *d*-amphetamine, cocaine, cocaine analogues, apomorphine and other drugs with conditioned reinforcement. *Psychopharmacology (Berlin), 80,* 113–19.
Rohrer, J. H. (1949). A motivational state resulting from nonreward. *Journal of Comparative and Physiological Psychology, 42,* 476–85.
Roitblat, H. L., Bever, T. G., & Terrace, H. S. (1984). *Animal cognition.* Hillsdale, NJ: Erlbaum.
Rose, G. (1983). Physiological and behavioral characteristics of dentate granule cells. In W. Seifert (Ed.), *Neurobiology of the hippocampus* (pp. 449–72). New York: Academic Press.
Rosellini, R. A., & Seligman, M. E. P. (1975). Frustration and learned helplessness. *Journal of Experimental Psychology: Animal Behavior Processes, 1,* 149–58.
Rosen, A. J., Glass, D. H., & Ison, J. R. (1967). Amobarbital sodium and instrumental performance following reward reduction. *Psychonomic Science, 9,* 129–30.
Rosenzweig, S. (1934). Types of reaction to frustration: An heuristic classification. *Journal of Abnormal and Social Psychology, 29,* 298–300.
　(1944). An outline of frustration theory. In J. Hunt (Ed.), *Personality and the behavior disorders.* New York: Ronald.
Rosenzweig, S., Mowrer, O. H., Haslerud, G. M., Curtis, Q. F., & Barker, R. G. (1938). Frustration as an experimental problem. *Character and Personality, 7,* 126–60.
Ross, R. R. (1964). Positive and negative partial-reinforcement extinction effects carried through continuous reinforcement, changed motivation, and changed response. *Journal of Experimental Psychology, 68,* 492–502.
Roussel, J. (1952). *Frustration effects as a function of repeated non-reinforcements and as a function of the consistency of reinforcement prior to introduction of non-reinforcement.* Unpublished master's thesis, Tulane University.
Rubel, E. W., & Rosenthal, M. H. (1975). The ontogeny of auditory frequency generalization in the chicken. *Journal of Experimental Psychology: Animal Behavior Processes, 1,* 287–97.

Ryan, T. J., & Watson, P. (1968). Frustrative nonreward theory applied to children's behavior. *Psychological Bulletin, 69*, 111–25.
Ryans, D. G. (1939). The measurement of persistence: An historical review. *Psychological Bulletin, 36*, 715–39.
Schachter, S., & Singer, J. E. (1962). Cognitive, social, and physiological determinants of emotional state. *Psychological Review, 69*, 379–99.
Schacter, D. L., & Moscovitch, M. (1984). Infants, amnesics, and dissociable memory systems. In M. Moscovitch (Ed.), *Infant memory* (pp. 173–216). New York: Plenum.
Scheff, S., Benardo, L., & Cotman, C. (1977). Progressive brain damage accelerates axon sprouting in the adult rat. *Science, 197*, 795–7.
Schlessinger, A. R., Cowan, W. M., and Gottlieb, D. I. (1975). An autoradiographic study of the time of origin and the pattern of granule cell migration in the dentate gyrus of the rat. *Journal of Comparative Neurology, 159*, 149–76.
Schmajuk, N. A., Segura, E. T., & Ruidiaz, A. C. (1981). Reward downshift in the toad. *Behavioral and Neural Biology, 33*, 519–23.
Schneirla, T. C. (1959). An evolutionary and developmental theory of biphasic processes underlying approach and withdrawal. In M. R. Jones (Ed.), *Nebraska symposium on motivation* (pp. 1–42). Lincoln: University of Nebraska Press.
Schwartz, B., & Gamzu, E. (1977). Pavlovian control of operant behavior. In W. K. Honig & J. E. R. Staddon (Eds.), *Handbook of operant behavior* (pp. 53–97). Englewood Cliffs, NJ: Prentice-Hall.
Scobie, S. R., Gold, D. C., & Fallon, D. (1974). Reward and reward omission: Time-dependent aftereffects in rats and fish. *Bulletin of the Psychonomic Society, 3*, 452–4.
Scull, J. (1971). Effects of shock and noise on running after training with partial or continuous reward. *Psychonomic Science, 23*, 325–6.
  (1973). The Amsel frustration effect: Interpretations and research. *Psychological Bulletin, 79*, 352–61.
Scull, J., Davies, K., & Amsel, A. (1970). Behavioral contrast and frustration effect in multiple and mixed fixed-interval schedules. *Journal of Comparative and Physiological Psychology, 71*, 478–83.
Seligman, M. E. P. (1970). On the generality of the laws of learning. *Psychological Review, 77*, 408–18.
Senf, G. M., & Miller, N. E. (1967). Evidence for positive induction in discrimination learning. *Journal of Comparative and Physiological Psychology, 64*, 121–7.
Senkowski, P. C. (1978). Variables affecting the overtraining extinction effect in discrete-trial lever pressing. *Journal of Experimental Psychology: Animal Behavior Processes, 4*, 131–43.
Seward, J. P. (1951). Experimental evidence for the motivating function of reward. *Psychological Bulletin, 58*, 130–49.
Seward, J. P., Pereboom, A. C., Butler, B., & Jones, R. B. (1957). The role of prefeeding in an apparent frustration effect. *Journal of Experimental Psychology, 54*, 445–50.
Sgro, J. A., Glotfelty, R. A., & Moore, B. D. (1970). Delay of reward in the double alleyway: A within-subjects versus between-groups comparison. *Journal of Experimental Psychology, 84*, 82–7.
Sgro, J. A., Showalter, J. R., & Cohn, N. H. (1971). Frustration effect following training with continuous and partial delay of reward. *Journal of Experimental Psychology, 87*, 320–5.
Shanab, M. E. (1971). Positive transfer between nonreward and delay. *Journal of Experimental Psychology, 91*, 98–102.

Sheffield, F. D. (1965). Relation between classical and instrumental learning. In W. F. Prokasy (Ed.), *Classical Conditioning* (pp. 302–22). New York: Appleton-Century-Crofts.

Sheffield, V. F. (1949). Extinction as a function of partial reinforcement and distribution of practice. *Journal of Experimental Psychology, 39,* 511–26.

Shinoda, A., & Bitterman, M. E. (1987). Analysis of the overlearning-extinction effect in honeybees. *Animal Learning & Behavior, 15,* 93–6.

Shoemaker, H. A. (1953). *The relative efficiency of rewarded and non-rewarded training in a black–white discrimination problem.* Unpublished doctoral dissertation, University of Colorado.

Simonov, P. V. (1969). Studies of emotional behavior of humans and animals by Soviet physiologists. *Annals of the New York Academy of Science, 159,* 1112–21.

(1974). On the role of the hippocampus in the integrative activity of the brain. *Acta Neurobiologicae Experimentalis, 34,* 33–41.

Sinden, J. D., Jarrard, L. E., & Gray, J. A. (1988). The effects of intra-subicular ibotenate on resistance to extinction after continuous or partial reinforcement. *Experimental Brain Research, 73,* 315–19.

Skinner, B. F. (1938). *The behavior of organisms.* New York: Appleton-Century-Crofts.

(1950). Are theories of learning necessary? *Psychological Review, 57,* 193–216.

Solomon, R. L., & Corbit, J. C. (1974). An opponent-process theory of motivation: I. Temporal dynamics of affect. *Psychological Review, 81,* 119–45.

Spear, N. E., & Pavlik, W. B. (1966). Percentage of reinforcement and reward magnitude effects in a T-maze: Between and within subjects. *Journal of Experimental Psychology, 71,* 521–8.

Spence, K. W. (1936). The nature of discrimination learning in animals. *Psychological Review, 43,* 427–49.

(1937). The differential response in animals to stimuli varying within a single dimension. *Psychological Review, 44,* 430–44.

(1940). Continuous versus non-continuous interpretations of discrimination learning. *Psychological Review, 47,* 271–88.

(1945). An experimental test of the continuity and non-continuity theories of discrimination learning. *Journal of Experimental Psychology, 35,* 253–66.

(1956). *Behavior theory and conditioning.* New Haven, CN: Yale University Press.

(1958). A theory of emotionally based drive (D) and its relation to performance in simple learning situations. *American Psychologist, 13,* 131–41.

(1960). *Behavior theory and learning.* Englewood Cliffs, NJ: Prentice-Hall.

(1961). Discussion. In N. S. Kline (Ed.), *Pavlovian conference on higher nervous activity.* New York: Annals of the New York Academy of Sciences, 92, 813–1198.

(1966). Cognitive and drive factors in the extinction of the conditioned eye-blink in human subjects. *Psychological Review, 73,* 445–58.

Spence, K. W., & Taylor, J. A. (1951). Anxiety and strength of the US as determiners of the amount of eyelid conditioning. *Journal of Experimental Psychology, 42,* 183–8.

Squire, L. R. (1982). The neuropsychology of human memory. *Annual Reviews of Neuroscience, 5,* 241–73.

(1987). *Memory and brain.* New York: Oxford University Press.

Staddon, J. E. R. (1983). *Adaptive behavior and learning.* Cambridge University Press.

Staddon, J. E. R., & Innis, N. K. (1969). Reinforcement omission on fixed-interval schedules. *Journal of the Experimental Analysis of Behavior, 12,* 689–700.

Stanton, M. E. (1982). Performance of 11- and 14-day-old rats on a working-memory problem. *Behavioral and Neural Biology, 36,* 304–10.

(1983). Dissociation of patterned alternation learning and the partial reinforcement extinction effect in preweanling rats. *Animal Learning & Behavior, 11,* 415–23.

Stanton, M. E., & Amsel, A. (1980). Adjustment to reward reduction (but no negative contrast) in rats 11, 14, and 16 days of age. *Journal of Comparative and Physiological Psychology, 94,* 446–58.

Stanton, M., Dailey, W., & Amsel, A. (1980). Patterned (single) alternation in 11- and 14-day-old rats under various reward conditions. *Journal of Comparative and Physiological Psychology, 94,* 459–71.

Stanton, M., Lobaugh, N., & Amsel, A. (1984). Age of first appearance of simultaneous and successive negative contrast in infant rats. *Journal of Experimental Psychology: Animal Behavior Processes, 10,* 376–89.

Stein, D. G., Finger, S., & Hart, T. (1983). Brain damage and recovery. *Behavioral and Neural Biology, 37,* 185–222.

Stein, L. (1966). Habituation and stimulus novelty: A model based on classical conditioning. *Psychological Review, 73,* 352–6.

Stimmel, D. T., & Adams, P. C. (1969). The magnitude of the frustration effect as a function of the number of previously reinforced trials. *Psychonomic Science, 16,* 31–2.

Stoltenburg-Didinger, G., & Spohr, H. L. (1983). Fetal alcohol syndrome and mental retardation: Spine distribution of pyramidal cells in prenatal alcohol-exposed rat cerebral cortex: A Golgi study. *Developmental Brain Research, 11,* 119–23.

Surridge, C. T., & Amsel, A. (1966). Acquisition and extinction under single alternation and random partial-reinforcement conditions with a 24-hour intertrial interval. *Journal of Experimental Psychology, 72,* 361–8.

Surridge, C. T., Boehnert, J., & Amsel, A. (1966). Effect of interpolated extinction on the reacquisition of partially and continuously rewarded responses. *Journal of Experimental Psychology, 72,* 564–70.

Surridge, C. T., Mock, K., & Amsel, A. (1968). Effect of interpolated extinction after partial delay of reward training on subsequent reacquisition and extinction. *Quarterly Journal of Experimental Psychology, 20,* 321–8.

Sutherland, N. S. (1966). Partial reinforcement and breadth of learning. *Quarterly Journal of Experimental Psychology, 18,* 289–302.

Sutherland, N. S., & Mackintosh, N. J. (1971). *Mechanisms of animal discrimination learning.* New York: Academic Press.

Swanson, A. M., & Isaacson, R. L. (1967). Hippocampal ablation and performance during withdrawal of reinforcement. *Journal of Comparative and Physiological Psychology, 64,* 30–5.

Taylor, J. A. (1953). A personality scale of manifest anxiety. *Journal of Abnormal and Social Psychology, 48,* 285–90.

Taylor, J. A., & Spence, K. W. (1952). The relationship of anxiety level to performance in serial learning. *Journal of Experimental Psychology, 44,* 61–4.

Taylor, J. R., & Robbins, T. W. (1984). Enhanced behavioral control by conditioned reinforcers following microinjections of *d*-amphetamine into the nucleus accumbens. *Psychopharmacology (Berlin), 84,* 405–12.

(1986). 6-Hydroxydopamine lesions of the caudate nucleus attenuate enhanced responding with reward-related stimuli produced by intra-accumbens *d*-amphetamine. *Psychopharmacology (Berlin), 90,* 390–7.

Teitelbaum, P., Cheng, M. F., & Rozin, P. (1969). Development of feeding parallels its recovery after hypothalamic damage. *Journal of Comparative and Physiological Psychology, 67,* 430–41.

Terrace, H. S. (1963a). Discrimination learning with and without "errors." *Journal of the Experimental Analysis of Behavior, 6,* 1–27.

(1963b). Errorless discrimination learning in the pigeon: Effects of chlorpromazine and imipramine. *Science, 340,* 318–19.

(1963c). Errorless transfer of a discrimination across two continua. *Journal of the Experimental Analysis of Behavior, 6,* 223–32.

(1966a). Stimulus control. In W. K. Honig (Ed.), *Operant behavior: Areas of research and application* (pp. 271–344). New York: Appleton-Century-Crofts.

(1966b). Discrimination learning and inhibition. *Science, 154,* 1677–80.

(1968). Discrimination learning, the peak shift, and behavioral contrast. *Journal of the Experimental Analysis of Behavior, 11,* 727–41.

(1972). By-products of discrimination learning. In G. H. Bower (Ed.), *The psychology of learning and motivation* (Vol. 5, pp. 195–265). New York: Academic Press.

Terris, W., German, D., & Enzie, R. (1969). Transsituational resistance to the effects of aversive stimulation. *Journal of Comparative and Physiological Psychology, 67,* 264–8.

Terris, W., & Wechkin, S. (1967). Learning to resist the effects of punishment. *Psychonomic Science, 7,* 169–70.

Theios, J. (1962). The partial reinforcement effect sustained through blocks of continuous reinforcement. *Journal of Experimental Psychology, 64,* 1–6.

Theios, J., & Brelsford, J. (1964). Overlearning-extinction effect as an incentive phenomenon. *Journal of Experimental Psychology, 67,* 463–7.

Thomas, G. J. (1984). Memory: Time finding in organisms. In L. R. Squire & N. Butlers (Eds.), *Neuropsychology of memory* (pp. 374–84). New York: Guilford.

Thompson, R. F. (1983). Neuronal substrate of simple associative learning: Classical conditioning. *Trends in Neuroscience, 6,* 270–2.

Thompson, R. F., Berger, T. W., Berry, S. D., Hoehler, F. K., Kettner, R. E., & Weisz, D. J. (1980). Hippocampal substrate of classical conditioning. *Physiological Psychology, 8,* 262–79.

Thompson, R. F., & Spencer, W. A. (1966). Habituation: A model phenomenon for the study of neuronal substrates of behavior. *Psychological Review, 73,* 16–43.

Tolman, E. C. (1938). The determiners of behavior at a choice point. *Psychological Review, 45,* 1–41.

Tominga, M., & Imada, H. (1968). Relationship between G1 reward value and the frustration effect in a double runway situation. *Japanese Journal of Psychology, 39,* 33–7.

Trapold, M. A., & Overmier, J. B. (1972). The second learning process in instrumental learning. In A. H. Black & W. F. Prokasy (Eds.), *Classical conditioning: II. Current research and theory* (pp. 427–52). New York: Appleton-Century-Crofts.

Traupmann, K. L., Amsel, A., & Wong, P. T. P. (1973). Persistence early and late in extinction as a function of number of continuous reinforcements preceding partial reinforcement training. *Animal Learning & Behavior, 3,* 219–22.

Traupmann, K. L., Wong, P. T. P., & Amsel, A. (1971). Durability of persistence as a function of number of partially reinforced trials. *Journal of Experimental Psychology, 88,* 372–5.

Tulving, E. (1972). Episodic and semantic memory. In E. Tulving & W. Donaldson (Eds.), *Organization of memory* (pp. 381–403). New York: Academic Press.

(1985). On the classification problem in learning and memory. In L.-G. Nilsson & T. Archer (Eds.), *Perspectives on learning and memory* (pp. 67–94). Hillsdale, NJ: Erlbaum.

Tulving, E., & Schacter, D. L. (1990). Priming and human memory systems. *Science, 247,* 301–6.

Tyler, D. W. (1956). Extinction following partial reinforcement with control of stimulus generalization and secondary reinforcement. *American Journal of Psychology, 69,* 359–68.

Tyler, D. W., Marx, M. H., & Collier, G. (1959). Frustration stimuli in discrimination. *Journal of Experimental Psychology, 58*, 295–301.
Tyler, D. W., Wortz, E. C., & Bitterman, M. E. (1953). The effect of random and alternating partial reinforcement on resistance to extinction in the rat. *American Journal of Psychology, 66*, 57–65.
Vanderwolf, C. H. (1971). Limbic–diencephalic mechanisms of voluntary movement. *Psychological Review, 78*, 83–113.
Vinogradova, O. S. (1975). Functional organization of the limbic system in the process of registration of information: Facts and hypotheses. In R. L. Isaacson & K. H. Pribram (Eds.), *The hippocampus: Vol. 2. Neurophysiology and behavior* (pp. 1–70). New York: Plenum.
Voorhees, J. W., & Remley, N. R. (1981). Mitral cell responses to the odors of reward and nonreward. *Physiological Psychology, 9*, 164–70.
Wagner, A. R. (1959). The role of reinforcement and nonreinforcement in an "apparent frustration effect." *Journal of Experimental Psychology, 57*, 130–6.
  (1961). Effects of amount and percentage of reinforcement and number of acquisition trials on conditioning and extinction. *Journal of Experimental Psychology, 62*, 234–42.
  (1963). Sodium amytal and partially reinforced runway performance. *Journal of Experimental Psychology, 65*, 474–7.
  (1966). Frustration and punishment. In R. N. Haber (Ed.), *Current research in motivation* (pp. 229–38). New York: Holt, Rinehart & Winston.
  (1969). Frustrative nonreward: A variety of punishment? In B. A. Campbell & R. M. Church (Eds.), *Punishment and aversive behavior* (pp. 157–81). New York: Appleton-Century-Crofts.
  (1981). SOP: A model of automatic memory processing in animal behavior. In N. E. Spear & R. R. Miller (Eds.), *Information processing in animals: Memory mechanisms* (pp. 5–47). Hillsdale, NJ: Erlbaum.
Wakerley, J. B., & Lincoln, D. W. (1971). Intermittent release of oxytocin during suckling in the rat. *Nature, 233*, 180–1.
Warrington, E. K., & Weiskrantz, L. (1982). Amnesia: A disconnection syndrome. *Neuropsychologia, 20*, 233–48.
Watson, J. B. (1913). Psychology as the behaviorist views it. *Psychological Review, 20*, 158–77.
  (1919). *Psychology from the standpoint of a behaviorist.* Philadelphia: Lippincott (revised 1924).
  (1925). *Behaviorism.* New York: Norton (revised 1930).
Watson, J. B., & Rayner, R. (1920). Conditioned emotional reactions. *Journal of Experimental Psychology, 3*, 1–14.
Weiner, I., Bercovitz, H., Lubow, R. E., & Feldon, J. (1985). The abolition of the partial reinforcement extinction effect (PREE) by amphetamine. *Psychopharmacology, 86*, 318–23.
Weinstock, S. (1954). Resistance to extinction of a running response following partial reinforcement under widely spaced trials. *Journal of Comparative and Physiological Psychology, 47*, 318–22.
  (1958). Acquisition and extinction of a partially reinforced running response at a 24-hour intertrial interval. *Journal of Experimental Psychology, 56*, 151–8.
Weiskrantz, L., & Warrington, E. K. (1979). Conditioning in amnesic patients. *Neuropsychologia, 17*, 187–94.
Wenrich, W. W., Eckman, G. E., Moore, J. J., & Houston, D. F. (1967). A trans-response effect of partial reinforcement. *Psychonomic Science, 9*, 247–8.

West, J. R., Hodges, C. A., & Black, A. C. (1981). Prenatal exposure to ethanol alters the organization of hippocampal mossy fibers in rats. *Science, 211,* 957–9.

West, J. R., & Hodges-Savola, C. A. (1983). Permanent hippocampal mossy fiber hyperdevelopment following prenatal ethanol exposure. *Neurobehavioral Toxicology and Teratology, 5,* 139–50.

Wickelgren, W. A. (1979). Chunking and consolidation: A theoretical synthesis of semantic networks, configuring in conditioning, S–R versus cognitive learning, normal forgetting, the amnesic syndrome, and the hippocampal arousal system. *Psychological Review, 86,* 44–60.

Wigal, T., & Amsel, A. (1990). Behavioral and neuroanatomical effects of prenatal, postnatal, or combined exposure to ethanol. *Behavioral Neuroscience.* 104, 116–26.

Wigal, T., Greene, P. L., & Amsel, A. (1988). Effects on the partial reinforcement extinction effect and on physical and reflex development of short-term in utero exposure to ethanol at different periods of gestation. *Behavioral Neuroscience, 102,* 51–3.

Wigal, T., Lobaugh, N. J., Wigal, S. B., Greene, P. L., & Amsel, A. (1988). Sparing of patterned alternation but not partial reinforcement extinction effect after prenatal chronic exposure to ethanol in infant rats. *Behavioral Neuroscience, 102,* 43–50.

Williams, J., Gray, J. A., Snape, M., & Holt, L. (1989). Long-term effects of septohippocampal stimulation on behavioral responses to anxiogenic stimuli. In P. Tyrer (Ed.), *Psychopharmacology of Anxiety* (pp. 80–108). Oxford: Oxford University Press.

Williams, J. M., Hamilton, L. W., & Carlton, P. L. (1974). Pharmacological and anatomical dissociation of two types of habituation. *Journal of Comparative and Physiological Psychology, 87,* 724–32.

(1975). Ontogenetic dissociation of two classes of habituation. *Journal of Comparative and Physiological Psychology, 89,* 733–7.

Williams, S. B., & Williams, E. (1943). Barrier-frustration and extinction in instrumental learning. *American Journal of Psychology, 56,* 247–61.

Wilson, W., Weiss, E. J., & Amsel, A. (1955). Two tests of the Sheffield hypothesis concerning resistance to extinction, partial reinforcement, and distribution of practice. *Journal of Experimental Psychology, 50,* 51–60.

Wilton, R. (1967). On frustration and the PREE. *Psychological Review, 74,* 149–50.

Wilton, R. N., & Clements, R. O. (1972). A failure to demonstrate behavioral contrast when the S+ and S− components of a discrimination schedule are separated by about 23 hours. *Psychonomic Science, 28,* 137–9.

Winocur, G., & Mills, J. A. (1969). Hippocampus and septum in response inhibition. *Journal of Comparative and Physiological Psychology, 67,* 352–7.

Wong, P. T. P. (1971a). Coerced approach to shock and resistance to punishment suppression and extinction in the rat. *Journal of Comparative and Physiological Psychology, 75,* 82–91.

(1971b). Coerced approach to shock, punishment of competing responses, and resistance to extinction in the rat. *Journal of Comparative and Physiological Psychology, 76,* 275–81.

(1977). A behavioral field approach to instrumental learning in the rat: I. Partial reinforcement effects and sex differences. *Animal Learning & Behavior, 5,* 5–13.

(1978). A behavioral field approach to instrumental learning in the rat: II. Training parameters and a stage model of extinction. *Animal Learning & Behavior, 6,* 82–93.

Wong, P. T. P., & Amsel, A. (1976). Prior fixed-ratio training and durable persistence in rats. *Animal Learning & Behavior, 4,* 461–6.

Wong, P. T. P., Lee, C. T., & Novier, F. H. (1971). The partial reinforcement effect (PRE) sustained through extinction and continuous reinforcement in two strains of inbred mice. *Psychonomic Science, 22,* 141–3.

Wong, P. T. P., Scull, J., & Amsel, A. (1970). The effect of partial "quinine" reward on acquisition and extinction. *Psychonomic Science, 18,* 48–9.

Wong, P. T. P., Traupmann, K. L., & Brake, S. (1974). Does delay of reinforcement produce durable persistence? *Quarterly Journal of Experimental Psychology, 26,* 218–28.

Woodworth, R. S. (1918). *Dynamic psychology.* New York: Columbia University Press.

Yellen, D. (1969). Magnitude of the frustration effect and number of training trials. *Psychonomic Science, 15,* 137–8.

Zander, A. F. (1944). A study of experimental frustration. *Psychological Monographs, 56,* (3, Whole No. 256).

Zolman, J. F., Sahley, C. L., & Mattingly, B. A. (1978). The development of habituation in the domestic chick. *Psychological Record, 28,* 69–83.

Name index

Abel, E. L., 197
Abramson, C. I., 168
Adams, P. C., 233
Adelman, H. M., 45, 49, 131, 233
Aderman, M., 166
Adrien, J., 182
Advokat, C., 171
Ahlskog, J. E., 234
Albin, R., 104
Allen, J. D., 237
Altman, J., 149, 152, 170, 176–7, 179f, 180, 212–14, 224–6
Altomari, T. S., 49, 131, 233
Ammon, D., 168
Amsel, A., 3–5, 9–11, 26–8, 30–1, 35–47, 49–50, 52–7, 62f, 63, 65–8, 70, 71f, 72–3, 76n, 79–83, 84f, 85, 86f, 87, 89–92, 94–100, 105f, 106–9, 110f, 112f, 113f, 114, 115f, 116f, 117f, 118–20, 123–7, 129, 131–3, 136–9, 140t, 141–2, 144, 145f, 146f, 147f, 148–9, 150f, 151f, 152f, 155–6, 157f, 158–60, 161f, 162, 163f, 164f, 165–6, 169, 171–2, 175–6, 180–3, 185, 187, 190, 194, 195f, 196–8, 199f, 200–2, 203f, 208, 210–11, 214, 218–19, 224, 227, 230, 233–7
Angevine, J. B., Jr., 176, 214
Anisman, H., 92
Anokhin, P. K., 42n
Araujo-Silva, M. T., 182, 191, 237
Armus, H. L., 123, 134, 235
Askew, H. R., 234
Aston-Jones, G., 20
Azrin, N. H., 45, 49, 130

Bachevalier, J., 139, 142, 196
Bacon, F., 176
Banks, R. K., 40, 54, 56, 81, 169, 236
Barker, R. G., 36, 37n, 58
Barnes, D. E., 181, 197
Barry, H., III, 99, 236
Bass, B., 236
Bayer, S. A., 152, 170, 175, 177, 180, 214
Becker, H. C., 180, 182
Becker, J. T., 185, 214
Behrend, E. R., 164
Beier, E. M., 124, 162, 234
Bell, R. W., 156
Benardo, L., 202

Bennett, T. L., 180
Bercovitz, H., 181
Berger, D. F., 48
Berger, T. W., 139
Bergman, G., 4
Berlyne, D. E., 10, 14, 20, 21f, 22
Bernard, C., 13
Berry, S. D., 139
Bertsch, G. J., 45
Bever, T. G., 6
Bilder, B. H., 149, 152, 171
Bindra, D., 15, 16t, 24–5
Birch, D., 126
Bitterman, M. E., 61, 63, 124, 137, 139, 148, 159, 161–2, 164–8, 181, 212, 234
Black, A. C., 197
Black, R. W., 101, 235
Blass, E., 153
Bloom, F. E., 20
Bloomfield, T. M., 101, 235
Boehnert, J. B., 129, 234
Bolles, R. C., 19, 24, 28–9, 153
Bootin, M., 190, 194, 195f, 196
Bower, G. H., 46, 48, 235
Boyagian, L. G., 92
Boyd, T. L., 92
Brake, S., 65, 140t, 144, 147f, 148
Brelsford, J., 235
Breuning, S. E., 166
Brodal, A., 20
Bronstein, P. M., 149, 152, 171
Brown, J. S., 6, 27–8, 36–8, 236
Brown, R. T., 54, 57, 81, 89–90, 133, 169, 219, 234, 236
Brown, T. S., 179, 180, 196
Brown, W. L., 63, 124, 162, 234
Bruce, R. H., 47
Brunner, R. L., 152, 170, 175, 181, 214
Burdette, D. R., 140t, 144, 147f, 148, 155, 158, 175
Burns, R., 166
Bush, R. R., 123, 141
Butler, B., 46

Cador, M., 230
Campbell, B. A., 144, 149, 151, 160, 171
Cannon, W. B., 13
Capaldi, E. J., 9, 52, 55, 61, 67, 75, 118,

126–7, 128, 132, 134, 141, 164–5, 174–5, 204, 236
Capobianco, S., 180
Carlson, A. D., 153
Carlson, J., 83, 85
Carlton, P. L., 89–90, 170–1, 234
Champlin, G., 235
Chen, J.-S., 82, 83f, 87, 91–2, 140t, 144, 145f, 146f, 147f, 148–9, 151f, 152f, 158–60, 161f, 162, 163f, 164f, 165, 169, 171, 172f, 183, 210, 219, 234, 236
Cheng, M. F., 193
Child, I. L., 35–6
Chomsky, N., 14
Choquette, K. A., 233
Clark, C. V. H., 180
Clayton, K. N., 81
Clements, R. O., 235
Coffey, P. J., 188, 190–2
Cogan, D. C., 179
Cohen, J. M., 234
Cohn, N. H., 46
Cole, K. F., 31
Collier, G., 50, 125, 234
Cook, P. E., 234
Cooney, J. B., 67
Corbit, J. C., 77
Corkin, S., 138
Cotman, C. W., 176–7, 202, 214
Cotter, J. R., 197
Coulter, X., 144, 151, 160
Couvillon, P. A., 166–8, 181
Cowan, W. M., 177
Cramer, C. P., 153
Crespi, L. P., 37, 53, 235
Crum, J., 63, 124, 162, 234
Curtis, Q. F., 37n

Dachowski, L., 44, 49, 131, 233
Dailey, W., 70, 140t, 156, 157f, 159, 175
Daly, H. B., 22, 48–50, 124, 131, 142, 233–7
Daly, J. T., 22, 124, 142, 237
D'Amato, M. R., 52, 125
Darwin, C., 6, 12–13
Das, G. D., 149, 177, 179f
Davenport, J. W., 44, 135, 233
Davies, K., 100n, 235
Davis, H., 45, 233
Davis, M., 78–9, 81, 192
Davison, C., 237
Deadwyler, S. A., 184
deCatanzaro, F., 92
Dembo, T., 36, 58
Dempster, J. P., 149
Dericco, D., 149
Deza, L., 149
Diaz, J., 201

Diaz-Granados, J. L., 180–2, 196–7, 200–1, 201f, 202, 203f, 211
Diercks, J. K., 165
Dodson, J. D., 22
Dollard, J., 5, 18, 19f, 36–7, 45, 49
Donenfeld, I., 45, 233
Donin, J. A., 62f, 63, 80, 127, 234
Doob, L. W., 36–7, 45, 49
Douglas, D. P., 149, 171
Douglas, R. J., 149, 152, 171, 176, 185, 204, 214
Douglas, V. I., 226
Drewett, R. F., 153
Dudderidge, H., 134, 236, 237
Dunlap, W. P., 49, 131, 233
Dworkin, T., 149, 152, 171
Dyrud, J. P., 44, 233

Eckman, G. E., 83
Eidelberg, E., 149
Einstein, A., 10
Eirick, R., 237
Eisenberger, R., 83, 85
Ellen, P., 179, 196
Elliott, M. H., 37, 53, 235
Endsely, R. C., 46
Eninger, M. U., 95
Ennis, C., 230
Enzie, R., 54
Ernhart, C. B., 108, 233
Ernst, A. J., 149
Eskin, R. M., 165
Estes, W. K., 22, 23f, 31, 53, 67, 72, 75, 123, 141
Evenden, J. L., 230
Everitt, B. J., 230

Falk, J. L., 130
Fallon, D., 54, 81, 166, 169, 233, 236
Farber, I. E., 36–8
Fedderson, W. E., 61, 165
Feigley, D. A., 149, 152, 170
Feldon, J., 180–1, 185, 188–92
Ferry, M., 166
Festinger, L., 47, 55, 67, 124–6
Fibiger, H. C., 149, 171
Fields, P. E., 47
File, S. E., 170
Finch, G., 233
Finger, S., 193
Fitzwater, M. E., 95
Flaherty, C. F., 44, 166, 180, 182, 233, 235
Franchina, J. J., 179–80, 196
Frank, M., 85
Freibergs, V., 226
Freud, S., 6, 13–14
Frey, P. W., 235

Name index

Galbraith, K. J., 104, 105f, 236
Galbrecht, C. R., 108, 233
Galileo, 10
Gallup, G. G., 45, 49, 131, 233
Gamzu, E., 100n
Garcia, J., 82
Gartrell, K. E., 67
Gaskin, M., 230
Gazzara, R. A., 181
German, D., 54
Gillespie, L. A., 159, 182
Glass, D. H., 234, 236, 237
Glazer, H. I., 81, 93, 169, 182–3, 190, 214, 236–7
Glotfelty, R. A., 46
Gold, D. C., 166
Goldman, H. M., 95
Gonzalez, R. C., 161, 164, 166, 212, 235
Goodrich, K. P., 53, 102, 119–20, 129, 135, 162, 234
Gottlieb, D. I., 177
Graham, M., 131, 155, 156f, 166, 198
Grant, M., 181, 196–7, 201f, 208
Gray, J. A., 9, 54, 88, 93, 134, 136, 143, 173, 176, 177n, 180–91, 196, 204, 206, 208, 211, 213–15, 223, 225, 236–7
Greene, P. L., 180–2, 196–8, 199f, 200–1, 201f, 202, 203f, 208, 211
Gregg, B., 140t, 158, 175
Grice, G. R., 95
Grigson, P. S., 182
Gross, K., 140t, 160, 161f, 162, 163f, 164f, 165, 234
Grosslight, J. H., 35
Grove, G. R., 95
Grusec, T., 48
Guile, M., 83
Guthrie, E. R., 22, 53, 141
Guttman, N., 235

Haggard, D. F., 53, 102, 119–20, 129, 135, 162, 234
Haggbloom, S. J., 181
Hake, D. F., 45, 49
Hall, W. G., 153, 155, 201
Hallgren, S. O., 54, 234, 236
Hamilton, L. W., 149, 152, 170–1, 180
Hammer, R. P., 197
Hancock, W., 107–8, 118, 233
Handlemann, G. E., 185, 214
Hanson, H. M., 235
Harlow, H. F., 5, 126
Hart, T., 193
Hartshorne, M., 93n
Haslerud, G. M., 37n
Hawking, S. W., 122
Hebb, D. O., 18–21, 137, 140, 167
Henderson, K., 102, 103f, 119, 236

Henderson, T. B., 166
Henke, P. G., 45, 180, 191–2, 196, 237
Hill, A., 154
Hill, W. F., 131–3
Hiscock, M., 226
Hitchcock, J. M., 192
Hodges-Savola, C. A., 197
Hoehler, F. K., 139
Hoffman, D., 80
Holder, E. E., 125
Holder, W. B., 125
Holinka, C. F., 153
Holland, P. S., 81
Holt, L., 93, 183–4
Honig, W. K., 235
Hooper, R., 236
Horie, K., 46
Houston, D. F., 83
Hug, J. J., 44, 47, 132–3, 233, 237
Hughes, L. F., 44, 49, 131, 233
Hughes, R. A., 89–90
Hull, C. L., 3–5, 9–10, 12, 14–16, 16t, 17–18, 20–2, 25–9, 31, 36–9, 41, 48, 50, 52, 94, 123–5, 127, 129, 130, 132, 137, 139, 140–1
Hulse, S. H., Jr., 123–4, 134, 149, 160, 165, 234–5
Humphreys, L. G., 61, 124, 132, 234
Hutchinson, R. R., 45, 49

Imada, H., 233
Innis, N. K., 44, 48, 233
Isaacson, R. L., 45, 176, 179–80, 196, 214, 237
Ison, J. R., 104, 123, 126, 234–7

Jackson, J. H., 152
James, H., 85
Jarrard, L. E., 179, 188–9, 191, 196
Jenkins, H. M., 61, 63, 65–8, 127, 144, 234
Jensen, C., 233
Jernstedt, G. C., 234
Johanson, I. B., 155
Johnson, R. N., 45, 237
Jones, H. M., 61, 124
Jones, R. B., 46

Katz, S., 52, 125
Keller, F. S., 59
Kello, J. E., 48
Kendler, H. H., 28, 52, 125
Kettner, R. E., 139
Killeen, P. R., 237
Kimble, D. P., 152, 176, 179, 196, 214
Kimble, G. A., 29
Kimble, R. J., 19f, 179, 196
Kincaid, W. D., 124, 162, 234
Kirkby, R. J., 149

Name index

Kittrell, M. E., 140t, 158, 175
Klee, J. B., 36
Klipec, W. D., 237
Koelling, R. A., 82
Kolb, B. E., 180
Kramer, T. J., 234
Krane, R. V., 104
Krippner, R. A., 46

Lachman, R., 9, 38
Lakey, J. R., 81, 169, 190, 214, 236
Lambert, W. W., 45
Lanfumey, L., 182
Latta, R. M., 233
Lawrence, D. H., 47, 55, 67, 125–6
Lawson, R., 35–7
Lee, C. T., 65
Leitenberg, H., 45
Leon, M., 153
Leonard, D. W., 104
Letz, R., 131, 140t, 155, 156f, 158, 166, 175, 198
Levine, M. J., 149, 152, 171
Levis, D. J., 92, 172
Levy, R. S., 45, 237
Lewin, K., 15, 16t, 36, 58
Lewis, J. H., 49, 131, 233, 236
Lilliquist, M., 163n
Lincoln, D. W., 153–4
Linden, D. R., 54, 234, 236
Lindner, M., 79, 159
Little, L., 235
Livesey, P. J., 137
Lloyd, D., 38
Lobaugh, N. J., 166, 181, 190, 194–8, 200, 201f, 208, 211
Lobdell, P., 45, 237
Logan, F. A., 28, 47, 57, 59, 68, 89, 90, 106n, 124, 133, 162–3, 220, 234, 236
Lord, J., 235
Lowes, G., 166
Lubow, R. E., 181
Ludvigson, H. W., 46, 233, 237
Lynch, G., 177, 184
Lyons, J., 237
Lytle, L. D., 149, 171

Maatsch, J. L., 49, 131, 233
Mabry, P. D., 45, 149, 192, 237
MacKinnon, J. R., 57, 89, 90, 91f, 99, 102, 104f, 119, 133, 233–6
Mackintosh, N. J., 47, 55, 127–8, 166, 235
Magoun, H. W., 19
Maier, N. R. F., 36
Maier, S. F., 92
Malamut, B., 139, 196
Maller, J. B., 93n
Maltzman, I., 31, 47

Mandler, J. M., 142, 235
Manto, P. G., 89–90
Marazzo, M. J., 149
Martin, B., 40, 56
Marx, M. H., 37, 42, 50, 125, 234
Marzocco, F. N., 233
Massad, P., 217
Mathis, D. A., 233
Mattingly, B. A., 171
Maxwell, D., 45
May, M. A., 93n
McCain, G., 45
McCleary, R. A., 153
McCroskery, J. H., 49, 233
McCuller, T., 81, 83, 84f, 85, 169, 190, 214, 236
McDougall, W., 13
McFarland, D. J., 130
McHose, J. H., 46, 107, 233
McNaughton, N., 177n, 180, 185, 188, 190, 196, 214, 225, 236n
McVean, G., 45
Mellgren, R. L., 80, 87, 219
Miles, W. R., 58
Miller, D. J., 58, 128, 175
Miller, N. E., 3, 5–6, 9–10, 18, 19f, 28, 36–8, 45, 49, 54, 56, 58, 75, 81, 99, 104, 111, 223, 228, 229f, 230, 234–6
Mills, J. A., 179, 196
Mishkin, M., 3, 139, 142, 192, 196–7, 208, 211
Mitchell, S., 188, 191
Mock, K., 65
Moltz, H., 153
Moore, B. D., 46
Moore, J. J., 83
Moore, J. W., 61n
Morgan, C. T., 47
Morris, R. G. M., 214
Morruzi, G., 19
Moscovitch, M., 142
Mosteller, F., 123, 141
Mowrer, O. H., 5–6, 10, 27–8, 36–8, 40, 45, 49, 56, 59, 61, 72, 124, 143, 223
Murillo, N. R., 165
Mustaca, A. E., 167
Myers, D. B., 46

Nadel, L., 142, 178f, 182, 214–15
Nanson, J. L., 226
Nation, J. R., 67, 80, 87, 92, 217, 219
Nawrocki, T. M., 128, 175
Neiman, H., 149, 152, 171
Neimann, J., 149
Nevin, J. A., 100, 235
Newton, I., 10
Nick, T., 181, 196–7, 201f, 208
Noirot, E., 156

Name index

Nonneman, A. J., 180
North, A. J., 123, 134–5
Novier, F. H., 65

O'Brien, T. J., 49, 131, 233
O'Keefe, J., 178f, 182, 214–15
Olds, J., 125
Olton, D. S., 185, 208, 214
Overmier, J. B., 9, 38
Owen, S., 185, 190

Panksepp, J., 45, 237
Papini, M. R., 167
Parry, P. A., 226
Parsons, P. A., 149, 152, 170
Patten, R. L., 46, 233
Pavlik, W. B., 89, 90, 234, 236
Pavlov, I. P., 1, 6, 9–10, 25, 38–9, 42–3n, 74, 81–2, 100–1, 104, 124, 139, 140, 142, 185, 231, 235
Peckham, R. H., 46–7, 233
Peeler, D. F., 45, 192, 237
Pennes, E. S., 236
Pentney, R. J., 197
Pereboom, E. C., 46
Perry, S. L., 61n
Pert, A., 166
Peterson, J. J., 149, 171
Petri, H. L., 3, 139, 192, 197, 208, 211
Pitcoff, K., 166
Planck, M., 10
Pliskoff, S. S., 52, 125
Pohorecky, L., 180
Posey, T. B., 179
Potts, A., 166
Pratt, C. C., 4
Powers, A. S., 166
Premack, D., 101
Pribram, K. H., 185
Prouty, D., 49, 234
Pubols, B. H., Jr., 126, 236

Quintao, L., 182, 191, 237

Radek, C., 131, 155, 156f, 166, 198
Ramallo, P., 167
Rashotte, M. E., 5, 10, 52, 54, 57, 63, 64f, 67–8, 70, 71f, 80, 83, 89–91, 99, 119, 127, 129, 131–6
Ratliff, R. G., 81
Rawlins, J. N. P., 180, 185, 188–92, 197, 208, 211
Raymond, B., 166
Rayner, R., 31
Razran, G., 124
Reeves, J. L., 179
Reid, L. S., 126, 236

Remington, G., 92
Remley, N. R., 237
Rescorla, R. A., 9, 22, 28, 38, 81, 123–4, 141–2
Reynolds, G. S., 100–1, 235
Riccio, D. C., 149
Richter, C., 13
Rickert, E. J., 180
Rilling, M. E., 234
Robbins, T. W., 230
Rohrer, J. H., 41
Roitblat, H. L., 6
Rose, G., 183
Rosellini, R. A., 169, 236
Rosen, A. J., 236–7
Rosen, J. B., 192
Rosenbaum, G., 45
Rosenthal, M. H., 171
Rosenzweig, S., 35, 37n
Ross, L. E., 235
Ross, R. R., 54, 65–6, 66t, 67, 80, 90, 133, 219, 236
Roussel, J., 27f, 41, 43, 45, 49, 107, 118, 233
Rowan, G. A., 182
Rozin, P., 193
Rubel, E. W., 171
Ruidiaz, A. C., 166
Russotti, J. S., 235
Ryan, T. J., 44
Ryans, D. G., 93n

Sahley, C. L., 171
Samson, H. H., 201
Samuelson, R. J., 214
Schachter, S., 15, 16t
Schacter, D. L., 138, 142
Scheff, S., 202
Scheibel, A. B., 197
Schlessinger, A. R., 177
Schmajuk, N. A., 166
Schneirla, T. C., 137–9, 155, 176, 197, 208, 211–13, 215
Schoenfeld, W. N., 59
Schwartz, B., 100n
Scobie, S. R., 166
Scott, E. M., 170
Scull, J., 44–5, 81, 82f, 85, 100n, 131, 235
Sears, R. R., 36–7, 45, 49
Segura, E. T., 166
Seligman, M. E. P., 14, 92, 169, 236
Senf, G. M., 104, 235
Senkowski, P. C., 235n
Seward, J. P., 26, 28, 38, 46
Sgro, J. A., 46
Shanab, M. E., 80
Shapiro, M. L., 214
Shapiro, N., 83

Name index

Sheffield, F. D., 22
Sheffield, V. F., 50, 52, 75, 124, 127, 132, 164, 234
Sherrington, C. S., 2, 9
Shettleworth, S. J., 100, 235
Shinoda, A, 168
Shoemaker, H. A., 95, 108, 110
Showalter, J. R., 46
Simonov, P. V., 15, 16t, 204, 213–14
Sinden, J. D., 188–9, 191
Singer, J. E., 15
Skinner, B. F., 4–5, 9, 31–2, 37, 53, 61, 68, 100–1, 124
Smith, P. T., 136, 237
Snape, M., 93, 183
Solomon, R. L., 9, 28, 38, 45, 77
Sparling, D. L., 134, 236
Spear, N. E., 149, 152, 170, 234, 236
Spence, K. W., 4–6, 10, 16t, 25, 28–30, 37–9, 95, 101, 105–6, 125–6, 138, 211
Spencer, W. A., 153
Sperling, S. E., 126
Spivey, M., 87, 219
Spohr, H. L., 197
Squire, L. R., 2, 3t, 138–9, 142
Staddon, J. E. R., 44, 48, 100n, 233
Stanley, W. C., 124
Stanton, M. E., 11, 137, 139, 140t, 156, 157f, 160, 164f, 165–6, 175–6
Statham, C., 153
Stehouwer, D. J., 171
Stein, D. G., 193
Stein, L., 76–8, 93
Steinsultz, G., 237
Stevens, S. S., 4
Stevenson, H. W., 126
Stevenson, S. S., 37, 58, 234
Stimmel, D. T., 123, 134, 233, 235
Stoltenburg-Didinger, G., 197
Surridge, C. T., 50, 52, 62f, 63, 64f, 65, 80, 119, 127, 129, 175, 233–4, 236
Sutherland, N. S., 55, 67, 127–8, 130
Swanson, A. M., 45, 237
Sytsma, D., 233

Taylor, D., 177
Taylor, J. A., 30
Taylor, J. R., 230
Teitelbaum, P., 193
Terborg, R., 83
Terrace, H. S., 6, 99–101, 105–6, 134–5, 221, 234–5, 237
Terris, W., 54, 169
Theios, J., 61, 63, 65–8, 127, 144, 234–5
Thomas, D. R., 235
Thomas, G. J., 1, 2
Thompson, C. I., 44
Thompson, R. F., 139, 153

Thorndike, E. L., 1, 38–9, 81, 137, 140–1
Tolman, E. C., 3–6, 14–15, 16t, 25, 139, 141
Tominga, M., 233
Tonkiss, J., 188, 190–2
Torney, D., 236
Trapold, M. A., 9, 38
Traupmann, K. L., 65, 76n, 79, 132, 218, 235
Trowill, J., 45, 237
Tucker, R. S., 46
Tulving, E., 138–9
Tyler, D. W., 50, 61, 124, 165, 234

Ursin, H., 190

Vanderwolf, C. H., 182
Verry, D. R., 128, 175
Vinogradova, O. S., 204, 213–14
Voigt, J., 180
Voorhees, J. W., 237

Wagner, A. R., 22, 47, 54, 56–7, 76–9, 81, 99, 107, 118–20, 123–5, 134, 141–2, 149, 165, 169, 219, 233–7
Wakerley, J. B., 153–4, 160
Walker, D. W., 181, 197
Ward, J. S., 39–40f, 42, 44, 47, 49–50, 54–5, 96, 99, 108–9, 112–13f, 114, 115–17f, 120, 123, 132, 185, 233–5
Warrington, E. K., 138
Waterhouse, I. K., 36
Watson, B. A., 4, 230
Watson, J. B., 9, 31
Watson, P., 44
Wechkin, S., 169
Weimer, J., 104
Weiner, I., 181
Weinstock, S., 52–3, 75, 119, 124, 129, 135
Weiskrantz, L., 138
Weiss, E. J., 52, 125
Weisz, D. J., 139
Wenrich, W. W., 83
West, J. R., 197
West, M., 184
Wible, C. G., 214
Wickelgren, W. A., 137, 179, 196, 208
Wigal, T., 181, 198, 199f, 200–1, 202f, 211
Wigal, S. B., 198
Williams, E., 36
Williams, J., 93, 183
Williams, J. D., 80
Williams, J. M., 170–1
Williams, S. B., 36
Wilson, A. S., 179, 196
Wilson, W., 52, 125
Wilton, R., 93, 131–3, 235
Winocur, G., 179, 196
Wolach, A. H., 166

Wolkoff, F. D., 149, 152, 171
Wong, P. T. P., 65, 76n, 79, 81, 82f, 83, 84f, 85, 86f, 128–32, 169, 190, 214, 218–19, 235–6
Woodard, W. T., 166
Woods, D. J., 217
Woods, P. J., 153
Woodworth, R. S., 15, 16t
Wortz, E. C., 124

Wrather, D. M., 80, 87, 219

Yellen, D., 233
Yerkes, R. M., 22

Zander, A. F., 36
Zola-Morgan, S., 142
Zolman, J. F., 171

Subject index

adaptation, 12
aggression, 11, 45, 53, 216
amphetamine effects, 229f
animal cognitivism, 4, 6
arousal, 34, 215, 231
attention deficit–hyperactivity disorder, 11
   action of stimulants, 228–9
   conflict theory, 228–30
   deficit in intermediate-term memory, 213
   deficits in persistence, 212, 225
   fetal alcohol and X-irradiation, 224
   general persistence theory, 227
   hippocampal microneuronal hypoplasia, 224
   increased distractability, 225
   responsiveness to anticipatory frustration, 226
   stimulant effects, 228–30
auditory evoked potentials, 184

behavioral field theory, 128–31
behavioral habituation; see also behavioral inhibition system (BIS): general theory of persistence
   "active" accounts of, 76–9
   counterconditioning as mechanism, 76
behavioral inhibition system (BIS), Gray's
   and frustration theory, 185–8
   relation to early hypotheses of hippocampal function, 185
   subicular comparator, 185
   tests with subicular, hippocampal, and septal lesions, 188–91
behaviorism and behaviorisms, 4

Capaldi's sequential hypothesis, 52, 127–8
classical association theories, 123
cognitive dissonance, 47, 124–6
cognitive map, 3, 214
cognitive revolution, 3
conditioning model theory (Lachman), 9, 38–40
counterconditioning, 51, 75–6
   in behavioral habituation, 74
   in children with ADHD, 230–1

demotivation hypothesis, 46
desistance, 11

developmental psychobiology
   dispositional learning, 6, 8
   hippocampal formation, 7
   levels of functioning, 6
   limbic system, 7, 176–8, 178f
   ontogeny of reward-schedule effects, 140t, 141f
discontinuously negatively correlated reinforcement (Logan), 68, 163
discrimination learning
   dispositional effects of prediscrimination exposures, 106–7, 109t
   in double-runway frustration effect, 95–6, 96f
   and frustration theory, 94–5
   and generalization of inhibition, 102, 105f
   Pavlovian positive induction, 100–1
   and positive behavioral contrast, 100–1
   as partial-reinforcement training, 96–100, 97f
   within-subjects experiments, 101–4
dispositional learning
   and developmental psychobiology, 6–8
   and hippocampal formation, 177
   human applications, 11, 216–24
   reward-schedule effects, 1
dispositional memory, 1–2
double-runway apparatus, 27
durable persistence, 64–5, 160

emotion as intervening variable, 36
entorhinal cortex, 179, 186–7

fear–frustration commonality, 56–7; see also transfer of persistence
fractional anticipatory frustration (r_F–s_F), 40–1, 40f, 42n
   aversive counterpart of r_R, 43
   in behavioral inhibition system theory, 188
   counterconditioning of, 76
   as inhibition, 41
fractional anticipatory goal responses (r_G–s_g)
   four subtypes, 40–1
   Lachman's conditioning model theory, 9
   as Pavlovian conditioning, 38

Subject index

"pure stimulus acts," 9
schematic representation, 39f
in two-process theory, 9, 38
fractional anticipatory reward (r_R), 40–1, 40f
frustration, conceptualizations of, 36–8, 42, 45
frustration–aggression hypothesis, 45
frustration drive stimulus (S_F), 50
 in choice behavior, aggression, and escape responses, 49–50
frustration effect (FE); *see also* frustrative nonreward; primary frustration (R_F)
 blocking, 42
 competing response hypothesis, 46
 delay and reduced reward, 46
 demotivation hypothesis, 46–7
 double-runway apparatus, 27
 effortfulness, 48
 preparatory response hypothesis, 47
 summation of, 46
 as temporal inhibition, 48
 "try harder," 47
 ultrasounds as indicant, 156
 within-subjects effect, 46
frustration theory of the partial-reinforcement extinction effect; *see also* partial reinforcement extinction effect, (PREE), alternative theories of
 boundary conditions, 34–5
 categories of, 35–7
 and cognitively oriented behavior therapies, 217
 as conditioning-model theory, 38–40
 of discrimination learning, 94–100, 97f
 earlier versions in learning theory, 37–8
 four-stage theory, 50–1
 modification based on developmental, neurobiological work, 207–13
 other phenomena explained, 53
 revisions of, 131–3
 six-stage theory, 55
 special case of general persistence, 53, 72–5, 74f
 theory applied to punishment, 56
frustrative nonreward; *see also* frustration effect (FE); primary frustration (R_F)
 in arousal, suppression, and persistence, 34
 indexed by activity in open field and stabilimeter, 48–50
 indexed by the frustration effect, 44

general theory of persistence, 72–5, 74f; *see also* frustration theory; transfer of persistence
 as behavioral habituation, 76–9
 and learned helplessness, 91–2

ontogeny of habituation, 169–73
persistence training in psychotherapy, 217n
generalization of inhibition, 102, 105f
generalized partial-reinforcement extinction effect (PREE), 88–91; *see also* general theory of persistence
learned helplessness, 91–2
transfer of persistence, 57–8
in within-subjects experiments, 91

habit-family hierarchy, 31f; *see also* response-family hierarchy (Hull)
role of drive and drive stimuli, 32–2
hippocampal EEG in theta range
 age first seen and paradoxical effects, 182
 and general persistence theory, 183
 relation to anticipatory frustration, 182–3
hippocampal formation, 7
 cell structure, 177, 178f
 development of, 177
hippocampal function
 cognitive vs. associative explanations of, 215
 cognitive map, 214
 development of in the rat, 149
 dispositional learning, 214
 drug effects, 181–2
 ethanol effects: in adults, 180–1; in infants, 197–204
 frustration and persistence, 214
 gating function of, 181, 185
 internal inhibition, 185
 lesion effects: in adults, 179–80; in infants, 194–7
 neuromodulation, 214
 reactions to novelty, discrepancy, 214
 relation to specific experimental tests, 214–15
 suppression and inhibition, 152, 214
 X-irradiation effects: in adults, 180–1; in infants, 200–4
 working memory, 214
hippocampal granule cell development, 177–9
 as theta-producing cells, 183
homeostasis, 13–14
Hull–Sheffield hypothesis, 52

idiosyncratic response rituals, 67–71; *see also* persistence; transfer of persistence
incentive motivation, 40
infantile amnesia, 151, 160, 222–3
inhibition; *see also* behavioral inhibition system (BIS)
 and hippocampal development, 152
 reactive and conditioned (Hull), 41

learned helplessness, 91–2
levels of functioning
  Bitterman's distinction: carryover vs. reinstatement, 139
  and developmental neurobiology and neuroanatomy, 142–3
  and developmental psychobiology, 6–9
  dispositional learning and memory, 6–9, 177
  hippocampal maturity, 208, 209f
  in humans, 138–9
  Korsakoff patients, 138
  memory, development of, 142
  in phylogeny and ontogeny, 137–8
  reversion of function, 176–7
  Schneirla's distinction: approach–withdrawal vs. seeking–avoidance, 139
liberalized stimulus–response theory, 3
limbic system, 7, 176–8, 178f

magnitude of reinforcement extinction effect, 158, 163
memory; see also patterned alternation
  classifications of, 3t
  dispositional and representational, 1
  episodic-semantic, 139
  implicit–explicit, 138
  limbic-dependent memory system, 196
  non-matching to sample, 196
  ontogeny of, 142
  priming of, 138
  procedural-declarative, 138–9
  in relation to habit, 139
  working or carryover, 175
memory-based learning, 197, 201; see also patterned alternation
microneuronal hypoplasia, 212
  in ADHD, 224
  minimal brain dysfunction, learning deficits, 225
  relation to alcohol in pregnancy, 226
  relation to distractability, spontaneous alternation, persistence, 225
Miller's analysis of conflict, 111–12
  in analysis of prediscrimination effects, 106
  stimulant effects on hyperactivity, 228–30
minimal brain dysfunction, 225; see also microneuronal hypoplasia
motivation
  arousal and reticular activating system, 19–22, 20f, 21f
  associative and nonassociative mechanisms, 16ff, 29f
  associative theories of, 22–5
  central motive state, 24–5
  cue and drive functions, theories of, 17–22, 19f

  early conceptualizations, 12–14
  facilitating and interfering properties of, 30–3
  and the frustration effect, 27
  generalized and selective aspects, 28–30
  Hullian theories of, 17–18
  and the learning–performance distinction, 16t
  major historical figures, 12–14
  primary and acquired, 25–8
  stimulus sampling theory of, 23f

neobehaviorism, 3–6
  Hull's, 4, 9
  N. E. Miller's liberalized, 3, 9
neocognitivism
  animal, 6
  mentalism, 3
  a window on the mind, 6
neuromodulation, 214
non-matching to sample, 196; see also memory

omission effect, 48
ontogeny of habituation
  and generalized persistence, 169
  reflex and exploratory, 170
  to shock and appetitive persistence, 171–3
ontogeny of procedural and declarative memory, 142
ontogeny of reward-schedule effects, 140t, 141f, 143
  and developing hippocampal formation, 177, 179f
  first four postnatal weeks, 149–52
  ten to fifteen days of age, 152–64
  young adults, juveniles, weanlings, 144–9
ontogeny of suppression, 152–3
oral cannula, 154
oxytocin, 153

paradoxical and nonparadoxical effects; see also reward-schedule effects
  and classical learning theories, 123–4
  in development, 140–1
  and dispositional learning, 137, 140
  in fishes and turtles, 164–6
  mathematical models of, 141–2
  and ontogeny of suppression, 152–3
  theories based on results from adult and infant rats, 174–6
partial delay of reinforcement extinction effect, ontogeny of, 160–3
partial-reinforcement acquisition effect (PRAE), 53, 162–3
  between subjects, 120

Subject index

ontogeny of, 163
within subjects, 118–19
partial-reinforcement extinction effect (PREE); *see also* persistence
 behavioral field approach, 128–31
 cognitive dissonance theory of, 124–6
 different mechanisms in infants and adults, 213
 durability of, 63–5
 early work in humans, 93n
 first appearance in infancy: guinea pig, 159; rat, 158–9
 frustration theory of, 50–2
 frustrative and nonfrustrative persistence, 133–5
 generalized, 88–91
 in human eyeblink conditioning, 138
 inborn persistence, 210–11
 loss of persistence, 159–60, 163
 ontogeny of, 159–60, 163
 preservation and reinstatement of persistence, 210–11
 as regression, 59
 in relation to age in infancy and a variety of treatments, 208–10
 in relation to vigor and choice, 135
 sequential theory of, 127–8
 stimulus analyzer theory of, 128
 survival of, 61–3
 Theios–Jenkins experiment, 61
 transfer of, 57–8
partial-reinforcement extinction effect (PREE), alternative theories
 behavioral field approach, 128–31
 cognitive dissonance, 124–6
 early theories, 124
 extensions and revisions of frustration theory, 131–5
 sequential theory, 127–8
 stimulus analyzer theory, 128
passive avoidance learning, 152
patterned alternation, 156–7, 207–8; *see also* memory-based learning
 in infants, 156–7
Pavlovian positive induction, 101; *see also* positive behavioral contrast
perseveration in relation to persistence, 229n
persistence, 50–1, 54, 56, 231; *see also* regression; transfer of persistence
 durability of, 63–5
 frustrative and nonfrustrative, 133–4
 as idiosyncratic response rituals, 67–71
 learned disposition, 160
 preservation of, 210
 reinstatement of, 210
 survival of, 63
 temperamental characteristics, 160
 transfer of, 65–7, 218–21

persistence theory applied to humans, 217
physiologizing, 9–10
positive behavioral contrast; *see also* Pavlovian positive induction
 as primary frustration effect, 101
 Skinnerian explanations of, 100–1, see 100n
prediscrimination effects, analysis of facilitation and retardation of discrimination learning, 106–7, 113f, 115f, 116f
 in frustration effect in a double runway, 116–18, 117f
 and Miller's analysis of conflict, 111–12
 prediscrimination experience, 109t
prediscrimination exposure effects, 106ff
 in relation to conflict and displacement, 111
 in relation to the frustration effect, 117f
 and the within-subjects partial-reinforcement acquisition effect, 118–20
preparatory response hypothesis, 47
preparedness, 214
primary frustration (R_F), 38, 41–2, 44; *see also* frustration effect; frustrative nonreward
 and chlorpromazine, 134n
 demotivation hypothesis, 46
 early experimental studies of, 37n
 generalized drive, 46
process need (Maltzman), 47
psychotherapy, 217n
punishment and the partial-reinforcement extinction effect, 56
pure stimulus acts (Hull), 9

regression, 11, 215, 232; *see also* transfer of persistence
 as persistence, 58–9
 as transfer of persistence in humans, 220
response-family hierarchy (Hull), 130–1
reticular activating system, 18–22
reversion of function; *see also* levels of functioning
 application to humans after fetal alcohol and X-ray exposure, 223
 in dispositional learning and memory, 177
 infant as hippocampally damaged adult, 176
 tests with lesions, 194–7: fetal alcohol, 197–201; postnatal alcohol, X-irradiation, 200–4
 two kinds, 193–4
reward-schedule effects, xii–xiii, 1; *see also* paradoxical and nonparadoxical effects

across species, vertebrates, 164–7
and dispositional learning, 1, 137
in the honeybee, 167–9
ontogeny of, 139–41, 140t, 141f, 142n, 143
steps in the psychobiological study of, 205–7
theories applied to infants, 174–6
Ritalin, 230
runway apparatus for pups, 154f

sensory evoked potentials, 184–5
septum, 177, 180, 186–9
sequential theory (Capaldi), 52, 127–8, 141, 174–6
statistical association theories, 123
stimulus analyzer theory, 128
subicular comparator, 185; see also behavioral inhibition system (BIS)
subicular lesions, 188–9
successive negative contrast, ontogeny of, 160–1, 163
suppression, 34, 152–3, 215, 232
survival of persistence, 61–3

temperament, 216
temporal inhibition, 48
toughening-up (Miller), 183–4, 223
and dispositional learning and memory, 184
in relation to theta rhythm, 183–4
transfer of persistence; see also fear–frustration commonality
applications to humans, 218–20
to different responses and motivations, 65–7
to the general theory of persistence, 79ff
generalized partial-reinforcement extinction effect, 88–91
idiosyncratic rituals, 67–71
persistence as discontinuously negatively correlated reinforcement, 39
persistence as regression, 59
resistance to discrimination, 122
two-process theory, 9

ultrasounds, 155–6

variable magnitude of reinforcement extinction effect, ontogeny of, 160–3